EXPLORING THE UNIVERSE

A Festschrift in Honor of Riccardo Giacconi

ROMANA STVDIORVM VNIVERSITAS
"SAPIENTIA"

Advanced Series in Astrophysics and Cosmology – Vol. 13
Series Editors: Fang Li Zhi and Remo Ruffini

EXPLORING THE UNIVERSE

A Festschrift in Honor of Riccardo Giacconi

Editors

H. Gursky
Naval Research Laboratory, USA

R. Ruffini
University of Rome La Sapienza

L. Stella
Osservatorio Astronomico di Roma, Italy

World Scientific
Singapore • New Jersey • London • Hong Kong

Published by

World Scientific Publishing Co. Pte. Ltd.

P O Box 128, Farrer Road, Singapore 912805

USA office: Suite 1B, 1060 Main Street, River Edge, NJ 07661

UK office: 57 Shelton Street, Covent Garden, London WC2H 9HE

British Library Cataloguing-in-Publication Data
A catalogue record for this book is available from the British Library.

EXPLORING THE UNIVERSE: A FESTSCHRIFT IN HONOR OF RICARDO GIACCONI

ISBN 981-02-4423-1

Printed in Singapore by Uto-Print

Preface

Opening new fields of scientific research, reaching the sky with new instruments of unprecedented capabilities, obtaining fundamental results at the forefront of our understanding of the Universe, creating the infrastructures of the world's most advanced facilities to allow scientists to achieve progress in their cutting edge research: any one of these would have been a major achievement in the life of a great scientist. In his uniquely distinguished career, Riccardo Giacconi has achieved these goals five times, one after the other:

- By expanding the developing field of high energy astrophysics with the establishment of X ray astrophysics, and through the discovery of the first extrasolar X-ray source (Sco X-1) and a diffuse background of X-ray radiation during the historical rocket flight of 1962.

- By conceiving and managing the construction and exploitation of the first satellite for X-ray astronomy: the UHURU. This led to a number of crucial new findings. First the discovery of pulsating X-ray binaries, which proved that accretion of matter onto collapsed stars powers the X-ray emission of the brightest X-ray sources in the galaxy and which yielded the first neutron star mass measurements. Then the study of Cyg X-1 and the identification of its optical counterpart, which provided the first observational evidence for a black hole. And finally the discovery of a tenous and hot X-ray emitting plasma trapped in clusters of galaxies, containing an amount of matter comparable to that in stars and galaxies.

- By conceiving the X-ray mirror technology that was the key to the great success of the Einstein space telescope, the first X-ray satellite with imaging capabilities. A large project that Riccardo managed through all its stages, the Einstein telescope provided a major leap forward in the identification and study of thousands of new X-ray sources of many different classes, ranging from nearby stars, to clusters of galaxies, to active galactic nuclei at cosmological distances. With the Einstein telescope, the X-ray window became a new way of studying the Universe in addition to the traditional optical astronomy. The same optics were used by Riccardo and Giuseppe (Pippo) Vaiana to develop instruments for obtaining high quality images of the sun at x-ray wavelengths. Beginning with instruments on sounding rockets during the 1960s and long duration observing on NASA's Skylab Mission, such instruments have profoundly transformed solar physics.

- As the first director of the Space Telescope Science Institute, Riccardo oversaw its development into one of the world's largest and most successful astronomical institutions. He managed its operation for over a decade, from well before launch to the post-Costar era, when the telescope achieved its design performance. The Hubble Space Telescope has yielded innumerable new fundamental results. Especially relevant to Riccardo's favorite themes are the remarkable findings which revealed the presence of massive gravitationally collapsed objects in the nuclei of active galaxies and the cosmological evolution of galaxies and clusters of galaxies.

- By realizing the largest optical observatory ever, the European Southern Observatory, consisting of four 8 meter telescopes which will also be operated cooperatively as an interferometer with a baseline of a few hundred meters.

In October 1997 we celebrated Riccardo's sixty-fifth birthday on the occasion of his being awarded a "Laurea Honoris Causa" in physics at the University of Rome "La Sapienza". A close number of his past and present collaborators gathered in Rome, where a meeting was held in his honor. Presentations were given that reflected the state of the art of research on subjects related to Riccardo's scientific interests and achievements through the decades. This Festschrift is based upon the papers presented at that meeting. It is intended to provide a tribute to Riccardo Giacconi for his outstanding contributions towards opening up the entire field of high energy astrophysics. We wish Riccardo continued success both in his own career and in helping many other scientists participate in the quest for knowledge of the Universe.

Herbert Gursky
Remo Ruffini
Luigi Stella

University of Rome "La Sapienza"
Honorary Degree in Physics Awarded to Riccardo Giacconi

After having obtained the Doctorate in Physics at the University of Milan and having worked for a few years as a researcher in the area of particle physics at the University of Milan, of Indiana and at Princeton, Riccardo Giacconi became member of the American Science and Engineering (ASE) in 1959. This private company based in Cambridge, Massachusetts had twenty-eight employees at the time. Mr. Giacconi was given the task to take up space research supported by government funds. In 1970, the ASE Space Research Team led by Riccardo Giacconi had grown to 500 people. The activities of the team ranged from satellite projects and construction to data reduction, analysis and interpretation. These research activities were commissioned and sponsored by the US Department of Defense and NASA.

In 1962, Riccardo Giacconi, Herbert Gursky, Paolini and Bruno Rossi published an article in which they related the first observation of an X-ray source outside the Solar System. This observation marked the origin of X-ray Astrophysics. In 1963, Giacconi proposed a project for a satellite devoted to X-ray Astrophysics. This satellite was built between 1966 and 1970, and was launched at the San Marco platform of Rome University in 1970. To celebrate this Italian launch site, built in Kenya by Luigi Broglio, the satellite was called "Uhuru", which in swahili means freedom. This mission marked the great progress made by X-ray Astrophysics and gave the first evidence for the existence of black holes as well as neutron stars in binary systems in our Galaxy. The results given by the "Uhuru" satellite at the time set up a world record for the number of scientific quotations, not only in the area of Physics and Astronomy, but also in other areas of the Natural Sciences.

In those same years, a series of scientific contacts was established between Riccardo Giacconi and Remo Ruffini, who - at the time - was working at Princeton University. These interactions allowed a direct comparison between the observing results obtained in Cambridge by the "Uhuru" satellite and the theoretical predictions about neutron stars, critical mass and the consequences of black holes observations, as developed by Ruffini and his team at Princeton.

Meanwhile, the first space mission devoted to the Sun's X-ray emission was realized in collaboration with Giuseppe Vaiana, by means of a telescope on board the Skylab.

In 1970, Giacconi started to plan the first space X-ray observatory, which was eventually approved in 1973. This telescope was successfully launched in 1978, and took the name of Einstein Observatory.

In 1973, Riccardo Giacconi and other members of his team were invited by Harvard University to take part in one of the major acquisitions ever made by a US university, namely, the "High Energy Astrophysics" section, which would attract about a hundred researchers in a few years.

In 1975, the knowledge acquired both in theory and in observations by the "Physics and Astrophysics of Neutron Stars and Black Holes" were introduced and summarized during the Sixty-fifth "Enrico Fermi" Summer School organized by Giacconi and Ruffini at Varenna, near Lake Como.

From 1978 onwards, in addition to the observing program which was carried out by the HEA researchers of the Giacconi team, a guest observer program was created at the Einstein Observatory for scientists coming both from US and International universities. The research activity thus started soon engaged a number of participants equal to the staff of the major ground-based observatories. This mission led to a new accurate study of Galactic Sources, as well as the identification of Extra-galactic Sources and to the observation of diffuse X-ray emission from Clusters of Galaxies, thus introducing the issue of Galactic Systems' evolution and formation into the field of Cosmology.

In 1981, Riccardo Giacconi put an end to the first stage of his scientific activity - the birth and growth of X-Ray Astrophysics, in order to start his second stage: Space Optical Astronomy and Astrophysics. In fact, at the beginning of September 1981, Giacconi accepted the post of Director of the soon-to-be-established Space Telescope Science Institute of the Johns Hopkins University in Baltimore. With his well-known energy and management skills, Giacconi founded the Institute and chose the staff, which soon reached 350 people, about a hundred of whom were astronomers. The aim he achieved was to take upon himself the entire responsibility of the Space Telescope Science Institute for the maintenance and running of the telescope, as well as for the development of scientific research programs.

Even in the dramatic situation created by the serious error which occurred in the making of the Hubble Space Telescope mirror, Giacconi - together with his Institute - provided the necessary technical and managerial skills for the designing of the correcting instrument - called COSTAR - thus leading to the reinstatement of the planned scientific capabilities of the telescope. In a way, this work marked the beginning of a new era in space science research - namely, the era of astronauts intervention through a second carrier upon an orbiting astronomical satellite.

In 1985, Giacconi took part in the creation of an International Consortium for the promotion of an activity which would later become very important for Rome University "La Sapienza". This Consortium, called ICRA (International Center for Relativistic Astrophysics), included Rome University "La Sapienza", the Space Telescope Science Institute, Stanford University, Washington University, the ICTP and the TWAS of Trieste, the Vatican Observatory and the University of Hofei in China.

After having concluded this second stage of activity - spanning from the birth to the maturity of Space Optical Astronomy, and having deeply changed the development of Astronomy as a whole, as well as the human picture of the Universe itself - Giacconi faced yet another challenge in the latter period of his life. This consisted in the creation of the largest ground-based astronomical observatory with the largest and most technologically advanced instruments in the whole world.

In December 1992, Riccardo Giacconi was appointed General Director of the ESO (European Southern Observatory). The ESO project, which will be completed in the next few years, is the VLT (Very Large Telescope), a set of four eight-metre telescopes which can be used either separately or together as an interferometer.

Despite his manifold activity, which led to a world leadership in the creation and development of X-Ray Astronomy, and - later on - of Space Optical Astronomy, Riccardo Giacconi did not neglect an ongoing scientific work, which has given relevant contributions through more than 150 research articles published in International scientific journals. Moreover, his important activity as a teacher can be easily gauged in a number of manuals of X-Ray Astrophysics, which he published during his lifetime.

Finally, one cannot help mentioning three recent activities of Riccardo Giacconi:

In 1987/88 he directed the Donegani Institute of the Montedison Group;

In 1991, he was appointed Honorary Professor of Astronomy at the University of Milan;

In 1995, he was called to take part in a Committee created by the Italian Government and led by Carlo Rubbia, with the aim of re-organizing Space Research in Italy.

In conclusion, we are glad to propose the awarding of an Honorary Degree in Physics at Rome University "La Sapienza" to Professor Riccardo Giacconi for his unique contribution to the progress of our scientific knowledge of the Universe and for his unprecedented role in the research areas of stellar evolution and cosmological structures - as well as for his continued attention to the technological and cultural debates of Italy in general and inside "La Sapienza" University in particular.

C. Bernardini
G. Diambrini Palazzi
R. Ruffini

Table of Contents

X-RAY ASTRONOMY-THE FIRST DECADE

HERBERT GURSKY
Naval Research Laboratory
Washington, DC 20375 USA

The years 1960 to 1970 saw the emergence of x-ray astronomy. By 1960 plans were being made for cosmic x-ray observations by a number of groups around the world, based on extrapolations from solar and cosmic ray observations. Still the discovery of significant fluxes of x-rays was a complete surprise in 1962. Over the next several years, follow-up observations were made and many new groups entered the field. Notable discoveries included the identification of the Crab Nebula as an x-ray source and the identification of Sco X-1 with a 13th magnitude stellar object that had the characteristics of an old nova. By the end of the decade the Uhuru x-ray satellite was flying and planning was underway for the HEAO spacecraft, including the Einstein Observatory, and many smaller instruments.

1 INTRODUCTION

Giacconi and I go back a long way. In 1957 I was completing work as a graduate student at Princeton University, working in George Reynolds' Cosmic Ray Physics Laboratory. Giacconi had joined the group as a research associate, having completed a postdoctoral appointment at Indiana University after receiving his doctorate at the University of Milan. Reynolds dream, as it was for many cosmic ray researchers, was to find new particles and we spent the summer of 1958, with Fred Hendel, developing and operating a detector array to chase a mass 500 particle reported by Russian researchers, naturally without success (see figure 1). In hind sight it was a remarkable time and place to conduct research. Not only did we have virtually unlimited resources to pursue our research, but Reynolds had assembled an extraordinary research staff at the Laboratory, including Joe Ballam, Val Fitch, Pierre Piroue and Ronald Rau. That fall I went off to Columbia University as an instructor in the Physics Department; Giacconi remained at Princeton for another year and then joined American Science and Engineering, Inc. (AS&E), a small company in Cambridge, Massachusetts. I next encountered Giacconi at the winter meeting of the American Physical Society in New York in 1961 where I was looking for a job and he was recruiting scientists for AS&E. We had both somehow 'failed' at establishing academic careers and had gone off to industry to find positions.

In the next ten years, Giacconi, or at least the group of us at AS&E, was able to create a scientific discipline that had not existed until then. We performed many of the crucial experiments, developed new instruments and devised the basic space experiments. My purpose in preparing this material is to recount Giacconi's accomplishments during that decade. It also makes a good story.

I have not tried to make the material bibliographically correct, but fortunately a number of noteworthy books are available that go over the same ground in some detail. I particularly note works by Wallace Tucker and Riccardo Giacconi[1], Herbert Friedman[2] and Richard F. Hirsh[3]. Of more general interest regarding the development of space science in the US is the work of David H. DeVorkin[4]. The development of the British program of space science has been described by Sir Harrie Massie and M. O. Robins.[5]

2 THE SCIENTIFIC SETTING

The decade between 1960 and 1970 was one of extraordinary productivity for astronomy. The reasons are not hard to find. In the 1950s a broad range of new technology produced during World War II found wide application in the commercial world and in the laboratory. Scientists returning from wartime research found eager employers in universities and industry. Returning servicemen and women flocked to the universities with the help of government financing. More important was that the US accepted the need for federal support for the sciences. Vast sums of money for the research establishment flowed from the military services, the Office of Naval Research, the National Aeronautics and Space Administration and the Atomic Energy Commission. In my own case, I attended graduate school under a fellowship awarded by the Atomic Energy Commission and my Ph.D. thesis research was paid for by the Office of Naval Research. It was a fortunate time to begin a scientific career.

In astronomy, the time was ripe for discovery. The optical astronomers were exploiting large, high quality telescopes. Physicists were entering astronomy in large numbers via the new 'windows', cosmic rays and radio astronomy. Finally scientists were beginning to exploit a new technology, rockets, that allowed carrying instruments above the atmosphere and providing for a few minutes of observation without its absorption, the equivalent of a stack of about 20 lead bricks for energetic radiation such as x-rays.

Of most direct importance to x-ray astronomy was the work done on solar physics and on cosmic rays. Both areas were among the principal objectives of the first rocket scientists. Atmospheric scientists had long recognized that the sun needed to produce ionizing radiation in order to account for well documented effects in the ionosphere and the aurora. In the 1930s this lead to the speculation that the sun was a source of x-rays which was supported by the astounding discovery by Bengt Edlen that the sun's corona was at a tempera-

Figure 1: Riccardo Giacconi, Herbert Gursky, and Fred Hendel at the Cosmic Ray Physics Laboratory at Princeton University (1958)

ture of millions of degrees based on his identification of coronal emission lines with highly ionized metallic atoms.

By 1960, science from space borne instruments was entering its second generation. Even under wartime conditions, Fritz Regener had developed a scientific payload for the German V-2 rockets concentrating on atmospheric research. Regener's payload never did fly, but the use of captured V-2 rockets after the war in the US did result in a variety of scientific results in the areas of atmospheric science, solar physics and cosmic rays under the sponsorship of the US military services, principally the Navy and the Air Force. Perhaps more important, the use of these rockets attracted a number of talented scientists and engineers and established the value of space borne platforms for scientific research. The early rockets appear incredibly primitive viewed 50 years later. As an example Richard Tousey's first attempt to obtain ultraviolet spectra of the sun used a spectrograph mounted on the rocket tail and with film canisters made of tungsten so that it could survive the hard impact on the floor of the rocket range. (It didn't.) The scientists and engineers did not realize they could cause the rocket to tumble after reentry into the earth's atmosphere to reduce its speed at impact, much less did they have parachutes available to provide a soft landing.

More than technology emerged in the decade following the first successful scientific rocket investigations from above the atmosphere. The US committed itself to the development of orbiting satellites in support of the International Geophysical Year in 1957. This effort received an enormous impetus from the Soviet space program and with the recognition that Soviet developments in missile technology were superior to that of the US. Part of the US response was the creation of a civilian agency, NASA, that would be responsible for developing a comprehensive program of space science. The next several years saw the development of small, special purpose satellites and the emergence of the first 'observatories'-the Orbiting Solar Observatory (OSO), the first of which flew in 1962 and the Orbiting Astronomical Observatory (OAO), the first of which was scheduled to fly in 1964 and eventually did in 1968.

Despite the grand plans, astronomical information from space platforms in 1960 was extremely limited. Observations had established that the sun was a copious source of ultraviolet and x-radiation. By extension it could be argued that stars would also be ultraviolet and x-ray emitters. The NRL group under Herbert Friedman had initiated a program of ultraviolet surveys from Aerobee rockets that yielded information on night-sky Lyman alpha emission. NRL and other groups had obtained UV photometry of many stars and Fred Whipple at the Smithsonian Astrophysical Observatory and Lyman Spitzer at Princeton were preparing instruments for rockets of a kind that would become the payload for the first OAO. There was yet another avenue to x-rays, based on cosmic rays phenomena. By 1960, radio astronomers had established that energetic electrons were common in supernova remnants and in the galaxy as a whole. Such electrons produced radio emission via the synchrotron process; indeed, efforts were underway to find a gamma ray component to the cosmic rays that would result from the interactions of these electrons. In one object, the Crab Nebula, the synchrotron emission from energetic electrons was observed in the optical range as a power law extension from the radio. It was easy enough, even if unrealistic, to extrapolate the spectrum into the x-ray regime.

3 THE DISCOVERY OF COSMIC X-RAY EMISSION

By 1960, Giacconi was at AS&E developing a program of x-ray observations with Air Force and NASA support. AS&E was an ideal place from which to launch a program of x-ray astronomy. The principal focus of the company was x-ray phenomena, albeit associated with nuclear weapons. Directors of the company included George Clark and Bruno Rossi of MIT who were active cosmic ray physicists. Clark was in the midst of organizing a program of gamma ray astronomy and, with William Kraushaar of MIT, had developed an instrument for the Orbiting Solar Observatory to measure multi-mev gamma rays that could be produced in cosmic ray interactions in the galaxy. With Japanese collaborators, Clark also was working on an air shower experiment to be installed on Mount Chacaltaya in Bolivia to record 10^{14} ev gamma rays that could be produced by the same electrons responsible for the emission observed by radio astronomers.

Rossi suggested to Giacconi that he consider developing a program of x-ray observations. The story is told that Rossi had originally made the suggestion to his colleagues at MIT who naturally were too busy pursuing their cosmic ray and gamma ray research to take up the idea. Two projects emerged from this suggestion; one was the search for x-ray sources from above the atmosphere and the other was the development of x-ray optics.

The prospects were not especially encouraging. Although the sun was a powerful source of x-rays, the nearest solar-type stars are a million times more distant and there was no hope of detecting them with the instruments available in 1960. Although other objects, such as cataclysmic variables or bright radio sources, might be intrinsically brighter than the sun they were also rarer and much more distant. Fortunately, Giacconi and his colleagues came upon a more local candidate source of x-rays-the moon. In 1960, the solar wind was just being investigated and its intensity was estimated as being so large that bremstrahlung from solar wind electrons impinging on the moon would produce measurable fluxes of x-rays in the vicinity of the earth. Accordingly an experiment was designed and flown on an Asp rocket from the Eglin Air Force Base in the US in June 1960 to search for such emission from the moon. However a rocket malfunction prevented any observations; mercifully perhaps, since it was soon found that the electron fluxes were substantially lower than the early estimates. Giacconi and his colleagues then redesigned the experiment to achieve much greater sensitivity based on the possibility of seeing fluorescence of lunar surface material produced by incident solar x-rays. The new instrument was configured for the much larger Aerobee rocket to be flown from the White Sands Missile Range. The first flight took place in October 1961 not long after I joined the company. This one also returned no data because of a failure of doors to open at altitude. Fortunately the payload was recovered allowing for refurbishment and reflight, which took place in June 1962. It was this flight that led to the first evidence for cosmic x-rays.

The signal was so strong that it was evident on the paper roll that recorded the data as it was being sent from the rocket. Even one of the technicians who was present said, "There it is", thinking that the signal came from our primary target, the moon. We scientists knew better; the signal from the moon could have been revealed only after summing all the data. It was only after various technical possibilities were eliminated and after looking closely at the data themselves that we convinced ourselves that the results represented cosmic radiation from sources well distant from the earth. The principal results from the rocket flight were that a large peak of x-rays was observed from the general direction of the galactic center, a smaller excess was observed from the direction of Cygnus and a general background of radiation was present. Based on results from two different detectors with different entrance windows, the radiation was consistent with x-rays of several kilovolts energy originating from beyond the solar system. The paper describing these results was submitted to the Physical Review Letters and appeared in the December 1 issue.[6]

In trying to understand these results we were extremely fortunate in having good neighbors. MIT, under the direction of George Clark, was just beginning the operation of an air shower experiment in Bolivia and their Japanese collaborators, Minoru Oda and Satio Hayakawa, were visiting the US. Given the cosmic orientation of everyone's research interests it was natural that we adopted a cosmic origin. In fact; the air shower experiment was showing a significant excess of energetic gamma rays from the vicinity of the galactic center (4σ) that coincided with the general direction of the major emission peak of x-rays that we had observed from the rocket. These gamma rays were believed to originate in the interaction of equally energetic comic ray electrons. These same electrons could lose most of their energy in x-rays by synchrotron emission in the galactic magnetic field. Fortuitously the total energy flux in x-rays was comparable to that in gamma rays.

Perhaps not surprisingly the 4σ peak in the Bolivian air shower experiment disappeared. It was to be more than a decade and the development of Cerenkov air shower telescopes before very high energy gamma rays were discovered from cosmic sources.

Giacconi's group at AS&E was not alone in developing space instruments to study the sky in x-rays. Herbert Friedman's group at NRL had expanded their rocket observations to the night sky following their pioneering work on x-rays from the sun; indeed, Friedman had noted that during one of his solar rocket observations, a scintillation counter designed to record x-rays greater than 20 keV recorded an excess flux that could be interpreted as a cosmic background. Subsequently Friedman had employed small, uncollimated geiger counters to survey the sky for x-ray emission as an auxillary measurement to his work on surveying the sky for ultraviolet emission. Also NRL staff members, Jim Kupperian and John Lindsey, continued their interest in x-ray observations after joining NASA shortly after its formation in 1958. Kupperian supported observational x-ray research at Lockheed under Philip Fisher; Lindsay was to play an important role in the development of imaging of the sun using grazing incidence optics.

Fisher's work at Lockheed was well along when AS&E had its successful flight in June, 1962; in fact, he had a rocket flight that September to survey the sky in x-rays using thin window proportional counters. However he was not able to detect any significant night sky emission.[7] Friedman was more successful. After learning about our positive results he accelerated the development of a payload using a large array of collimated proportional counters, comprising about 65 cm_2, to survey the sky. His flight, in April 1963, was very successful and produced three noteworthy results. First it was able to localize a strong source in the constellation Scorpius which was undoubtedly the source found during the AS&E flight; secondly, it found a weaker source in Taurus, in the direction of the Crab Nebula, thirdly; it also detected a significant background of x-ray emission.[8] At AS&E we had additional flights in the fall of 1962 when the galactic center region was below the night time horizon and in June 1963 to duplicate as best we could the conditions of the June 1962 flight. We used the same payload as the discovery flight. In light of the NRL flight, we were too conservative; our objective was more to confirm our original results, which we did, rather than extending our knowledge.[9]

Thus by the end of 1963, x-ray emission as a feature of the cosmos had become an established fact and the stage was set for the 'second generation' of x-ray observations. The first generation of experiments was planned in terms of the most general strategy for searching. At AS&E we made a guess at what one target-the moon-might look like in x-rays, but otherwise there was little in the way of guidance from traditional astronomy. By the end of 1963, however, enough was known to allow actually planning experiments. In particular, the intensity of the brightest sources was known, as was the intensity of the diffuse background radiation. These facts, meager as they were, allowed the experimenters to design instruments appropriate to conducting x-ray observations and to plan rocket experiments designed to develop specific information.

4 THE SECOND GENERATION

It is important to recall the state of the art of rocket technology in the 1960s, at least what was available to the scientific community. The Aerobee rockets flown in the discovery flights through 1963 were provided with no active guidance beyond the stability inherent in a spinning object. Since the rockets were fired essentially straight up, the zenith orientation was determined by the latitude of the rocket range, the season of the year and the local time. Observations by AS&E allowed the rocket to retain its full spin of about 2 revolutions per second. Since the rocket would acquire a small precession angle of about 10 degrees, instruments looking out the side of the rocket repeatedly scanned a band in the sky of that same width. The NRL group despun the rocket after it reached altitude. This had the effect of increasing the precession angle significantly and the scan circles could cover a large portion of the visible sky. It also looked at the earth much of the time thus reducing the available observing time for celestial targets. It was a crude but effective means of achieving a scan of a desired piece of the sky; especially for the broadly collimated detectors of the times. A major technical advance in rocketry took place in the early 1960s that made it possible to point in selected directions. NASA introduced gyroscopically controlled rocket payloads using star trackers to determine inertial orientation. With these it was possible either to point steadily at single targets or to scan selected regions at prescribed rates, although the accuracy was only a few degrees. Pointing at a single object increased observing time 10 to 100 times compared to scanning and made possible new kinds of observations.

Two notable experiments were subsequently conducted using controlled rockets that lead to unambiguous identifications of x-ray sources. In July 1964, the NRL group flew a rocket just at the time that the moon was passing in front of the Crab Nebula. This technique had been exploited by radio astronomers to establish the characteristics of radio sources. The motion of the moon in the sky, about a half arc minute per minute, matched well with the angular size of the nebula and the five minute duration of the rocket flight. Friedman was hoping to see a sudden decrease in the recorded x-ray intensity as the edge of the moon crossed the nebula, which would be the signature of a neutron star.

The discovery of x-ray sources had elicited a significant amount of theoretical interest in their nature, among which was the idea that they represented the thermal emission from the hot surface of a neutron star. The idea was especially attractive for a source associated with the Crab Nebula, since it had been speculated for about 30 years that neutron stars were formed during supernova explosion The Crab itself had been searched optically for such an object by Minkowski and Baade in the 1930s. Unfortunately the x-ray data from the rocket flight revealed a gradual diminution of the signal, reflecting emission from a region comparable in size to that seen in optical and radio emission.[10] It was still a landmark result since it indicated that the synchrotron emission from the Crab Nebula, known to be responsible for its optical and radio emission, extended three more decades into the x-ray regime. Ironically, three years later, the NRL group measured the x-ray pulsations from the 33 millisecond Crab Nebula pulsar after it was discovered in the optical and radio regimes. Thus Friedman was in the unenviable position of having first demonstrated that the x-ray emission from the Crab was not originating from a neutron star and then demonstrating that the neutron star in the Crab was indeed an x-ray source. Friedman was of course unlucky. The strong emission from the nebula masked the much weaker emission from the neutron star during his occultation experiment.

The second notable experiment was the precise measurement of the location and angular size of the bright source in Scorpius, now known as Sco X-1. Following the discovery that at least one supernova remnant was an x-ray source, it was speculated that others might be supernovae remnants as well. However there did not appear to be bright radio emission from the few x-ray sources that had good positions at that time; nor did any of the sources coincide with the more prominent supernova remnants, other than the Crab. At AS&E we addressed the question of the nature of Sco X-1 with a combination of spectral, angular size and positional measurements. With sounding rocket data we were able to establish that the kilovolt x-rays from Sco X-1 exhibited an exponential spectrum characteristic of bremsstrahlung, rather than a power law characteristic of synchrotron radiation. This pointed to Sco X-1 being something other than a supernova remnant; furthermore it allowed for a prediction of the minimum optical brightness based on the extrapolation from the x-ray into the optical domain. Then Minoru Oda, then still visiting at MIT, invented a device, the modulation collimator, that provided fine angular resolution within a broad field of view and would allow a measurement of both angular extent and position. Oda came to Giacconi with the suggestion that MIT and AS&E join forces to fly his collimator, which he and I agreed to. The first flight still used a spinning rocket. It was successful in the sense that we were able to demonstrate that the source intensity was modulated as if it were a point source, but the limit on the angular size was not adequate to restrict the nature of the object.[11] The next year we planned a reflight and I redesigned the experiment with the assistance of Paul Gorenstein at AS&E and Oda, Hale Bradt and Gordon Garmire at MIT. The most important innovation was the use of a controlled rocket that could point continuously in the direction of a source. The next innovation was the use of the length of the rocket rather than its width as the principal axis of the experiment. Traditionally experiments looked out the side of a rocket in order to maximize the area available for detectors. For this experiment, where angular resolution was the key experimental parameter, we needed length more than area. Other innovations included multiple planes of grids in the modulation collimator, a vernier technique to eliminate ambiguities commonly associated with modulation cameras, star cameras to determine attitude and a fiducial system to tie together the axes of the modulation collimator and the star cameras. The rocket flight, which took place in June 1966, was a complete success. Then began the work of reconstructing an 'image' and locating the source on the sky. The data, even as it was recorded during the flight, indicated that the source had no appreciable angular extent. This told us that the source was most likely star-like in appearance, as opposed to being nebular.[12]

We recognized that the importance of the experiment lay in obtaining an optical counterpart rather than just reporting an x-ray source location. A simple extrapolation from the x-rays, indicated that the object could reveal itself as faint, blue object, perhaps of 13th magnitude. Giacconi and I contacted Margaret Burbidge to interest her searching for the object once we obtained a location. Burbidge pointed out that the thrust of her work was spectroscopy of known objects rather than searching, as we needed. She sent us to Allan Sandage, who agreed to work with us. Meanwhile Oda who returned to Japan arranged with his colleagues there to search for the counterpart. Once the location was obtained, the candidate identification by both Sandage and the Japanese followed almost immediately. The object had the appearance as an old nova. The model that emerged, a binary system with a collapsed star, powered by accretion, has remained intact since.[13]

The rocket payload never flew again; there simply was no candidate x-ray source bright enough or properly situated in the sky to warrant one. Some years later, the experiment, reincarnated in a satellite version, did fly as part of the HEAO-1 mission.

As a follow-on, the AS&E group decided on a simpler, more traditional experiment. We used large area proportional counters of the kind being developed for the Uhuru satellite and narrow mechanical collimators to scan regions with sources to obtain positions. The decision was made to concentrate on the Cygnus region which was known to contain bright x-ray sources. Cygnus was far to the north and allowed for good observing for North American astronomers. Also we felt that the region containing these sources would be less subject to optical obscuration than the region near the galactic center that contained other strong sources. Finally we wanted to confirm a report from the NRL group that the bright radio galaxy, Cygnus A, was seen to be an x-ray source. Instead of pointing steadily in single directions, this experiment was performed with a controlled rocket and slewed slowly between fixed points.

This experiment, which took place in the fall of 1967, was also completely successful and resulted in good positions for three sources in Cygnus, now known as Cyg X-1, Cyg X-2 and Cyg X-3.[14] Allan Sandage was able to find an optical candidate for Cyg X-2, which has held up in spite of the fact that the error uncertainty was about a quarter of a degree. He was not able to find candidates for Cyg X-1 or Cyg X-3. We did not see emission from Cygnus A. Ironically, the optical counterpart for Cyg X-1 was of 9th magnitude and the brightest object in the error box. Somehow it never occurred to us that once we invoked accretion from a normal star to a collapsed star as the model, that the optical emission from the normal star could dominate the optical emission.[15]

The Cygnus flight was the last one that the AS&E group used with traditional detector-collimator instruments. Subsequent flights concentrated on the development of grazing incidence optics for x-ray observations. Also the AS&E group was becoming more and more involved with the development of a small satellite for surveying the sky in x-rays for which it had proposed to NASA in 1964. In response, NASA began the Small Astronomy Satellite (SAS) program, of which the AS&E survey experiment was to be the first. It would eventually fly in December 1969 from the Italian launching site off the coast of Kenya and was given the name "Uhuru", the Swahili word for freedom.

5 THE DEVELOPMENT OF SOLAR IMAGING

The work on x-ray optics had been proceeding at AS&E more or less independently from the work on cosmic x-rays. Giacconi's initial thoughts on x-ray optics, as described in a paper he wrote with Bruno Rossi[16] considered only concentrators, paraboloidal reflecting elements that would produce a spot of x-rays and provide for improved signal to noise in detectors, compared to mechanically collimated detectors and describes how such a concentrator would be used in a satellite. The following excerpt is revealing of the state of thinking at the time. "Among the obvious applications are a detailed analysis of the distribution of x-ray sources on the solar disk and the solar corona, and a search for weak x-ray sources, for example in the Crab Nebula." In the article was also an estimate of what sensitivity could be achieved by a telescope in orbit. The flux limit they quoted, 10^{-5} photons/cm^2-sec, was precisely the value achieved by the Einstein Observatory when it flew in 1978.

The article does reference the work of Wolter that described combinations of surfaces that would produce genuine images. These include the paraboloid-hyperboloid combination, now known as Wolter type I, that has been used almost exclusively for x-ray astronomy. By 1961, with NASA sponsorship, AS&E had developed small grazing incidence reflectors from glass to demonstrate the principal of concentrating and focusing x-rays. From this point the technology proceeded to molded mirrors of epoxy of the Wolter type I, to electroformed mirrors of nickel, to kanigen coated aluminum or beryllium mirrors. Programmatically, the work proceeded from sounding rockets, to an experiment on the OSO-D spacecraft, to one on Skylab as part of the Solar Apollo Telescope Mount.

As I recall, the work on x-ray telescopes proceeded at a more leisurely pace than did the rocket program of cosmic x-rays. There was simply not the compelling science nor the intense competition that our discovery of significant x-ray emission from the night sky had inspired. An important element in the AS&E program was the decision, probably in 1962, to join with John Lindsay at NASA's Goddard Space Flight Center in a program of rocket observations of the sun. Lindsay continued an interest in imaging the sun that had begun at NRL. Friedman had carried out several notable rocket experiments that delineated solar x-ray features. In 1958 he mounted an expedition to the South Pacific and fired rockets to view the solar x-ray emission as the sun was being eclipsed by the moon. The moon acted as a shutter and the solar x-ray flux diminished as x-ray emitting features slid behind the disc of the moon. Significantly there was sizable emission even when the sun was completely covered, indicating that the x-ray emission was, in part, originating high in the solar atmosphere. In another rocket experiment in 1960, Friedman's group took a pinhole picture of the sun on film. It was the first time that an image of solar x-ray emission of any sort was obtained. But NRL never pursued x-ray optics, nor did Lindsay when he first went to NASA. Thus it was natural that Lindsay join with AS&E to employ the newly developed x-ray optics in a program of sounding rocket imaging of the sun. The first images were obtained in 1963 and others, with improved optics, in 1965.[17] Unfortunately, John Lindsay died suddenly in 1965, and the collaboration between AS&E and

GSFC ended, both groups going their own way with x-ray optics development and flight.

These rocket flights, as limited as they were, provided the impetus for ambitious satellite instruments. In contrast to the other areas of the space sciences, the solar community was already heavily into space research. Solar observations were among the first to be conducted from research rockets. The sun was bright and amenable to study by small instruments and, by present standards, unsophisticated detectors. Furthermore the sun was of recognized importance scientifically, so it was natural that NASA commit to developing a significant program of solar observing. Among the first spacecraft series that NASA developed were the Orbiting Solar Observatories. AS&E was successful in obtaining inclusion in the fourth OSO a set of x-ray telescopes of the kind flown in its early sounding rockets for systematically imaging the sun. The instrument flew successfully in 1967 and succeeded in returning about 1000 images of the sun, which doesn't sound like a lot. But considering our limited ability to handle data it was almost too many. I can recall Frank Paolini, the principal scientist involved in the program, creating images, pixel by pixel, using color pencils.

In the late 1960s developments ensued of the utmost significance to x-ray astronomy. One was the decision by NASA to pursue an advanced solar observatory in space. This was accompanied by a corresponding increase in the funding available for advanced technology. At AS&E, the impetus of this new mission potential lead the group to abandon electroforming as a method for fabricating x-ray telescopes. Instead they went to Diffraction Limited, an optics manufacturer in the Boston area, with a design for an x-ray telescope and ordered one to be built using more or less conventional materials and conventional figuring and polishing techniques, although for a grazing incidence surface rather than a normal incidence surface.

The resulting telescope was first flown in a sounding rocket in 1968 and yielded spectacular images compared to what had been achieved earlier.[18] The resolution was a few arc seconds, producing pictures similar to those now being routinely generated by x-ray telescopes on the Yohkoh and SOHO spacecraft. Subsequently, Ed Kellog, Leon Van Spoeybroeck and I devised and flew a sounding rocket experiment using the same mirror to image Sco X-1 successfully. Programmatically, NASA chose its SKYLAB mission to host an advanced solar observatory which carried two x-ray telescope experiments, one from GSFC and one from AS&E when in flew in 1973.

6 SUMMARY

Whenever I think back to the period of time 1960 to 1970, I am absolutely astonished at how much happened. In 1960, x-ray astronomy was recognized as something that could be done from space, but no direct observations existed and the means to conduct observations was very limited. By 1970 it was a full-fledged scientific discipline with large numbers of practitioners in many countries. Not only were there rocket and balloon measurements being carried out, but one satellite dedicated to x-ray observations was in orbit; others, including the ambitious HEAO program were well along in their development phases. To their credit, NASA and other organizations were willing to take enormous risks, committing substantial resources to space programs before the technology was fully developed. The payoff for the scientific community was tremendous. Instrument concepts developed by Giacconi and his colleagues in the 60s lead directly to the Uhuru and Einstein missions of the 70s. It's hard to think of two other space missions that have had such a large scientific impact.

This brief review does not do justice to the many groups that entered the field during the initial discovery phase. As an example of the breath of interest, Robert Gould, writing in 1967, listed 11 different institutions that had made spectral measurements of cosmic x-ray sources. The British had a strong program centered at the University College, London under R.L.F. Boyd and at Leicester University under Kenneth Pounds. Even in 1959 their National Committee for Space Research had noted that the equipment being developed for solar x-ray observations could be developed for stellar sources. Even before the discovery of x-ray sources in 1962, Boyd, Len Culhane and A.P. Willmore were designing paraboloid concentrators for satellite applications. Skylark rockets, flown from Woomera, Australia, provided a firm basis for the British program and was heavily oriented to solar observations, including pinhole imagery of the sun (beginning in 1964) and grazing incidence spectrometers. The first x-ray astronomy experiments were flown in 1967 by the Leicester group. The British developed the Ariel series of satellites as their national platform for space science. Ariel 4, flown in 1974, was the first to carry an x-ray astronomy payload.

In Japan, both Minoru Oda and Satio Hayakawa began x-ray astronomy programs. Hayakawa established a rocket program at his home institute in Nagoya and flew rockets concentrating on soft x-rays. Oda was responsible for establishing a national space satellite program at ISAS, associated with the University of Tokyo. The first satellite in the series flew in 1971 and the first one dedicated to x-ray astronomy, Hakucho, flew in 1975.

Aside from NASA, Department of Energy laboratories conducted notable x-ray astronomy experiments during the 1960s. At the Livermore laboratory, Hans Mark directed the development of instruments for x-ray observations on sub-orbital flights of DoE rockets. A notable result from this program was the observation of the first transient x-ray source, Cen X-2, that was discovered earlier by the Australian group headed by Kenneth McCracken. The Los Alamos National Laboratory made very effective use of the instruments on the Vela spacecraft that were part of the US program of nuclear test ban verification. In 1969, within a few weeks of the launch of the Vela 5 spacecraft they discovered a second transient source, also in the constellation Centaurus, and recorded the first gamma-ray bursts.

The pace of discovery had not diminished.

References

1. Wallace Tucker and Riccardo Giacconi, *The X-ray Universe* (Harvard University Press, 1985).
2. Herbert Friedman, *The Astronomer's Universe* (W.W. Norton and Company, 1990).
3. Richard F. Hirsh, *Glimpsing an Invisible Universe* (Cambridge University Press, Cambridge, 1983).
4. David H. DeVorkin, *Science With a Vengeance* (Springer-Verlag, New York, 1992).
5. Harrie Massey and M.O. Robins, *History of British Space Science* (Cambridge University Press, Cambridge, 1986).
6. R. Giacconi, H. Gursky, F. Paolini and B. Rossi, Phys. Rev. Lett, 9, 439 (1962).
7. P.C. Fisher and A.J. Meyerott, Astrophys. J., 139, 123 (1964).
8. S. Bowyer, E.T. Byram, T.A. Chubb, and H. Friedman, Nature, 201, 1307 (1964).
9. H. Gursky, R. Giacconi, F. Paolini and B. Rossi, Phys. Rev. Lett, 11, 530 (1963).
10. S. Bowyer, E.T. Byram, T.A. Chubb, and H. Friedman, Science, 146, 912 (1964).
11. M. Oda et al, Nature, 207, 554 (1965).
12. H. Gursky et al, Astrophy. J., **146**, 310 (1966).
13. A. Sandage et al, Astrophy. J., **146**, 314 (1966).
14. R. Giacconi, P. Gorenstein, H. Gursky and J. Waters, Astrophy. J., **148**, L119, 1967.
15. R. Giacconi et al, Astrophy. J., **148**, L129, (1967).

16. Riccardo Giacconi and Bruno Rossi, J. Geophys. Res., **65**, No. 2, 773 (1960).
17. R. Giacconi, et al, Astrophy. J., **142**, 1275 (1965).
18. G. Vaiana et al, Science, 161, **564** (1968).

X-RAY ASTRONOMY FROM UHURU TO EINSTEIN

W. H. Tucker and H. Tananbaum
Harvard-Smithsonian Center for Astrophysics

60 Garden St., Cambridge, MA 02138 USA
e-mail: wtucker@cfa.harvard.edu; ht@cfa.harvard.edu

1 Introduction

1.1 A Sensible World, Not a World On Paper

Everyone who participates in the endeavor we call science has in common at least one goal: to understand the universe a little better than we did before.

Those who worked with Riccardo Giacconi during the sixties and seventies and eighties were accorded a rare privilege. They participated in the birth and coming of age of an entirely new way of looking at the universe, x-ray astronomy, and they were involved first hand in a paradigm shift caused by the discovery of new cosmic phenomena. The contributions to this volume show the richness of the field, covering stellar coronae, neutron stars, black holes, supernova remnants, active galactic nuclei and quasars, galactic coronae, clusters of galaxies and the x-ray background radiation.

In his contribution, Herb Gursky discusses the birth of the field, when the group led by Riccardo discovered objects, presumably stars, emitting nearly all of their energy as x-radiation at luminosities ultimately shown to be thousands of times greater than the Sun. It was also apparent in the very first, discovery rocket flight that there is no dark night sky for x-ray astronomers. A strong, diffuse, isotropic background exists, and its photons bring information to us about the universe far away and long ago.

We will cover the coming of age: the exhilarating years of Uhuru and Einstein when the binary nature of x-ray stars was determined, when the first hard evidence for black holes was obtained, when a large fraction of the baryonic matter in the universe was discovered to be in the form of hot gas in clusters of galaxies, and when x-ray astronomy became part of the mainstream, i.e., more a part of the "standard" paradigm. We will close with some reflections about some of the changes brought on in part by the success of x-ray astronomy.

We both arrived at the converted milk truck garage on Carleton Street in Cambridge, MA which were the headquarters of American Science & Engineering in the late 60's. We were less than thirty years old, and it was a time and a place when "don't trust anyone over thirty" was often heard in the news, at the coffee houses, and on the streets.

We were diffident about the transition we were making to the corporate world from academia, the only world we had known, presumably from a milieu dedicated to the pursuit of ideas, to one dedicated to the pursuit of profits.

What a pleasant surprise it was to find that we had entered the force field of a visionary natural philosopher disguised as an executive vice-president!

Sure, most of the time was spent working on or discussing things such as collimators and count rates, pulse shape discrimination and proportional counters, progress reports and proposals, rockets and radiative processes, satellites and schedules - all the thousands of things that have to be checked and fixed in the course of space science projects.

But throughout it all, underlying it all, was a philosophy that inspired the group, prepared us to meet the challenges we faced, and enabled us to grow scientifically. This philosophy had three main components: scientific, technical and managerial, all of which involved a creative tension between skepticism and trust.

The philosophy of the science was motivated by Galileo's famous dictum, "Our subject is the sensible world, not a world on paper." Be skeptical of the world of paper, i.e., the existing paradigm. Trust that nature will be kind, that if you build it, the photons will come."

The approach to the technical mountains that had to be climbed was to be skeptical of any untried and untested technology. Take nothing for granted: plan for soft failure, i.e., build in redundancy as well as reduced modes of operation able to return much, even if not all, of the desired data. Also, plan for hard failure, i.e., design a data handling system to get as much data as quickly as possible. Trust in simplicity and elegance in design, and trust that if it can be done, we can do it.

This dictum flowed into the managerial philosophy. Be skeptical of talk about what a person is going to do, and appreciate the work which they did. Demand a lot from the people who work for you, but don't take them for granted. Give them freedom to be creative and grow. This tenet was epitomized by what became known as "the Uhuru spirit." It was because of this spirit that the philosophy transmitted by Riccardo was not engraved on a stone tablet (though sometimes he encouraged this belief to get the job done!), but actually embodied an evolving set of principles with no particular claim to authorship.

Rather than give a detailed survey of the Uhuru and Einstein years, which has been done elsewhere[1,2,3] we discuss how these principles shaped the course of these projects and with them, the recent history of astronomy.

2 The Origin of The Uhuru and Einstein Projects

Although the Uhuru satellite flew eight years before the Einstein observatory, plans for a focusing x-ray telescope observatory date from the birth of the field of x-ray astronomy. In late 1959, Giacconi met Bruno Rossi, who suggested that an attempt to detect x-rays beyond the solar system might be worthwhile. Rossi had no theoretical predictions on which to base this belief, nor were the pioneering solar experiments of Herbert Friedman and his colleagues at the Naval Research Laboratory (NRL) in Washington, D.C. encouraging. On the contrary, if all stars were as luminous as the Sun in x-rays, then detectors millions of times better than those used by Friedman's group would be needed. Flare stars, magnetic stars or supernova remnants might help by a factor of a thousand or so, but the prospects for success still looked bleak, implying that larger increase in capability would still be needed. So said the world of paper.

But Giacconi, who shared Rossi's "deep-seated faith in the boundless resourcefulness of nature, which so often leaves the most daring imagination of man far behind," was undaunted. In 1960, he and Rossi published a description of a grazing-incidence telescope with nested mirrors[4].

Although a modest experimental research program demonstrated the feasibility of the concept with several small prototypes by 1963, only a year after the discovery of Sco X-1, no funds were available to build a telescope with optically polished surfaces, nor to explore the preparation of surfaces beyond the state of the art.

Undeterred, Giacconi pressed forward. In 1963, with only Sco X-1 and the Crab Nebula as known sources, he presented to Nancy Roman, then Chief of the Astronomy Branch of NASA, a long-range program for x-ray astronomy. In it he described a program of sounding rockets, an instrument on the Orbiting Solar Observatory series of spacecraft, an Explorer satellite dedicated to x-ray observations, a small focusing x-ray telescope, and a large focusing x-ray telescope. The telescopes were judged premature, but the idea of an X-ray Explorer satellite intrigued Roman. She asked for more details and the AS&E group responded in April 1964 with a proposal entitled "An X-ray Explorer to Survey Galactic and Extragalactic Sources."

The proposal was a classic, combining a bold vision and innovative technical approaches. The proposed satellite mission would make a detailed survey of the sky for x-ray sources. In the proposal, the AS&E group broke ranks with the then prevalent thinking favoring satellites that housed experiments from various groups with diverse goals. Instead, they asked for a satellite dedicated to one purpose, to be built and operated by one group. They argued that the scientific return would be optimized by a satellite with an extremely slow spin rate, an equatorial orbit, and quick-look access to a significant fraction of the data on a daily basis. The slow spin rate would keep the sources in the field of view for a relatively long time, thus facilitating the determination of accurate positions for known sources and the discovery of weak sources. The equatorial orbit would permit the launching of a larger payload, and the insertion into an orbit with low charged particle background. The quick-look access to $\sim 20\%$ of the data would permit assessment of the scientific content on a daily basis and allow for modification of the observing program in the light of important discoveries. They proposed a launch in December 1965. In essence, on the basis of the discovery of few sources - recall that at most ten cosmic x-ray sources were known at this time [5] - they were saying, "Here is an exciting new field with great potential. Let us use proven technology to build the best satellite we can to explore it." Trust in the richness of nature.

To their credit, a NASA review committee of astronomers, chaired by Roman, endorsed the concept of an X-ray Explorer. Endorsement of a concept by a committee is one thing. Implementation of the project is quite another. NASA decided that AS&E could be entrusted with the construction of the scientific payload, but not with the construction of the spacecraft. Also, there was the question as to which of NASA's field centers would agree to manage the project. To keep a project alive through the vicissitudes of budget struggles within NASA and in Congress, it is essential for the staff of one of the field centers to be committed to the project. The response of NASA center managers to the X-ray Explorer project was lukewarm. After all, it would be a small program of interest to a small group of scientists outside NASA. Fortunately, the X-ray Explorer had a strong supporter in high places, John Naugle, then Associate Administrator for Science and Applications at NASA.

A small, innovative, project such as the X-ray Explorer stood little chance in getting through the NASA bureaucracy. However, a larger program based on what was clearly a good idea did. Two and a half years after the submission of the X-ray Explorer proposal, and a year after the proposed launch date (space scientists are always over optimistic about launch dates; they have to be in order to keep their spirits up), NASA announced that it would begin a program of Small Astronomy Satellites, of which the X-ray Explorer would be the first. Goddard Space Flight Center would manage the program under the direction of Marjorie Townsend. The Applied Physics Laboratory of Johns Hopkins University would design and build the spacecraft. In December 1966, the AS&E group submitted a modification of their origin proposal. It described in much greater detail the scientific payload and the planning and execution of the X-ray Explorer Mission. The launch date was set for late 1970.

Meanwhile, Giacconi had not dropped the idea of an observatory with a focusing x-ray telescope. A collaboration between Giacconi and his colleagues at AS&E and a group led by John Lindsay at Goddard produced an x-ray telescope about the size of a mailing tube that made telescopic x-ray images of the Sun for the first time in October of 1963 and again in March of 1965. Plans to put a much larger telescope on a manned spacecraft were deferred when Lindsay died of a heart attack in September of 1965.

Despite the successes of the solar x-ray telescope flights, NASA officials and scientific opinion-makers saw no urgent need for an x-ray telescope to study extrasolar x-ray sources. They preferred to wait and see if all-sky surveys with conventional instruments would show whether an x-ray telescope were really necessary. What if x-ray astronomy turned out to be the study of a few peculiar stars?

Armed with the success of the solar x-ray telescopes and with his belief that nature would provide the photons from sufficiently abundant and scientifically interesting sources, Giacconi argued the case of an x-ray telescope to whomever would listen at whatever forum, from personal conversations, to lectures, to presentations before scientific advisory committees. One of the most important of these was a meeting at Woods Hole, Massachusetts in the summer of 1965. The X-ray and Gamma-Ray Astronomy Panel of the Space Science Board, chaired by Herbert Friedman, had convened to study plans for future x-ray and gamma-ray experiments.

Frank McDonald of Goddard Space Flight Center was there. He was championing the idea of a cosmic ray, x- ray and gamma-ray program that would use a series of identical, large, and relatively inexpensive spacecraft built from hardware left over from the Apollo lunar exploration program. Giacconi pressed for inclusion of an x-ray telescope as a natural and essential follow up to a large survey type experiment proposed by Friedman. The panel agreed, as did NASA's Astronomy Missions Board two years later, and feasibility studies of the High Energy Astronomy Observatory concept began in 1968 at the Marshall Space Flight Center.

3 UHURU

Meanwhile, the field of x-ray astronomy was progressing rapidly. Groups from AS&E, Lawrence Livermore Laboratory, Lockheed, NRL, MIT and the University of Leicester had discovered more than thirty sources. About half a dozen were associated with supernova remnants, including the Crab Nebula pulsar, one or maybe two or three with extragalactic sources, and the remainder with some bizarre, x-ray emitting stars whose exact nature was a mystery. Technical advances were also occurring apace. The slow scan technique [6] and sophisticated collimators [7,8,9] had been developed, along with very thin beryllium window proportional counters (led by John Waters, at AS&E) and improved methods of rejecting background noise [10].

Together the scientists and engineers at AS&E examined new developments alert to how they could improve the design of the X-ray Explorer while staying on budget and schedule. The unusual nature of AS&E in the late '60s helped this process. The research staff could focus on the job at hand without the distractions of teaching, faculty meetings, and other academic duties. In fact, they had to focus on it, because there was no tenure system. A lack of competence or dedication could be quickly identified and even more quickly remedied. This is not to say that the scientists and engineers were afraid to express an opinion or to take risks in their research. On the contrary, the evolving management philosophy fostered creative exploration of new ideas and approaches, and the rapid growth of AS&E in those days provided an expanding pool of scientific and engineering talent that could be tapped to make a more capable and more reliable scientific payload.

How would a particular new technology increase scientific returns? How would the added complication increase the risk? Questions such as these were hotly debated as the design and construction proceeded. In general, a new technology was adopted only if it could be shown with confidence that it would not affect reliability and only if it were consistent with the governing principle of elegant simplicity of design. For example, the idea was advanced to make the proportional counter bodies from beryllium to match the thermal coefficient of the beryllium windows and thereby minimize stresses at the seals, which could result in disastrous leaks. Since the satellite launch was planned for the humid ocean conditions, off the coast of Kenya, the AS&E team decided to evaluate experimentally the effects of extended exposure of the counters to high temperature (90-100°F) and high humidity. After two weeks in such an environment, simulated in a laboratory chamber, the counter bodies developed large leaks (at the locations of impurities in the beryllium walls and windows). The team responded by deciding to flow dry air (or nitrogen) around the counters for the entire period at the launch site, requiring extra efforts on the part of the spacecraft contractor and NASA, but ultimately solving the problem of how to get an all-beryllium counter through the launch environment and into orbit.

The environment at AS&E, with the constant interactions between engineers and scientists, enabled the scientists to direct the design toward a more user-friendly satellite. For example, since the nominal lifetime of the satellite was only six months, the scientists wanted to be able to analyze the data quickly so they could modify the observational program while the satellite was still operational. This was a departure from normal satellite operations, and required major innovations, under the direction of Ed Kellogg and Harvey Tananbaum, in the development before launch of an automated software system for checking the performance of the spacecraft, determining the pointing direction, and facilitating the data reduction. The staff at Goddard agreed to cooperate fully and arranged to have $\sim 20\%$ of the data transmitted from Goddard to AS&E daily over telephone lines. The efficient data-handling system would prove its worth just nine days after launch when the quick-look data from Cygnus X-1 began to come in and the scientists began to puzzle over new information which set them on the path to solving the mystery of the energy source powering the x-ray stars.

When completed, the X-ray Explorer weighed 64 kilograms. The detection system consisted of two sets of 6 beryllium-window, beryllium-body, argon and carbon-dioxide proportional counters with mechanical collimators in front. One collimator had a field of view (FWHM) of 0.5 degrees by 5 degrees (to optimize source locations) while the other was 5 degrees by 5 degrees (to maximize sensitivity to faint sources). The effective area was about 800 square centimeters for each of the two systems. The spin rate of the satellite was one revolution every 12 minutes. A magnetic torquing system could be used to adjust the spin rate. The data were recorded by accumulating counts for a fixed time. The shortest accumulation time was just below one-tenth of a second. The data-recording system was designed and approved before the discovery of the Crab Nebula pulsar, with its 33 millisecond period.

Following the Crab pulsar discovery, the ensuing negotiations between AS&E and NASA over the data-recording system serve as a reminder that "you can't win them all." AS&E requested additional funds (about $250,000) to modify the system to study the Crab pulsar and other similar objects which might exist. The request was denied on the grounds that the funds were needed for further safeguards against temperature variations of the electronics in orbit, a precautionary measure which proved unnecessary during the actual mission. The lack of fast time response made it impossible to study the Crab Nebula pulsar, and greatly complicated the analysis of the time variable signals from Cygnus X-1. Fortunately, this was not a fatal flaw in that it did not preclude the major breakthrough of the mission, the discovery of binary x-ray pulsars.

On December 12, 1970, after a delay of a few days to replace the spacecraft battery, the X-ray Explorer was launched into a 550-km-high equatorial orbit by a Scout rocket from the San Marco Platform, an old oil-drilling rig anchored three miles offshore from a small Kenyan village. The oil rig had been modified for rocket launches by the Italian Space Agency. December 12 is Kenya's Independence Day so, in appreciation of the hospitality and cooperation of

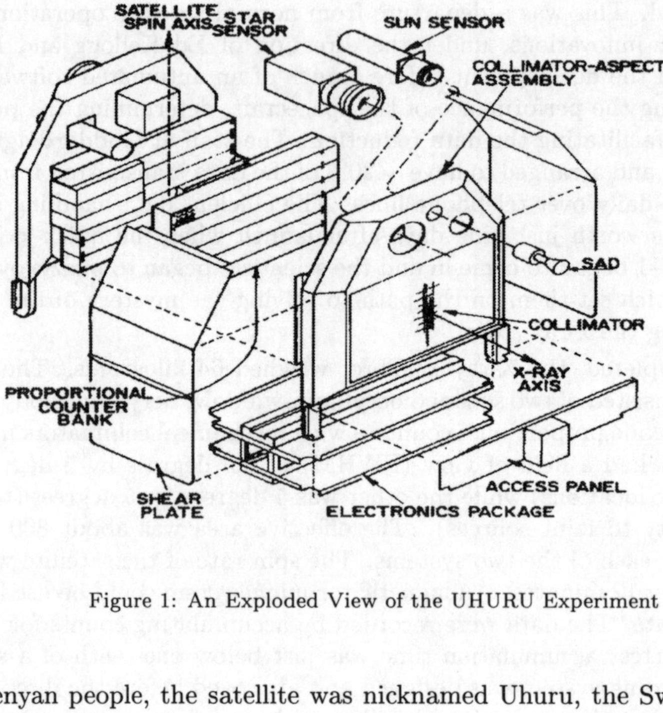

Figure 1: An Exploded View of the UHURU Experiment

the Kenyan people, the satellite was nicknamed Uhuru, the Swahili word for "freedom."

With Uhuru, it was possible for the first time to monitor x-ray stars for extended periods on time scales determined by the design of the satellite. Successive passes over a source were separated by 12 minutes based on the rotational period of the satellite (or 6 minutes from one band of detectors to the other). A source would be within the full-width field of view of the half-degree and 5 degree collimators for 2 and 20 seconds, respectively. Variability on timescales down to the 0.1 sec limit set by the data recording system could be studied by comparing the observed shape of the count rate as the source passed into and out of the field of view with the triangular response expected from a steady source.

A primary objective of the Uhuru mission was to resolve the issue of the variability of the x-ray stars. A prime target for starting this campaign was Cygnus X-1. It was one the brightest and most perplexing of a class of perplexing objects. Observations by several different rocket and balloon groups had yielded different fluxes and different spectra. At one time it had been argued (and many at AS&E agreed) that the differences were due to systematic observational errors and not variability. The x-ray spectrum, which was similar to the extended Crab Nebula supernova remnant, led many to believe that it was an unusual type of supernova remnant and not variable. Others thought that a better analogy was the pulsar in the Crab Nebula, in which case Cygnus X-1 would be variable. The absence of strong nonthermal radio emission confounded all model builders who were influenced primarily by the example of the Crab Nebula.

One of us (HT) had written his Ph.D. thesis, in large part, on balloon observations of a factor of two change in intensity of Cygnus X-1 over a timescale of $\sim$ 9 months, and argued forcefully about the importance of early observations of this object. The quest for scientific truth and the eagerness to let nature tell her story inspired by Giacconi, allowed for a wide latitude of inputs from all team members for possible targets to be scanned during the first days of the Uhuru mission. (Recall the philosophy of coping with possible hard failure meant getting the most important data as quickly as possible.)

The first observations of Cygnus X-1, in late December 1970 and early January 1971 immediately confirmed that Cygnus X-1 was variable on these longer time scales. Subsequent observations and analyses of the original Uhuru observations established that the intensity was varying much more rapidly (in less than 1 second) and possibly periodically. These results led the AS&E team to further exploit the capabilities of Uhuru by slowing down the spin rate to lengthen the exposure during individual scans across a source and thereby further investigate Cyg X-1 and to search for other variable sources. As described by Ethan Schreier and Harvey Tananbaum in their contribution, this technqiue led to the discovery of the periodicity in Cen X-3 and subsequently Her X-1. By the end of 1972, detailed studies of Cyg X-1, Cen X-3, and Her X-1 convinced astronomers that the riddle of x-ray stars had been solved. They are accreting neutron stars or black holes in close binary systems. It is important to emphasize that, besides the Uhuru observations, rocket flights of X-ray payloads with improved instrumentation for time resolution and source location, radio and optical observations, and theoretical work were all instrumental in arriving at this conclusion [11,12].

Giacconi's philosophy of how to do science played an important role throughout this turbulent, exhilarating period. This was exemplified by the "Uhuru meetings," of the AS&E scientists who were working on the Uhuru project. The meetings were held every Friday afternoon in Ed Kellogg's office. Part brainstorming session, part encounter group, and part support group, these meetings were the forums for discussion of the observation plan for the next week, the analysis and interpretation of the data, and any other Uhuru-related problem which anyone wanted to bring up.

The Uhuru group intensely debated the meaning of the short-time scale variability in Cygnus X-1, whether or not it could be a periodic signal (faster than the 0.1 sec sampling time, for example) and whether or not the data provided sufficient evidence to even mention the possibility of a black hole in the papers submitted to the Astrophysical Journal and the talks presented at various scientific meetings and colloquia. Another strenuous discussion ensued over the possibility of slowing the spin rate of the Uhuru satellite to much below the half-degree per second nominal value. There was the possibility that the spin rate would become so low that the satellite would begin to nutate or wobble excessively (like a top slowing down and eventually toppling over). On the other hand, there was so much which might be learned by slowing the spin by an order of magnitude and dwelling on the various bright Galactic sources for 100-200 seconds per pass rather than the nominal 20 seconds for the 5° collimator. As history notes, the team decided to see what nature might have to show, so the spin rate was slowed and the 4.8 second pulsation period in Cen X-3 was discovered in the next morning's quick-look data.

At the Uhuru meetings, personalities clashed, but almost always the clash was of an intellectual nature, as ideas, however wild, were floated with abandon and shot down remorselessly. Shocked newcomers and visitors could be forgiven for thinking that they were witnessing a free-for-all. Instead they were seeing the "Uhuru spirit" in action. Giacconi summarized best in an interview [13] years later:

"With the Uhuru spirit you absolutely let the other guy have it," he recalled. "But it was done in a spirit of searching for the truth, and of trust and love and help. People tended to see us as a tightly controlled monolith, but within the group there was always a lot of freedom. And most of all, it was a rational world. The validity of your ideas was what counted, not personalities or company politics."

Uhuru was a watershed in the history of astronomy. The discovery of radio pulsars in general and the Crab Nebula pulsar in particular had already established the existence of neutron stars. But, except for the Crab, which was clearly a short-lived special case, the energy generated by pulsars was relatively insignificant. The identification of x-ray sources with accreting neutron stars and black hole candidates demonstrated three important aspects of our universe: first, that our galaxy contains thousands of stellar objects whose emission is mostly in the x-ray band with luminosities on the order of a thousand solar luminosities or more, which can vary on a time scale of seconds or less; secondly, that accretion is an important process and source of power in the universe; and finally, that black holes very probably exist.

This last conclusion is of course what has captured the popular imagination, and the concept of black holes has become part of the popular literature as a metaphor for an irretrievable loss of material, time, money, information, etc. As anyone who has taught courses or given popular lectures on astronomy knows, there are three questions asked most frequently: How did the universe begin? Is there other life in the universe? What is it like to fall into a black hole? The concept of black holes did not start with Uhuru, but after Uhuru and the work of all those who made the case for Cygnus X-1 being a black hole, it gained tremendous legitimacy.

The role of accretion is of more consequence to practicing astrophysicists. The accretion process is invoked to explain a wide range of phenomena, from molecular jets to cataclysmic variables to quasars. After the Uhuru observations, and the theoretical work of Rashid Sunyaev and Martin Rees and their collaborators on accretion disks in x-ray binaries, it became possible to accept that the same process could explain active galactic nuclei and quasars. Ironically as x-ray binaries have become only one of many examples of the accretion process, their study has gained in significance. Because of their brightness and rapid variability, they are the best laboratories in which to test detailed theories of the accretion process.

The accomplishments of Uhuru were not confined to the densest states of matter in the universe. The discovery of x-ray emission from hot gas in clusters of galaxies also established x-ray observations as the primary means of studying perhaps the most abundant forms of baryonic matter in the universe. It would take the focusing power of the Einstein Observatory to fully explore the scientific implications of this discovery and to show that x-ray observations are vital for understanding dark matter, which is likely the most prevalent form of matter in the universe.

4 THE EINSTEIN X-RAY OBSERVATORY

Another, less appreciated, but very significant accomplishment of the Uhuru mission was that its successes were crucial to the preservation of the HEAO program. An economic recession prompted the Nixon Administration to put a ceiling on executive agency spending which led to a cut in NASA's budget at a time when it was staggering under the burden of cost overruns for the Mars Viking program. NASA had to find extra money to fund the Viking program to keep it on target for a Mars landing in the bicentennial year of 1976. The only program with that kind of money was HEAO. Thus, on January 2, 1973, HEAO was cancelled without warning.

Fortunately, HEAO had an able advocate in Jesse Mitchell, head of NASA's Physics and Astronomy Division, and a savvy veteran of NASA politics. He acted quickly to have the cancellation changed to an 18-month postponement while the NASA program managers and scientists scrambled to prepare a detailed, workable, alternative program costing half as much. Key to his case were images of the Sun made by the AS&E x-ray telescope on Skylab, which showed that the Sun behaved very differently from what had been conventionally thought (Surprise!) and that x-ray telescopes would work; and the discoveries by Uhuru which had electrified the astronomical community and prompted the American Astronomical Society to provide a strongly worded letter of support for HEAO.

Three missions emerged from the restructuring: HEAO-1, a scanning mission that would survey and map X-ray sources throughout the sky over a wide range of X-ray energies; HEAO-2, a focusing X-ray telescope, 1/2 the diameter originally proposed, which became known as the Einstein Observatory; and HEAO-3, which would study gamma rays and cosmic rays.

HEAO-1 and HEAO-3 each contained an array of instruments from different groups. Each group built its own instruments and each had exclusive rights to the data from those instruments. They thus shared little in common except space on the satellite, much like the owners of apartments in a condominium.

The proposed design of the restructured HEAO-2 called for a focusing x-ray telescope and four focal plane instruments, the high resolution imager (HRI), the imaging proportional counter (IPC), the focal plane crystal spectrometer (FPCS) and the solid state spectrometer (SSS). The HRI and the IPC were the responsibility of Giacconi's group, which had moved in 1973 from AS&E to the Harvard-Smithsonian Center for Astrophysics. An MIT group under the direction of George Clark had responsibility for the FPCS, and a Goddard group led by Elihu Boldt had responsibility for the SSS.

Although Giacconi knew that they could not achieve the elegant simplicity of Uhuru for an observatory with a focusing x-ray telescope and this complex set of instruments, he believed that they could at least have simplicity in the concept. They would take a unified scientific approach to the observatory. The entire complex would be integrated into a complete whole, led by Giacconi as Principal Investigator for the entire mission. The condominium approach of HEAO-1 and HEAO-3 would not work. The groups would have to live together like a family, and like a good family, they would have to deal with their crises together, sacrificing individual interests for the good of the whole as technological and monetary problems arose and were resolved. In return, all members would share equally in the scientific returns. Thus, the design philosophy of elegant simplicity and the overall philosophy exemplified by the "Uhuru spirit" evolved to adapt to the new environment of a larger, more complex satellite. The dedication to the family or consortium idea was tested

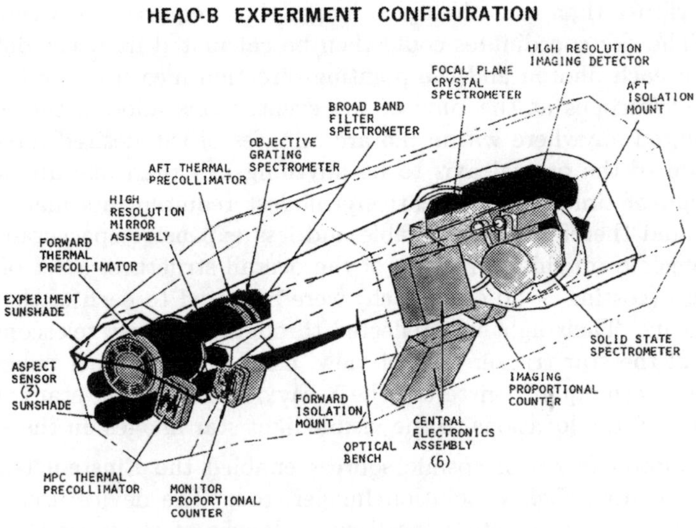

Figure 2: A Schematic of the HEAO-B Observatory

very early in the process. When the Columbia University polarimeter was dropped in the restructuring, they were reassured that they would retain their rights to a share of the data as members of the consortium.

The philosophy of elegant simplicity in design had other implications for the observatory. One was to keep the spacecraft requirements as simple as possible. It was decided to use only instruments that detect individual photons and to exclude those that integrate over extended periods, either in the detector or in onboard memory elements. This approach allowed for the removal of previously unanticipated errors. For example, after launch, an aspect distortion problem caused by residual magnetic fields was degrading the accuracy of the HRI source locations. However, since the data were reduced and the image was generated on the ground, the effect could be removed by calibration of the star trackers coupled with knowledge of the satellite location and orientation with respect to the earth's magnetic field.

The restriction to detectors that count individual photons made it possible to relax the pointing accuracy and stability requirements significantly, up to a fraction of the size of the useful field of view. It was necessary for the motion to be sufficiently small only during the time required for an aspect determination

(~ 1 second) rather than over the much longer time required for a complete observation. The sky coordinates could then be calculated from the detector coordinates for each photon and the pointing direction measured by the star trackers and gyroscopes at the time of the event. This allowed the optical axis to be pointed anywhere within 0.5 arc minutes of the desired direction, and the motion of the optical axis to be anything less than one arc second during a second of time. These relatively modest requirements made for a less complex, and therefore more reliable and less expensive spacecraft. For monitoring longer term shifts or drifts of the overall structure, a set of light emitting diodes (costing ~ 50 cents each) were mounted to each of the focal plane instruments. Their light was reflected through the X-ray telescope and into a corner of the star tracker field of view, using simple corner cubes and periscopes. This technique "registered" the X-ray detectors to the star trackers and thereby, the X-ray locations to the visible light star images on the sky.

The low photon fluxes of cosmic sources enabled the Einstein team to design a more reliable High Resolution Imager, too. The device used a microchannel plate array to create a localized avalanche of electrons that produced a signal when it hit a crossed grid of wires. After attempts by industry to construct a finely spaced grid failed, Giacconi and Steve Murray recalled previous work by the Leicester University X-ray group and realized that they could determine the centroid of the charge cloud produced by each individual X-ray photon, to get positions accurate to two arc seconds. This could be done with wires spaced a few tenths of a millimeter apart, a tolerance achieved in the Einstein team's own laboratory with very simple tools. Again, a gain in simplicity and reliability was achieved by designing with the nature of the phenomena they were trying to detect in mind.

Reliability is the one of the goals of any design, and simplicity is one of the surest ways to achieve reliability. When simplicity was not possible, the Einstein designers used another principle, that of "soft" failure. Whenever possible they put a redundancy into the design such that a single failure could not affect the mission too severely. The obvious application of this principal was the use of more than one focal plane detector, and redundant spacecraft components. More subtle examples include the use of low expansion materials with matched thermal coefficients in the optical train and the choice of orientation for the star trackers. The use of thermally matched optical components allowed for a simpler active thermal control system and provided for reasonable performance even if the control system failed. Rather than position the three star trackers in a traditional way along three orthogonal axes, the team chose to co-align one tracker with the optical axis, with the others offset from the optical axis by about 3 degrees in opposite directions. This arrangement, motivated by the recognition that the X-ray target would normally be well away from the solar direction and from the bright Earth, gave a probability of at least 95% for finding suitable guide stars. At the same time, it still allowed for attitude control and determination in case of failure of one star tracker. This proved crucial when one star tracker did indeed fail fairly early in the Einstein mission.

Just as with Uhuru, the Einstein scientific team also did not "win them all". After the HEAO program was restructured in 1973, the baseline spacecraft design involved the use of propellant to off-load momentum accumulated in the reaction wheels due to perturbing torques. Due to weight and volume constraints, the finite amount of propellant which was planned for HEAO-B implied a lifetime of $\sim$ 1 year after which useful operations on-orbit would cease. The science team sought to increase the amount of propellant (with modest success), and then proposed to use an innovative budget of $400,000 allocated by NASA for "science-enhancement" to add magnetic torquers to the spacecraft for unloading momentum and extending the lifetime indefinitely. NASA ultimately decided against this request and applied the $400,000 to cover cost overruns by various HEAO-B subcontractors (including the mirror vendor). Beaten, but not defeated, the Einstein team (with the assistance of MSFC engineer Tom Guffin, and a few dedicated engineers at the spacecraft contractor, TRW), turned to simultaneous operation of all four reaction wheels instead of the nominal three to reduce the frequency of momentum unloading, and to pairing of targets on-orbit to largely cancel momentum build-up in the wheels. Through these innovative approaches, Einstein achieved an on-orbit operating life of 2-1/2 years instead of the nominal one year. Perhaps this vignette illustrates another aspect of the philosophy of planning for "soft failure" (or getting the most possible out of the mission), while demonstrating the resiliency, ingenuity, and perseverance of the team to maximize opportunities to do good science.

The four pairs of nested, co-aligned x-ray mirrors were the crucial observatory components for which the Einstein team could not design for soft failure. No redundancy was possible; not even a test mirror was allowed by the tight budget and schedule. The mirrors had to be ground, polished and coated to a small-scale tolerance better than 30 A, on the first try. The primary responsibility for this awesome task was entrusted to Leon Van Speybroeck, who had worked with Giacconi, Giuseppe Vaiana and others to develop the x-ray mirror assembly which had worked so well on Skylab. In his contribution to this volume, Van Speybroeck describes the many technical and other problems that had to be overcome in the fabrication of the mirrors. What he does not say is that it is questionable that it could have been done by anyone other than him. He literally slept with the mirrors during crucial periods at Perkin-Elmer (the contractor for the mirror), and his insistence on floating the mirrors in mercury to obtain gravity-free measurements of the figure of the mirrors was in the venerable tradition established by Giacconi and his team of taking nothing for granted.

This skepticism extended to every phase of the project, preventing countless small mistakes and several large ones. One of the more dramatic examples occurred during the shipping of the mirrors. The crate containing the mirror assembly was to be loaded by a crane onto a truck. The ever cautious Einstein team insisted that the loading process be tested in a dry run, with a crate containing ballast. On the first test run, the crane dropped the crate! After much checking and rechecking of the loading procedure, the mirror assembly was safely loaded and shipped to MSFC for X-ray calibration.

The dry run of the loading of the mirror assembly was of course only one of more than a thousand tests that were eventually performed on the observatory. Most of these were done in August and September of 1977, when, because of schedule slippage, six months of calibration were compressed into thirty breathless days of round-the-clock testing. One valuable by-product of this crash effort was the development of an efficient data-handling system. Their experience with the discovery of x-ray binaries with Uhuru had convinced Giacconi and his collaborators that investment in software to process the data would yield great scientific rewards when Einstein began its observations. As a result, discovery of copious X-ray emission from quasars and from stellar coronae very early in the Einstein mission led to immediate updates and revisions to the observing plans to study the nature of these sources in a more coherent and statistically appropriate manner.

Twenty-two minutes after midnight on November 13, 1978 the Einstein Observatory was launched into orbit. Four days later, the HRI was moved into the focus and the telescope was pointed toward Cygnus X-1. Giacconi watched on a monitor as the telescope scanned toward Cygnus X-1. X-ray images of stars streaked across the field of view and then the telescope coasted to a stop and locked onto Cygnus X-1. A point like image formed on the screen. Eighteen years after the publication of the article with Rossi on the feasibility of an X-ray telescope, Giacconi knew he had a working x-ray telescope in orbit. "It was an unbelievable sensation," he said. "A profoundly moving moment."

Eighteen years might seem like a long time in the career of a scientist, but from the perspective of what had been accomplished, it was an astonishingly brief time. Within that time (only sixteen years, if you count from the discovery of Sco X-1) x-ray astronomers had, under Giacconi's leadership, achieved a million-fold increase in sensitivity. This is equivalent to the 340-year

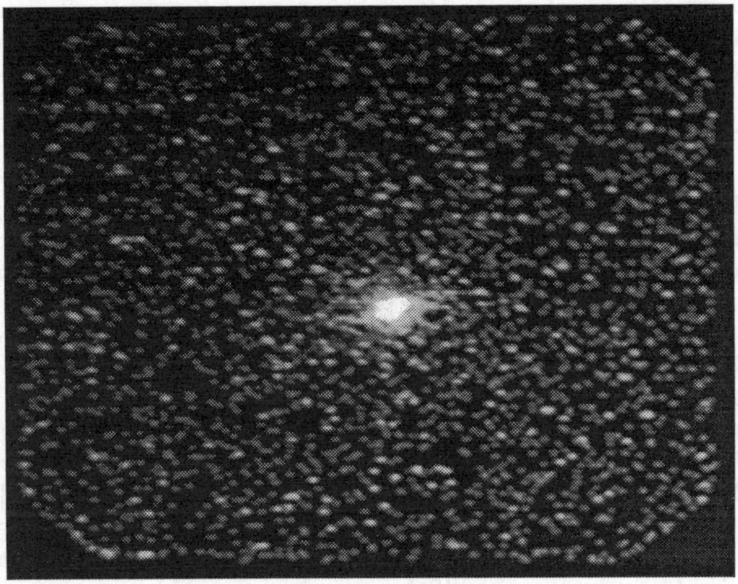

Figure 3: Cygnus X-1 First Light

progression from Galileo's telescope to the 200-inch Mt. Palomar telescope.

The Einstein Observatory brought the x-ray sky into focus, on an equal footing with optical and radio astronomy. Almost immediately, with the spectacular x-ray images of supernova remnants and clusters of galaxies, we had an immensely richer and more colorful view of the universe. X-radiation from stellar coronae, though weak, was detected from hundreds of stars of virtually every spectral type. These observations showed that a simple acoustical heating model will not work, that rotation and magnetic fields play an important role as well, accounting for the surprisingly strong x-ray luminosity of very young stars. At the other end of the evolutionary sequence, Einstein spectral observations of supernova remnants provided clear evidence that heavy elements are ejected into space by supernovae, while making us aware that the gas dynamics of the explosions are very complex.

Early type galaxies were found to have extensive hot coronae, which established the case for the existence of dark matter in galactic halos. The observations of M87 and the Virgo Cluster demonstrated the value of x-ray surface brightness and temperature data as a probe of the distribution of dark matter in clusters of galaxies for the first time. Radiative cooling in clusters with cD galaxies was shown to be a widespread and energetically important. This phenomenon is still not understood, nor do we understand the evolution of clusters as a whole, but the Einstein observations provided convincing proof that we live in the epoch when clusters are being assembled through mergers of sub-clusters.

Although, the x-ray imaging of optical and radio jets from active galaxies and quasars has done little more than whet our appetite for the sub-arc-second resolution of AXAF, the Einstein x-ray data on the intensity and variability of hundreds of active galactic nuclei (AGN) and quasars have had a profound impact. The data on individual AGN and quasars has been instrumental in establishing the black hole accretion disk model for this phenomenon. The Einstein deep surveys[14] and the Extended Medium Sensitivity Survey[15] made giant steps toward understanding the enduring mystery of the origin of the x- ray background radiation. These works showed that the background is primarily, and possibly completely composed of discrete objects and that a large fraction of these discrete objects are AGN and quasars. Here again, we must await AXAF for what we hope will be the total resolution of the mystery, but we should not be surprised if the x-ray background has more surprises in store.

The Einstein Observatory brought x-ray astronomy into the mainstream of astronomy through its increased sensitivity. Moreover, through its guest observer program, proposed by Giacconi from the onset of the HEAO program, it brought the mainstream of astronomy into x-ray astronomy. The program set standards for subsequent missions by voluntarily turning over a significant fraction of the observing time for use by guest observers, starting with 20% shortly after launch and reaching 50% by the end of the mission in 1981. The Einstein team also established a data archive, through which all of the Einstein data were eventually made available to the entire astronomy community. While this may all seem rather common-place now in the late '90s, it was quite revolutionary for a Principal Investigator-led space science mission in the late '70's. Consequently, many astronomers from outside the field of x-ray astronomy began making x-ray observations and using x-ray data for the first time, along with their observations at other wavelengths. This cross-fertilization or unification of various sub-disciplines contributed to a reinvigoration of astronomy as a whole, and changed the way in which most astronomers approach their research. Sensitive observations across the electromagnetic spectrum, from the ground and from space, are required and now obtained to develop a full understanding of most astronomical objects and systems.

5 THAT WAS THEN, THIS IS NOW

As larger and even more complex facilities such as the Advanced X-ray Astrophysics Facility (AXAF) are developed, there is the possibility that the concept of a unified scientific approach will be replaced by plans which depend on consensus from a hierarchy of large committees. For example, for the last cycle of Hubble Space Telescope (HST) observations, there was a top-level Telescope Allocation Committee of 6 scientists supported by 15 panels composed of 8 scientists each. And this is just to decide who shall get observing time on a telescope, let alone what might be required to build one! The competition for time on the telescopes is intense and many young scientists depend on the grants which come with the allocation of time for financial support. Under such conditions the pressure to conform is tremendous, and the risk-reward ratio for innovation may become unacceptably high. Moreover, peer review committees may try to please as many proposers as possible which may fragment the approach to the science. HST dealt with this issue, in part, by establishing key projects allocating larger blocks of observing time through a separate peer review process. Interestingly, but not surprising, Giacconi was Director of the Space Telescope Science Institute when key projects were conceived and initiated.

This progression towards larger teams is illustrated by the following numbers for Uhuru, Einstein, AXAF, and the proposed High Throughput X-ray Spectroscopy mission (HTXS). Uhuru had the PI (Giacconi) plus 4 Co-I's from AS& E and one from MIT for a total of 6. Einstein had the PI (Giacconi) and a total of 22 scientists from 6 different institutions. At the time of Instrument Selection for AXAF, 4 Instrument Principle Investigators, 1 Telescope Scientist, and 5 Inter-Disciplinary Scientists were selected to form the AXAF Science Working Group chaired by the NASA Project Scientist. Furthermore, as an example, the team for the AXAF High Resolution Camera consisted of 12 scientists from 4 separate institutions. Now, 15 months beore launch, the AXAF scientific team involves approximately 100 contributing scientists. The HTXS Science Working group which has undertaken an early-phase study for a facility following AXAF, is comprised of 58 scientists from 14 institutions.

How can a philosophy as clear and forceful and adaptive as Riccardo Giacconi's, with his vision and intuitive trust in the abundance of nature, emerge and thrive in such an environment? This is indeed a challenge for those who lead these present and future projects. We cannot go back to "the good old days." Large facilities are necessary if we are to answer the difficult questions posed by the earlier, highly successful missions. The complex nature of the universe demands it.

Perhaps the experiences during the recent AXAF science calibration indicate that many of Riccardo's teachings are still very much alive (and in fact thriving). The Project Scientist, Mission Support, Science Center, and Telescope and Instrument Principal Investigator teams worked as a unit to plan and execute a most successful integrated calibration (true, they had nearly five months compared to the 30 days utilized for Einstein calibration). The team worked around late deliveries of focal plane instruments, calibrated transfer or surrogate detectors during an available interval to provide a fall-back (soft failure equivalent), and reworked the options almost daily, ultimately achieving an action-packed, science-filled, deadline-beating 10 days of calibration for the flight AXAF CCD Imaging Spectrometer (ACIS) with the flight mirrors and gratings. The High Resolution Camera was calibrated a few weeks previously with the full flight system as well. So, perhaps many of us who worked with Riccardo during the heady Uhuru and Einstein years not only absorbed some of his methods and principles, but also were trained well enough to apply his philosophies under somewhat different arrangements. Senior members of the team took regular shifts, helping to cover 24 hours per day, 7 days per week. Since some of the people involved were now well over 30, the dictum of "not trusting anyone over 30" has clearly evolved over time for these members of the team.

Given the challenges we face, we must remember how we got here and somehow communicate the essence to the next generation of instrument builders and astronomers, building into the system means for encouraging innovation, insisting on elegant simplicity, encouraging individuals to contribute while working as a team, and most importantly trusting in the beauty and richness of Nature. Because that, more than an x-ray telescope, is the true legacy of Riccardo Giacconi.

References

1. R. Giacconi and H. Gursky, eds. *X-ray Astronomy*, (D. Reidel Publishing Co. Dordrecht, 1974)
2. R. Giacconi *et al*, "The Einstein Observatory and Future X-ray Telescopes", in *Telescopes for the 1980s*, ed. G. Burbidge and A. Hewitt, (1981): 195
3. W. Tucker and R. Giacconi *The X-ray Universe*, (Harvard U. Press, Cambridge, 1985)
4. R. Giacconi and B. Rossi, *J. Geophys. Res.* **65**, 773 (1960)
5. C. Bowyer *et al*, *Science* **152**, 66 (1965)
6. P. Fisher *et al*, Astrophys. J. **143**, 203 (1996)

7. M. Oda, *Appl. Opt.* **4**, 143 (1965)
8. M. Oda *et al*, *Nature* **205**, 554 (1966)
9. H. Gursky *et al*, Astrophys. J. Lett **146**, 310 (1966)
10. P. Gorenstein and S. Mickiewicz, *Rev. Sci. Instr.* **39**, 816 (1968)
11. H. Tananbaum and W. Tucker, "Compact X-ray Sources" in *X-ray Astronomy*, ed. R. Giacconi and H. Gursky, (D.Reidel Publishing Co. Dordrecht, 1974): 207
12. W. Tucker and R. Giacconi *The X-ray Universe*, (Harvard U. Press, Cambridge, 1985)
13. W. Tucker and K. Tucker *The Cosmic Inquirers*, (Harvard U. Press, Cambridge, 1986)
14. R. Giacconi *et al*, Astrophys. J. **234**, L1 (1979)
15. T. Maccacaro *et al*, Astrophys. J. **374**, 117 (1991)

HISTORICAL DEVELOPMENT OF X–RAY OPTICS FOR ASTRONOMY

L. P. VAN SPEYBROECK
Smithsonian Astrophysical Observatory, MS-6
60 Garden Street
Cambridge, MA 01238, USA
E-mail: lvs@cfa.harvard.edu

The history of high resolution X–ray optics for astronomical observations will be reviewed. This may be divided into roughly three eras, the early conceptual and invention phase, an intermediate period in which the technology for X–ray optics was defined by the traditional methods and experience in the optical industry, and the present era in which both adequate metrology and quantitative fabrication techniques have resulted in the superb mirrors of ROSAT and AXAF. The physics of these optics and their design and fabrication will be discussed. The progress in the field will be illustrated by examples of historical and recent X–ray mirrors. Some speculations on future missions and progress will be made.

1 Introduction and Early History

Compton[7] showed that X–rays reflect from a polished surface at small glancing or grazing angles in 1922, almost thirty years after the discovery of X–rays by Röentgen. However, it was another twenty five years before Kirkpatrick and Baez[20] achieved two dimensional X–ray imaging using orthogonally crossed cylindrical mirrors. Mirrors of this general configuration continue to be considered today for applications in which good angular resolution is less important than a large collecting area. The high resolution optics which are the subject of this article are figures of revolution about an optical axis; Wolter[40] studied such designs for the purposes of X–ray microscopy, and showed that imaging over an extended field can not be achieved with a single reflection[a].

Giacconi and Rossi[12] first suggested the use of reflecting optics for solar and extra–solar X–ray astronomy in 1960, about two years *before* the first discovery of an X–ray source outside of our solar system. In this seminal article they concentrated upon single reflection systems, which are useful as collectors but do not provide useful images of extended fields; however, they were aware of Wolter's work, and stated that they were considering the possibilities of using "multiple total reflections to construct image–forming X–ray telescopes". Giacconi, in private conversations, said that he initially was inspired to seek

[a] It often has been stated that an even number of reflections is required for good imaging over an extended field, but actually designs with an odd number of reflections are possible if there is an internal ray crossing between two of the reflections[21].

image concentration because of his frustration at waiting for something to happen during low event rate cosmic ray experiments. The first solar observations utilizing focusing X–ray optics were performed by Giacconi and his colleagues at American Science and Engineering, Inc. (AS&E) together with scientists from the NASA Goddard Space Flight Center (GSFC) in 1963, only three years after this technique was first suggested, and scientifically useful solar results were obtained by Giacconi and colleagues [11] two years later in 1965. Subsequently solar observations were conducted by Underwood and Muney [29] of GSFC, and by Vaiana and his colleagues at AS&E[31]. Summaries of some of this early work can be found in the 1968 review by Giacconi *et al*[10]. Some of these early mirrors are shown in figure 1. The three telescopes in the upper left panel of this figure were among the first X–ray telescopes, have diameters of about 75 mm, and were made using conventional machining and epoxy casting; other telescopes from this era were manufactured using electroformed nickel on polished mandrels, techniques which are still being developed and used today. The development and testing of some of these early mirrors is described by Giacconi *et al*[9]; resolutions of order a few arc minutes were demonstrated, but these early mirrors also had severe image losses due to scattering. The other mirrors in figure 1, going clockwise, are a 225 mm diameter kanigen coated beryllium mirror used in many solar rocket flights by Vaiana *et al*, a 300 mm diameter quartz rocket mirror, and the first nested system, the two (225 mm and 300 mm diameters) pair mirror for the S-054 experiment on Skylab [32]. These later mirrors had point response function widths of order two arc seconds, but only a very small fraction of the imaged energy was contained in the central image core because of severe scattering losses.

The mirrors discussed above are examples of the early conceptual phase, in which the people developing this new invention initially were not familiar with many standard optical fabrication techniques, or with much of the experience available in the optical industry. However, they became increasingly aware of this resource, especially during the development of the mirrors for the Einstein Observatory. The remainder of this chapter will include a brief review of the optics of X–ray mirrors, a brief discussion of the Einstein and ROSAT (Röentgensatellit) mirrors, a somewhat more thorough discussion of the AXAF (Advanced X–ray Astrophysics Facility) mirrors, and some speculations on future trends. All of these examples are missions with an emphasis upon high resolution rather than collecting area. Aschenbach [1] has reviewed this subject, including much of the ROSAT development. Pounds [22] discusses the EXOSAT and other programs, including these, from a European perspective elsewhere in this volume, and Gorenstein [15] discusses techniques for missions with an emphasis upon a large collecting area.

Figure 1: The top left panel shows some of the first experimental telescopes made at AS&E[11]; these mirrors have diameters of about 75 mm, and resolutions of a few arc minutes were achieved. The other mirrors in this photograph, proceeding clockwise, are an early kanigen coated beryllium mirror, used by Vaiana and colleagues on many solar rocket flights, a quartz rocket mirror, and the mirror for the S-054 experiment on Skylab, which was the earliest example of a nested mirror system.

2 Physics of X–ray Optics

Three factors must be considered for X–ray optics: first, X–rays will reflect efficiently over a broad bandwidth only at grazing incidence; second, grazing incidence optics have a somewhat unusual configuration; and finally, surface scattering is very important. These factors are discussed in the following sections.

2.1 Reflection Efficiency

We are interested in photon energies which are well above the resonance energies associated with most electrons in physical materials; as a result of this, the real part of the index of refraction is near to, but slightly less than unity. It is conventional to express the index of refraction n as:

$$n = 1 - \delta - \imath\beta \tag{1}$$

where the real part δ describes the phase change due to the interaction with the material, and the imaginary part, β, is the absorption term. Following Henke *et al*[18], we also define the following quantities[b]:

$$a = 2\delta - \delta^2 + \beta^2 \approx 2\delta \quad (2)$$
$$\gamma = 2\beta(1-\delta) \approx 2\beta \quad (3)$$
$$b = \sin^2\alpha - a \quad (4)$$
$$\eta^2 = (1/2)\left(b + \sqrt{b^2+\gamma^2}\right) \quad (5)$$

where α is the grazing angle. In terms of these quantities, the reflection probabilities ϵ_σ and ϵ_π, the cases in which the electric vector is perpendicular or parallel to the plane of reflection respectively, are given by:

$$\epsilon_\sigma = \frac{4\eta^2(\sin\alpha - \eta)^2 + \gamma^2}{4\eta^2(\sin\alpha + \eta)^2 + \gamma^2} \quad (6)$$
$$\frac{\epsilon_\pi}{\epsilon_\sigma} = \frac{4\eta^2(\eta - \cos\alpha\cot\alpha)^2 + \gamma^2}{4\eta^2(\eta + \cos\alpha\cot\alpha)^2 + \gamma^2} \quad (7)$$

The latter ratio is very near to unity for grazing incidence reflections.

Some examples of the reflection efficiency for non-polarized radiation by iridium and gold are given in figures 2–4; the grazing angles illustrated in figure 4 are typical for the four AXAF mirrors. The optical constants used for these calculations were obtained from Henke *et al*[17]. Relatively good efficiencies are obtained below the critical angle θ_c, which is given approximately by

$$\theta_c = \sqrt{2\delta}$$

Aschenbach[1] gives the following useful approximations for X–ray wavelengths sufficiently apart from absorption edges:

$$\delta \approx N_0 \frac{Zr_e}{A2\pi}\rho\lambda^2$$

where N_0 is Avogadros number, r_e is the classical electron radius, Z is the atomic number, A is the atomic weight, ρ is the mass density, and λ is the wavelength. For heavy elements, where $Z/A \approx 1/2$, he obtains $\theta_c \approx 5.6\lambda\sqrt{\rho}$ where λ is in Å, the density ρ is in g cm^{-3}, and θ_c is given in arc minutes.

Finally, the thickness of coating required for efficient reflection is quite small; the skindepth for energy flow into the medium is given by $d = \lambda\eta/(2\pi\gamma)$, the intensity falling off as $\exp(-z/d)$, where z is measured normal to the surface. The skindepth typically is well under 100Å when efficient reflection is obtained.

---

[b]Our a, α, and η are Henke's α, θ, and ρ respectively.

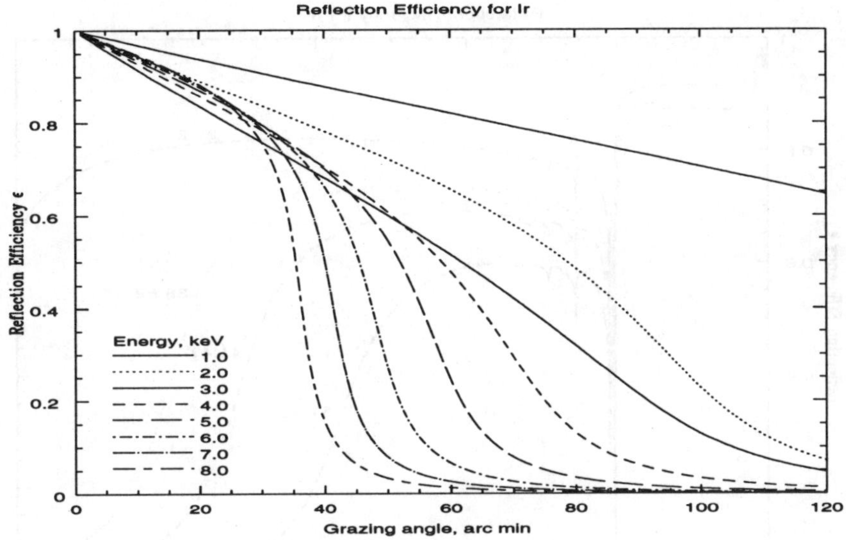

Figure 2: Reflection probability for iridium as a function of angle for various energies.

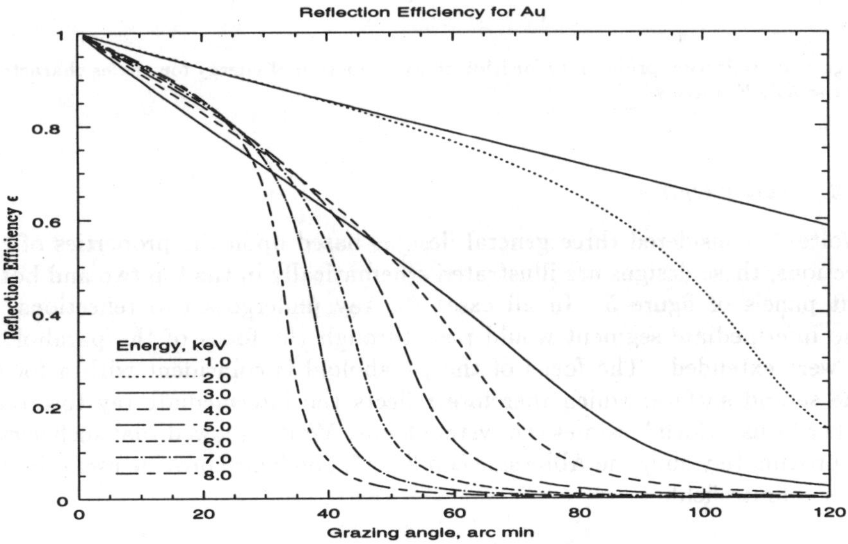

Figure 3: Reflection probability for gold as a function of angle for various energies.

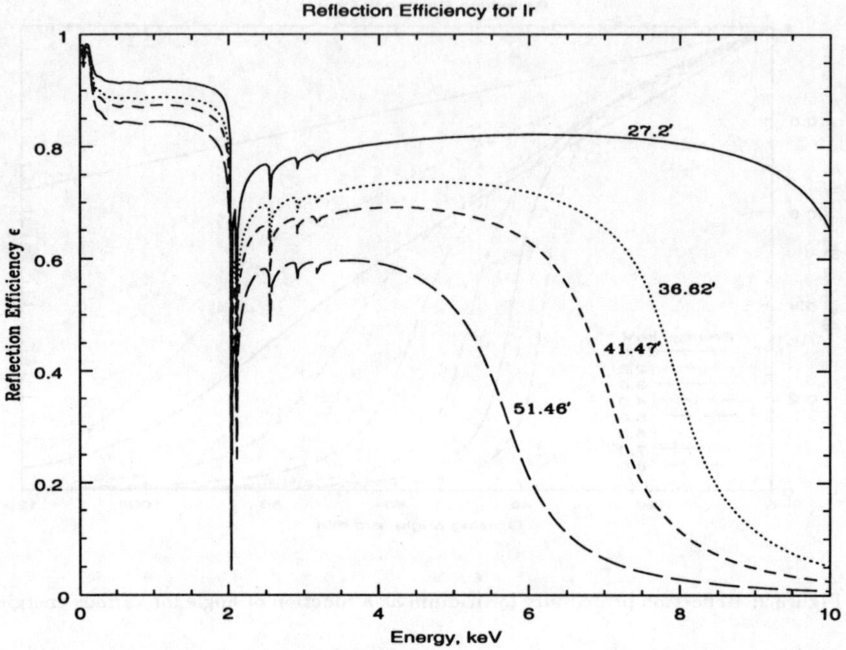

Figure 4: Reflection probability for iridium as a function of energy for angles characteristic of the AXAF mirrors.

2.2 Wolter Optics

Wolter [40] considered three general designs based upon the properties of conic sections; these designs are illustrated schematically in the top two and bottom left panels of figure 5. In all cases the ray undergoes two reflections, and the intermediate segment would pass through the focus of the paraboloid if it were extended. The focus of the paraboloid is coincident with a focus of the second surface, which therefore reflects the intermediate ray towards its other focus, which becomes the system focus. Wolter showed that such systems approximately obey the Abbe sine condition[c], and hence provide useful imaging over a finite field.

---

[c]The geometrical form of the Abbe sine condition for these systems is that the intersections of the entering and exiting segments of a ray path must lie on a sphere.

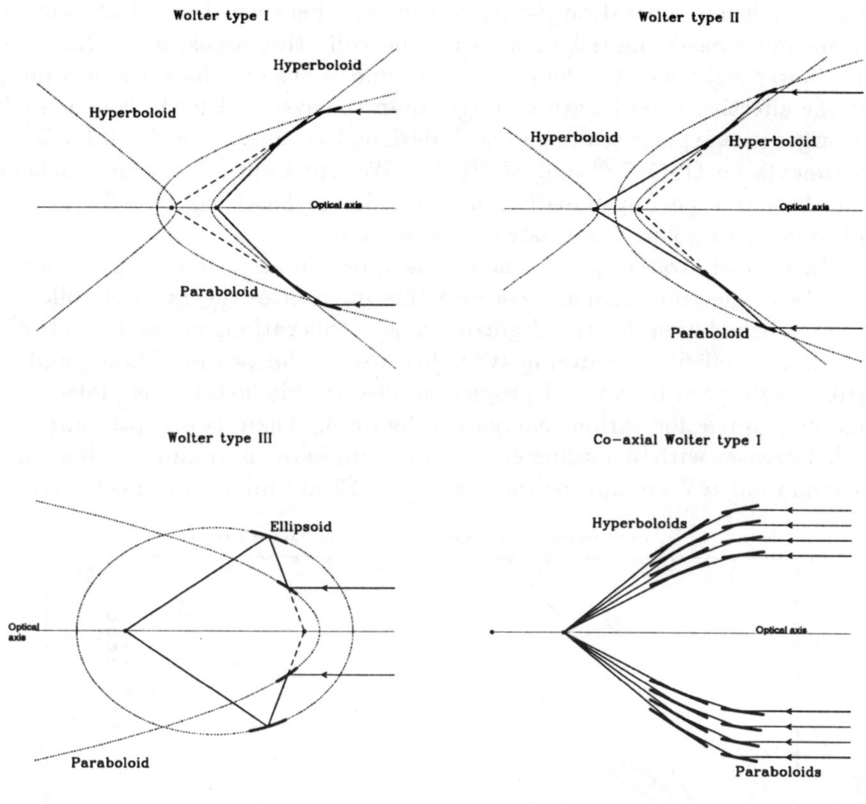

Figure 5: The top two and the bottom left panels show the grazing incidence mirror designs of Wolter[40]. Axial rays are reflected from a paraboloid. The reflected ray would pass through the focus of the paraboloid if it were extended. The focus of the paraboloid is coincident with a focus of the second surface, which therefore reflects the intermediate ray towards its other focus, which becomes the system focus. The lower right panel illustrates co–axial nesting of Wolter type I pairs to obtain a larger collecting area.

The reflection angles for Wolter type I systems are in the same direction, which results in a shorter system for a given grazing angle than the other designs. The type I systems are mechanically simpler than the other designs because there is less separation (none, in principle) between the optical elements, and are more easily nested to increase the collecting areas, as is illustrated in the lower right panel of figure 5. The type II systems have the advantage that the effective focal length is longer than the system length, which results in a larger image plate scale, and such designs have been used for solar X–ray experiments on OSO–7 [30] and SOHO [16]. We will limit our discussion below to the Wolter type I systems because this design has been used for all high resolution extra–solar X–ray astronomy missions.

The typical Wolter type I system has approximately equal grazing angles at the two reflecting elements because this maximizes the ratio of collecting area to polished area; for equal grazing angles this ratio is approximately $\epsilon^2\alpha$, the reflection efficiency entering twice because of the two reflections, and the grazing angle once because of projection effects; this quantity is plotted as a function of angle for various energies in figure 6. There is an optimum angle which decreases with increasing energy; for iridium the optimum grazing angles at 2.0 and 8.0 keV are approximately 42 and 27 arc minutes respectively.

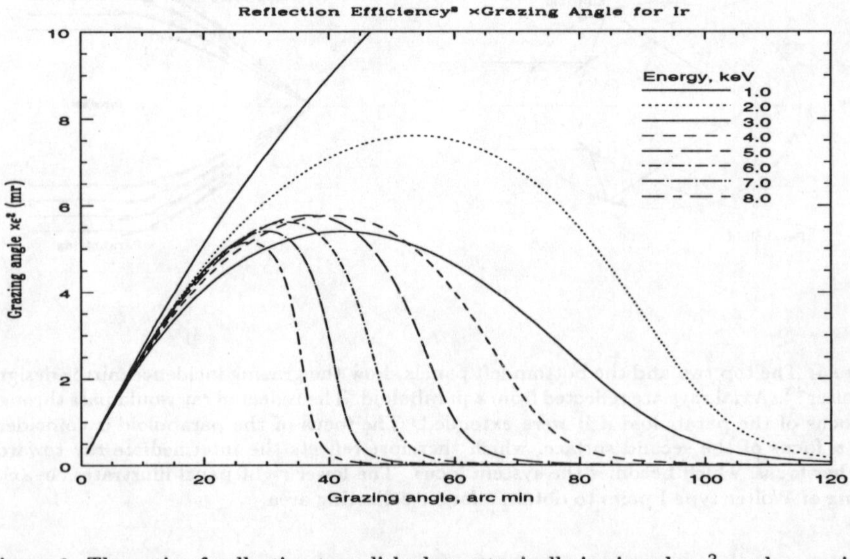

Figure 6: The ratio of collecting to polished area typically is given by $\epsilon^2\alpha$, where ϵ is the reflection efficiency and α is the grazing angle; this quantity is plotted as a function of angle for various energies.

The resolution of these systems is approximately proportional to θ^2, where θ is the incident angle with respect to the optical axis. The resolution degrades linearly with L/D, where L is the length of the element, and D is the diameter [34,24]. The empirical fit to the resolution also includes a small term proportional to $\tan\theta\tan^2\alpha$, where α is the grazing angle for axial rays. This term arises because the conic section geometry does not exactly satisfy the Abbe sine condition. Wolter [41] later obtained designs which satisfy the Abbe sine rule exactly; the difference in performance between these Wolter-Schwarzschild designs and the designs based upon conic sections is small for small grazing angles, but becomes appreciable for mirrors with grazing angles of more than about 1.5 degrees [5]. The mirror in the lower right panel of figure 1 and the SOHO mirrors [16] actually follow the Wolter-Schwarzschild prescriptions.

2.3 Scattering

The performances of all X–ray mirrors flown before ROSAT were dominated by surface scattering, which can be severe because of the short X–ray wavelengths. Beckmann and Spizzichino [3] have treated the scattering with the approximations that the surface can be considered stationary and that a scaler model can be used for the electric field. The small grazing angle α typical of X–ray mirrors results in the in–plane scattering dominating the out of plane scattering by a factor of order $1/\alpha \approx 100$, and so it is convenient to use the one dimensional form of the equations, as developed by Glenn [14]. The scattered intensity from a single surface in the one dimensional approximation can be obtained from:

$$v_z = -(\frac{4\pi}{\lambda})\sin\alpha \tag{8}$$

$$p_{v_z}(x) \equiv (1-\frac{|x|}{L})(e^{-v_z^2\sigma^2(1-\rho)} - e^{-v_z^2\sigma^2}) \tag{9}$$

$$\psi(\theta) = \psi_0(\theta) + \psi_1(\theta) \tag{10}$$

$$\psi_0(\theta) = \mathcal{S}\frac{L\sin\alpha}{\lambda}\left(\mathrm{sinc}(\frac{\pi L\sin\alpha\theta}{\lambda})\right)^2 \tag{11}$$

$$\psi_1(\theta) = \frac{\sin\alpha}{\lambda}\int_{-L}^{L} dx e^{\frac{i2\pi\sin\alpha\theta}{\lambda}x} p_{v_z}(x) \tag{12}$$

where θ is the angle measured from the specular image, L is the length of the optical element, ρ is the surface error correlation function, $\psi_0(\theta)$ is the one dimensional intensity distribution for smooth surfaces due to diffraction,

$\psi_1(\theta)$ represents the surface scattering, and the Strehl factor $\mathcal{S} \equiv \exp(-v_z^2\sigma^2)$, where σ is the rms surface error. The quantity $\psi(\theta)$ represents the probability of scattering into a *radial* increment $d\theta$, and in this approximation, $\psi(\theta) = \psi(-\theta)$. The normalization integrals are:

$$\int_{-\infty}^{\infty} d\theta\psi_0(\theta) \;=\; e^{-v_z^2\sigma^2} = \mathcal{S} \tag{13}$$

$$\int_{-\infty}^{\infty} d\theta\psi_1(\theta) \;=\; 1 - e^{-v_z^2\sigma^2} \;=\; 1 - \mathcal{S} \tag{14}$$

$$\int_{-\infty}^{\infty} d\theta\psi(\theta) \;=\; 1 \tag{15}$$

The smooth surface limits of the above equations are much simpler, and usually adequate for all but very narrow angle scattering. Ignoring the finite length of the glass, and assuming that the surface roughness σ is small, we can express the scattered distribution in terms of the surface roughness power spectral density (PSD), which essentially is the Fourier transform of $\rho\sigma^2$. We define the power spectral density per surface to be $2W_1(f)$, where the factor of 2 is conventional[d]. In the smooth surface limit, the surface frequency f and the scattering angle θ are related by:

$$f \sim \frac{\theta \sin\alpha}{\lambda} \tag{16}$$

The relative surface brightness for wide angle scattering is given by:

$$\psi(\theta) \;=\; 2 \times \frac{8\pi}{f}\left(\frac{\sin\alpha}{\lambda}\right)^4 (2W_1(f)) \tag{17}$$

The first factor of 2 comes from the fact that there are two reflections, whereas the power spectral density $2W_1(f)$ represents the average of the two surfaces. This analysis allows us to relate measurements taken at different X–ray energies, and to estimate scattering for other energies. However, the linear approximations are not valid at lower frequencies, where the surface errors times the grazing angle can exceed the wavelength.

3 The Einstein and ROSAT Observatories

The first sensitive and high resolution extra–solar observations were performed with the Einstein Observatory, which was launched in 1978 and had an angular

[d] The integral of power spectral densities to calculate surface roughness typically is limited to positive frequencies, whereas the definition of $W_1(f)$ includes negative frequencies; however, $W_1(-f) = W_1(f)$ for real frequencies.

resolution of about 3.5 arc seconds; these observations showed that the X–ray emission is an important characteristic of many of the most interesting astronomical systems. Tucker and Tananbaum [28] discuss the Einstein program and some of the more discoveries made with that observatory. The success of the Einstein Observatory has resulted in high resolution optics becoming the technique of choice for most soft X–ray astronomical observations. The mirrors of the Einstein Observatory are shown in the top two panels of figure 7; the bottom left panel shows the telescope of the Einstein Observatory being placed in an X–ray test chamber at the NASA Marshall Space Flight Center (MSFC).

These mirrors can be said to represent the culmination of the artisan approach to X–ray mirror fabrication based upon the traditional methods and experience in the optical industry. The performance of the Einstein Observatory mirrors greatly exceeded that of their predecessors, and yet remained dominated by surface scattering; for example, the half power diameters at 0.28, 1.49, 2.98, and 4.51 keV were approximately 9, 10, 17, and 30 arc seconds respectively [33].

Important advances were made in the German ROSAT program; this satellite has large mirror surfaces with very low scatter, and has provided sensitive observations since 1990. The ROSAT program has been discussed by Trumper [26,27] and by Aschenbach [1,2]. These were the first astronomical X–ray mirrors which were fabricated using adequate metrology and a quantitative understanding of the polishing process, and also were the first large X–ray mirrors which yielded performance consistent with expectations. The program utilized improvements in metrology which had become available at that time, particularly the WYKO instruments which made surface roughness measurements both accurate and convenient. The top left panel of figure 8 shows a ROSAT mirror element being ground to essentially final figure tolerances in a precision vertical grinding machine; the optical element was supported by an external cage with multiple silicone rubber attachments to the exterior of the optical element during this process. The upper right hand panel of this figure shows an element being smoothed with a large polishing lap, and the lower left panel shows the critical surface roughness measurements. The ROSAT telescope assembly is shown in the lower right hand panel of the figure. The ROSAT program included iteration of fabrication and X–ray testing, which demonstrated that the polishing methodology would yield satisfactory surfaces; final roughnesses of 2-3 Å in the WYKO bandwidth were achieved. This group deserves much credit for a systematic and successful approach to obtaining very smooth surfaces. The contributions of B. Aschenbach and H. Bräuninger of the Max-Planck-Institute für Extraterrestrische Physik, and of K. Beckstette and E. Heynacher of Carl Zeiss, are noteworthy.

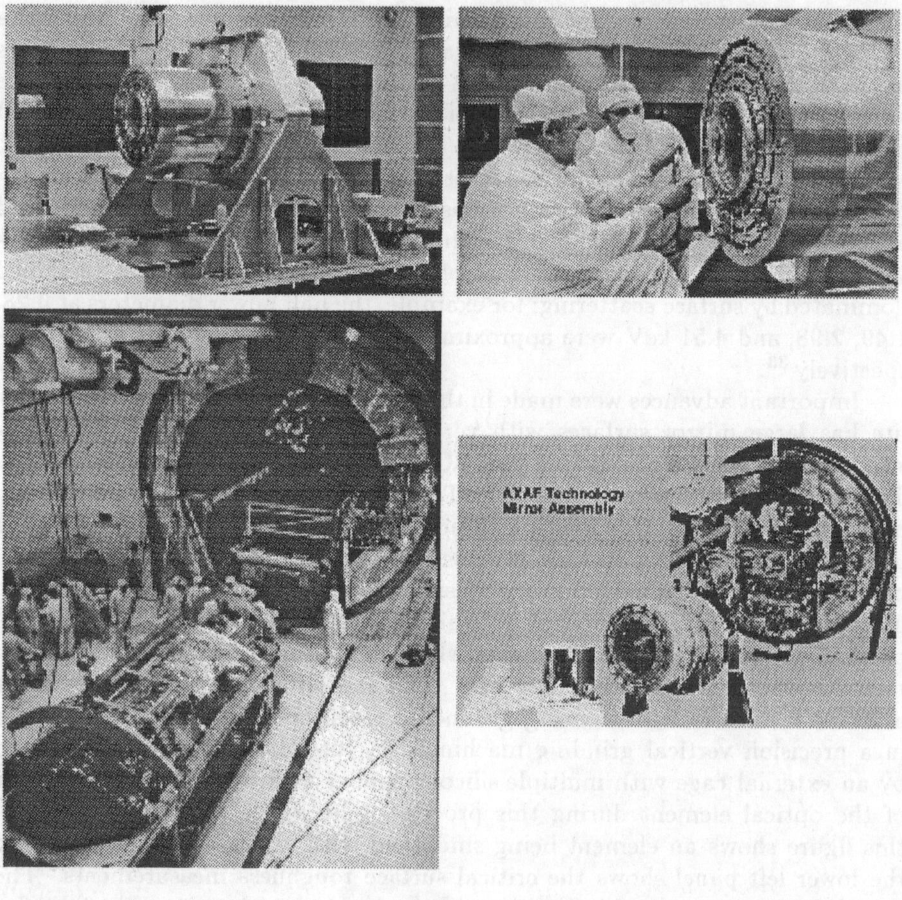

Figure 7: The top two photographs show the mirrors of the Einstein Observatory in the assembly clean room at Perkin-Elmer (now HDOS). The two lower panels show the telescope of the Einstein Observatory and the AXAF TMA being placed in an X–ray test chamber at MSFC.

Figure 8: The top two photographs show the ROSAT mirrors being ground and polished at the Carl Zeiss facility in Oberkochen, Germany. The bottom left photograph shows the measurement of the surface roughness of one of the elements; these were the first X-ray mirrors to achieve satisfactory surface finishes. The ROSAT telescope assembly is shown in the bottom right photograph.

4 AXAF

The AXAF is one of NASA's Great Observatories, and is expected to provide the highest resolution X–ray images yet obtained from astronomical sources outside of our solar system. Weisskopf[38,39] has described the overall mission. The project is managed by MSFC with technical support from the Smithsonian Astrophysical Observatory (SAO). TRW is the prime contractor, and is responsible for the entire mission. The AXAF High Resolution Mirror Assembly (HRMA) consists of four concentric Wolter-I parabola-hyperbola pairs with diameters between approximately 0.6 and 1.2 meters. The element axial lengths are 842.2 mm, wall thicknesses are between about 15.6 mm and about 22.5 mm. The focal lengths of all pairs are approximately 10 meters. The mirror pairs are numbered 1 (largest), 3, 4, and 6 (smallest) for historical reasons. The mirror fabrication program included an earlier, smaller mirror, the Technology Mirror Assembly (TMA), which was built to develop and evaluate the fabrication process; the insertion of this mirror into the MSFC test facility is shown in the lower right panel of figure 7. The final TMA performance was excellent, and in close agreement with predictions made before the beginning of X–ray testing [36]. The largest AXAF flight mirror pair was made before the other AXAF mirror elements to prove that the manufacture of the larger optics could be accomplished satisfactorily; this pair was tested with X–rays[8] as a further proof of the fabrication process, and also was used to test and improve the alignment and assembly procedures. The successful TMA and mirror pair 1 (MP1) results allowed the AXAF mirror fabrication to proceed based upon optical and mechanical metrology without the necessity of expensive and time consuming iterations of X–ray testing.

4.1 AXAF HRMA Fabrication

The optical elements are made of Zerodur™ from the Schott Glasswerke; these were ground and polished to the desired figure by Hughes Danbury Optical Systems, Inc. (HDOS). The primary metrology used at HDOS during the final mirror fabrication stages consisted of three instruments, the Precision Metrology Station (PMS), the Circularity and Inner Diameter Station (CIDS) and the Micro Phase Measuring Interferometer (MPMI). Some feeling for the scale of the operation is provided by the photograph in figure 9. A brief summary of the HDOS fabrication results is given below; further details have been given by Reid [23].

Figure 9: AXAF mirror element being placed in an environmentally controlled metrology enclosure at HDOS.

The PMS was used to obtain axial profiles for fixed azimuthal angles; the absolute accuracy of the PMS was estimated to be better than 20Å rms in the spatial frequency bandwidth 0.0006 to 1.0 mm^{-1}. Examples of the final axial figure errors for the first (P1) and last (H3) elements completed at HDOS are given in figures 10 and 11. The P1 mirror has a relatively large error in average curvature which has little effect upon performance if the elements are optimally spaced; the average curvature error has been removed in the flatter curve of the P1 figure. There was a substantial improvement between the first and last optical elements; Actually, all other elements are much more like H3 than P1; for example, the average curvature error in P1 is 5.88 times that of H1, the next worst element. The ends of the optical elements are not polished as well as the middle sections; this is evident in figures 10–11, and also in the surface roughness data to be discussed below.

The CIDS data were used to determine absolute diameters and also roundness errors at two axial stations near the ends of the elements. These errors are less important than the axial figure and surface roughness errors for the HRMA, and will not be discussed here. They are included in the performance predictions. The PMS and CIDS data were combined to form an error map and used generate a control file for computer controlled small tool polishing.

The mirror elements were smoothed with a large lap after satisfactory figure errors were obtained. The high frequency surface roughness errors were measured with the MPMI, an adaptation of a WYCO. Samples were taken at eight azimuths, and at axial intervals of about 100 mm in the middles and at about 10 mm near the ends of the of the elements. No significant azimuthal variations were found. The large, middle sections of the elements also have no significant variations, and these PSD data could be averaged without loss of information. The average PSDs for the middle 92% sections of P1 and H3 are shown in figures 12 and 12. The lighter lines show the PSDs obtained from the PMS data and from the MPMI with different magnifications, and hence different spatial frequency bandwidths. The dashed vertical line at 0.02 mm^{-1} indicates the approximate lower frequency boundary of the frequency interval used in the analysis. The calculated PSD from each instrument is suppressed at the low end of its range because average curvature has been removed from the data. This is done in the case of the PMS because this term is treated elsewhere in the analysis, and in the case of the MPMI data because it is contaminated by errors in aligning the instrument with the mirror axis. The high frequency PMS data are degraded by instrument noise.

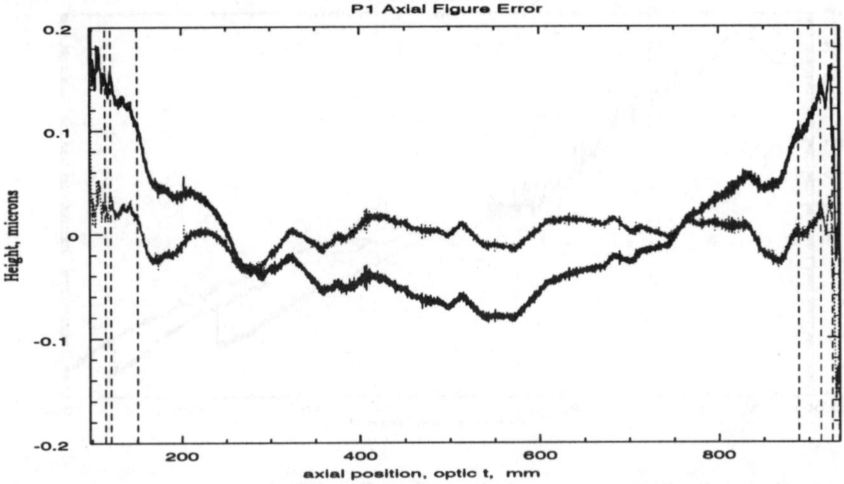

Figure 10: Axial profile errors for P1, the first completed AXAF mirror element. The flatter line in has average curvature error removed; this error has little effect upon performance if the element is positioned correctly. The vertical dashed lines are the segment boundaries used in the analysis.

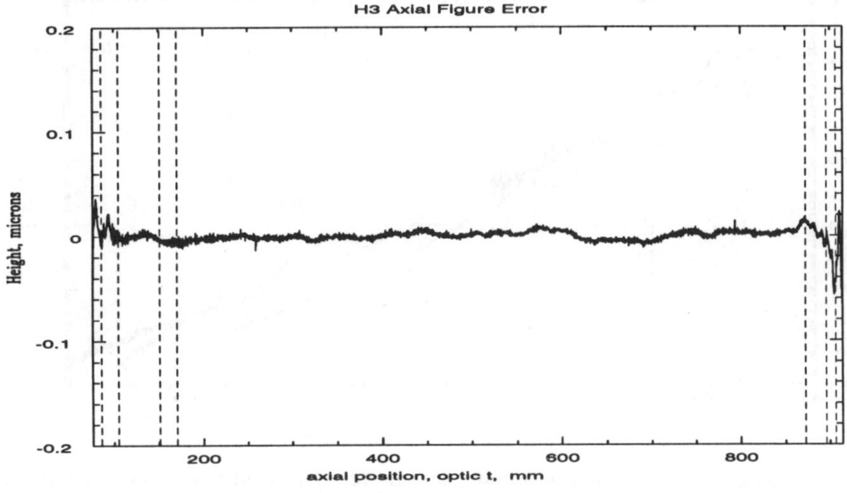

Figure 11: Axial profile errors for H3, the last completed AXAF mirror element. The vertical dashed lines are the segment boundaries used in the analysis. All elements other elements resemble H3 more than P1.

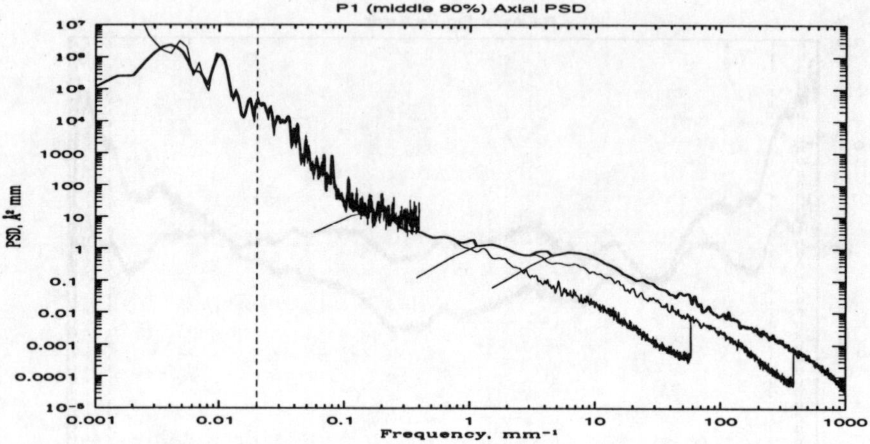

Figure 12: Composite PSDs for the middle 92% of P1. The bold lines were used in the performance predictions. The lighter lines correspond to data from the PMS and from the MPMI with magnifications of 1.5, 10.0, and 40.0 respectively. The dashed vertical line at 0.02 mm^{-1} indicates the approximate lower frequency boundary of the frequency interval used in the analysis.

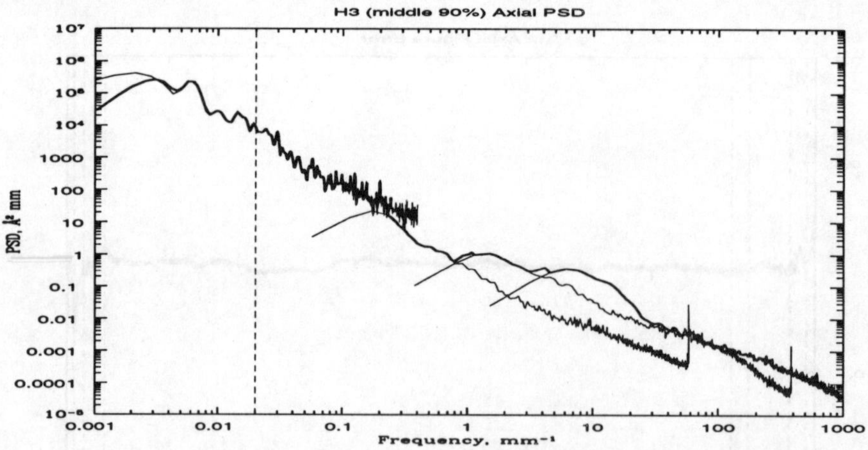

Figure 13: Composite PSDs for the middle 92% of H3. The bold lines were used in the performance predictions. The lighter lines correspond to data from the PMS and from the MPMI with magnifications of 1.5, 10.0, and 40.0 respectively. The dashed vertical line at 0.02 mm^{-1} indicates the approximate lower frequency boundary of the frequency interval used in the analysis.

The agreement between PSDs resulting from different magnifications of the MPMI in regions where the bandwidths overlap is not good; this is believed to be due to optic surface anisotropy and the MPMI optical transfer function [23]. We selected the highest frequency device in regions of overlap because this is expected to represent the surface most accurately. The measured integrated roughness in the 1-1000 mm^{-1} bands are given for the middle segments of all elements in table 1. The surface roughness increases near the edges of the elements where the smoothing by the large polishing laps was less effective; it also varies more rapidly with position, and consequently is not well determined. The last column of table 1 gives an estimate of the average roughness per surface over the entire length of the glass based upon MPMI data; the outer 2-4% of the glass at each end contributes significantly to the total integrated roughness, and hence to the wide angle X–ray scattering.

Table 1: Surface Roughness Summary

	Roughness 1 to 1000 mm^{-1} Å			
	Middle 770 of 832 mm			Estimated rms per
Pair	Paraboloid	Hyperboloid	rms average	surface, including ends
1	3.36	3.12	3.24	5.96
3	1.70	2.02	1.87	2.05
4	2.20	1.98	2.09	3.00
6	3.23	1.47	2.51	3.59

The mirror elements were coated with an underlayer of chromium and a reflecting surface of iridium by Optical Coating Laboratory, Inc. (OCLI) of Santa Rosa, California. The mirror thermal and mechanical design, and its alignment and assembly were performed by the Eastman Kodak Corporation (EKC) in Rochester, New York. Cohen [6] has given a brief description of the mechanical design and its expected stability. The mirror alignment and assembly were done in a clean environment with a vertical optical axis. The elements were supported in a relatively stress free manner, and manipulated to achieve a satisfactory focus location and to minimize coma errors. The alignment parameters were determined with a visible light instrument which is described more completely by Glenn [13] and Waldman [37]. This system allowed the measurement of alignment parameters to an equivalent image radius accuracy of order 0.01 arc seconds, and also detected important support induced strain terms. Figure 14 shows the mirror being aligned at EKC. The visible light measurements of the alignment parameters shown in figure 15 were confirmed to an accuracy of order 0.1-0.2 arc seconds by X–ray measurements.

Figure 14: Alignment of the AXAF mirrors at EKC. H3 has been aligned, and H1, the final element, can be seen in the background.

4.2 AXAF HRMA Calibration

The measurements performed during fabrication and assembly at HDOS and EKC were used to make performance predictions before the beginning of X–ray calibration. The X–ray tests of the mirror were performed in a special facility at MSFC which was built to support this mission. Some of the principal results are summarized below; more details can be found in Van Speybroeck *et al*[35].

Effective area measurements were performed with spectral line and with continuum dominated X–ray sources. In general, the line measurements were taken with proportional counters and the continuum measurements were taken with solid state detectors (SSDs). In both cases the area is determined by the ratio of counts observed in the focal plane to those observed through a known aperture forward of the mirror. The two techniques are complementary, and also provide a consistency check. The continuum source measurements provide a dense sampling of the effective areas with energy, but are more sensitive to the spectral resolution of the detectors in regions where either the incident flux or mirror transmission are changing rapidly with energy.

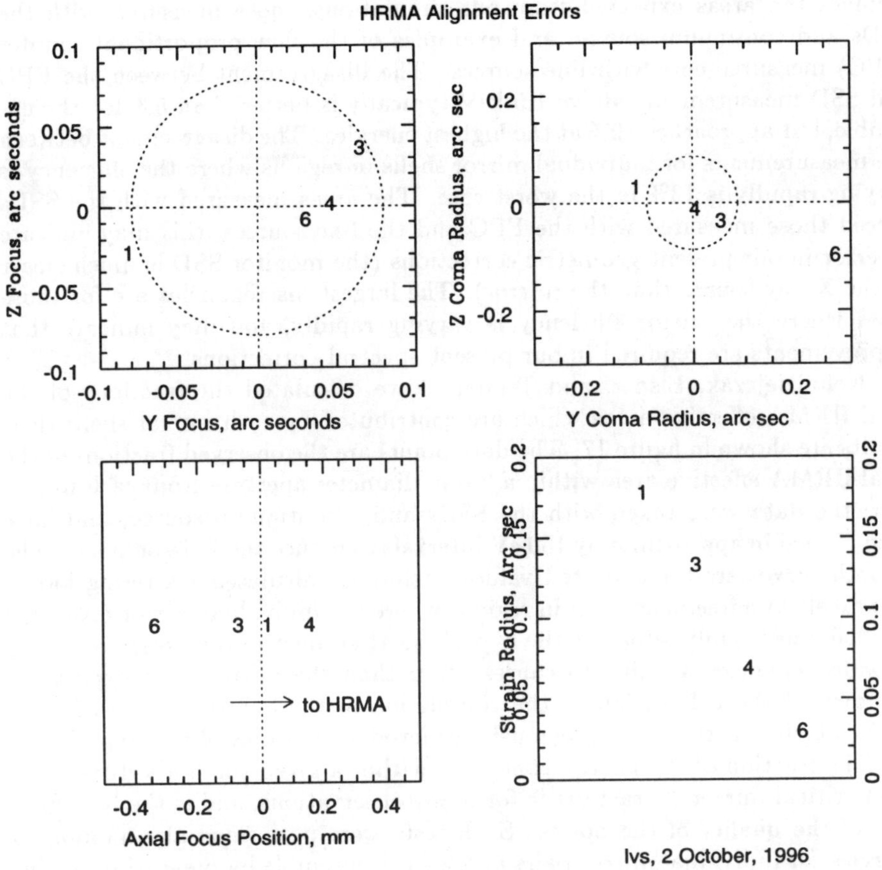

Figure 15: HRMA alignment errors measured in visible light at EKC. The upper left panel shows the lateral focus offsets from the best system focus, the upper right panel contains the optically measured coma radius for each pair. The dashed circles indicate the budgeted allowances for these terms; the only significant error is the coma term for mirror pair 6, which resulted from an asymmetry in the bonding procedure that was corrected for the later pairs. The lower two panels show the axial focus offsets from best focus and an estimate of the support induced strain terms respectively.

Preliminary effective areas for the ensemble measured within a 2 mm diameter (approximately 40 arc seconds) aperture are given in figure 16. The plot includes the areas expected from our simulations, those measured with the SSDs and continuum source, and examples of the flow proportional counter (FPC) measurements with line sources. The disagreement between the FPC and SSD measurements above 1.5 keV typically is better than 5% for the ensemble, but approaches 10% at the highest energies. The disagreement between the measurements for individual mirror shells in regions where the efficiency is varying rapidly is 13% in the worst case. The areas measured with the SSDs exceed those measured with the FPC and the line sources; this may indicate an error in our present geometric corrections (the monitor SSD is much closer to the X–ray source than the mirror). The largest discrepancies are for those cases where the mirror efficiency is varying rapidly, and may indicate that improvements are required in our present spectral corrections.

Kolodziejczak, Elsner, and Tennant have calculated the fractions of the total HRMA effective areas which are contributed by each mirror shell; their results are shown in figure 17. The data points are the observed fractions of the total HRMA effective area within a 2 mm diameter aperture from each mirror pair; the data were taken with the SSDs and a continuum source, and have been binned in approximately 150 eV intervals to reduce the Poisson noise. The smooth curves are the expected values, including calculated scattering losses. The excellent agreement, even in regions where the individual mirror responses are changing rapidly, supports the hypothesis that much of the present effective area discrepancies are due to causes other than the mirror. We expect the accuracy of the calibration at the completion of the analysis to be of order 1-2%, which is better than previously achieved with optics of this type.

The fraction of the energy contained within a given small aperture is the most critical mirror characteristic for many observations, and is the best measure of the quality of the optics. Such tests were performed at a number of energies for individual mirror pairs and for the ensemble by measuring the image power contained within circular apertures of varying sizes. Some examples of these results for on-axis tests of individual pairs are given in figures 18–21, and for the HRMA ensemble in figures 22–23. These curves are normalized to unity at an aperture diameter of 35 mm (about 12 arc minutes); integration of wide angle scattering data shows that there is never more than 4.0% of the image energy beyond this diameter, and typically much less, for example 0.4% at 1.486 keV. There are three important characteristics of these results; first, the fractions obtained within representative apertures for AXAF are high; second, the results do not vary greatly with energy; and finally, the HRMA model describes the data accurately.

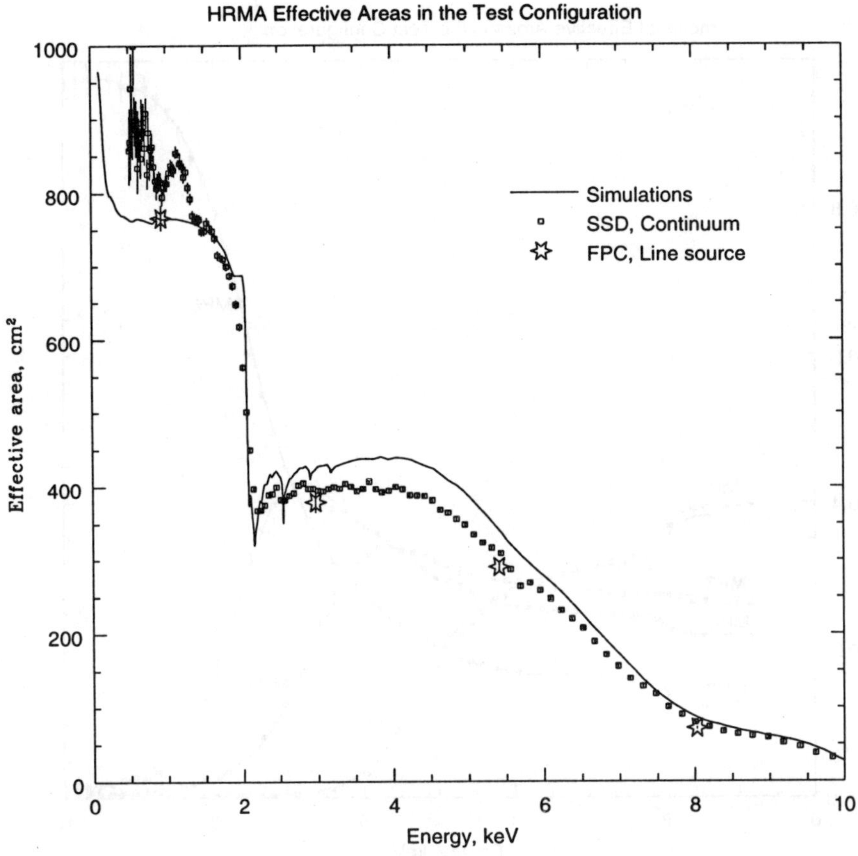

Figure 16: The dashed line and points give the effective areas within a 40 arc second diameter aperture which were measured at the XRCF using a continuum dominated source and the SSDs. The four isolated points are equivalent measurements taken with a line dominated source and the FPC detectors. The solid line is a plot of the effective areas expected for these measurements from simulations.

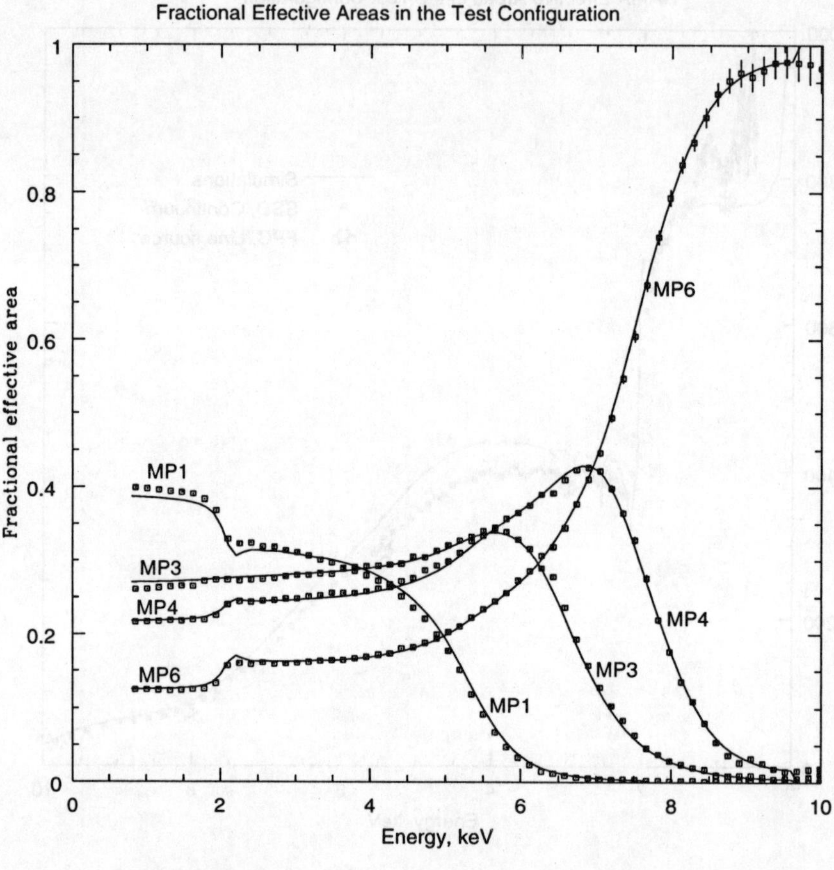

Figure 17: The smooth curves are plots of the fraction of the HRMA effective area expected from each mirror pair as a function of energy. The points are the fractions measured using the SSD and a continuum source. The data reduction was provided by J. Kolodziejczak, R. Elsner, and A. Tennant.

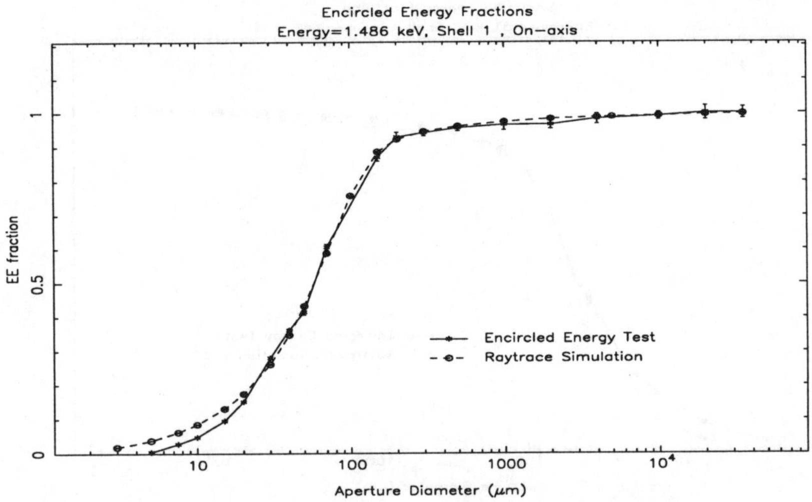

Figure 18: Measured encircled energy fractions for mirror pair 1 on axis at 1.486 keV. The smooth curves are the predicted encircled energy fractions for the model of the HRMA in the test configuration.

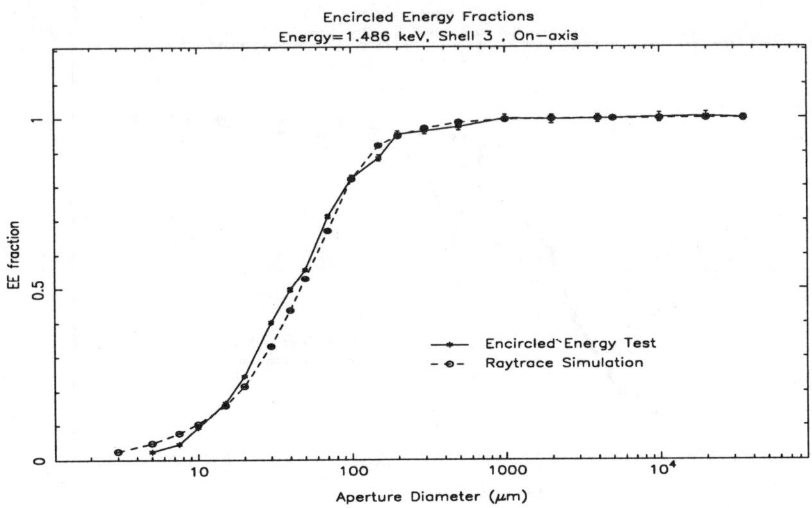

Figure 19: Measured encircled energy fractions for mirror pair 3 on axis at 1.486 keV. The smooth curves are the predicted encircled energy fractions for the model of the HRMA in the test configuration.

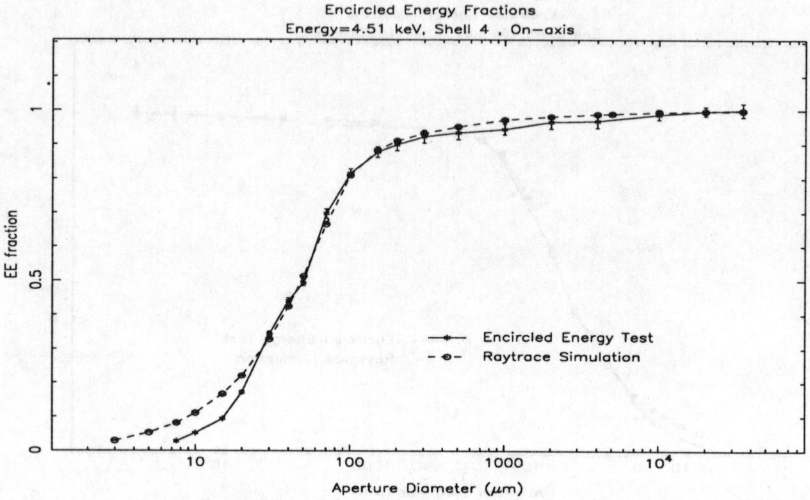

Figure 20: Measured encircled energy fractions for mirror pair 4 on axis at 4.51 keV. The smooth curves are the predicted encircled energy fractions for the model of the HRMA in the test configuration.

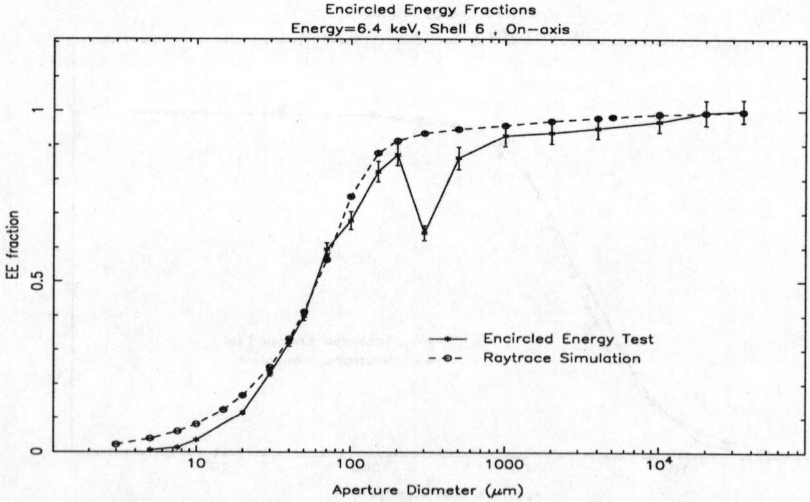

Figure 21: Measured encircled energy fractions for mirror pair 6 on axis at 6.4 keV. The obviously spurious point is being investigated. The smooth curves are the predicted encircled energy fractions for the model of the HRMA in the test configuration.

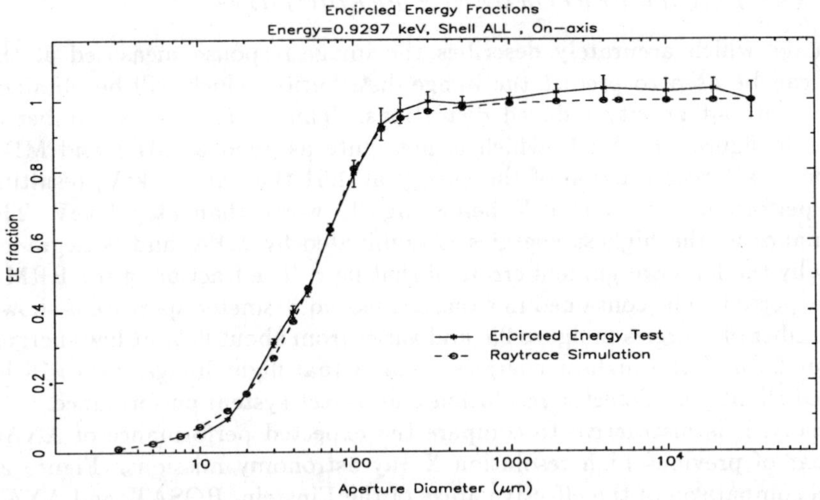

Figure 22: Measured encircled energy fractions for the mirror ensemble on axis at 0.9297 keV. The smooth curve is the predicted encircled energy fraction for the model of the HRMA in the test configuration.

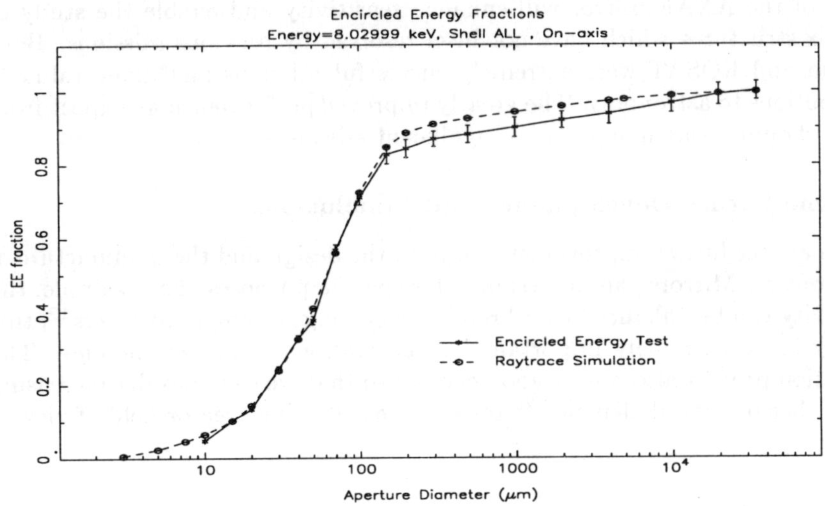

Figure 23: Measured encircled energy fraction for the mirror ensemble on axis at 8.02999 keV. The smooth curve is the predicted encircled energy fraction for the model of the HRMA in the test configuration.

4.3 AXAF FLIGHT PERFORMANCE PREDICTIONS

The model which accurately describes the image response measured at the XRCF can be used to predict the image distribution which will be obtained in flight without gravity induced distortions. This is done for a number of energies in figure 24. MP1, which is not quite as good as MP3 and MP4, contributes a larger fraction of the energy at 4.51 than at 6.4 keV, resulting in the performance at 4.51 keV being slightly worse than at 6.4 keV. The performance at the highest energies is dominated by MP6, and is degraded slightly by the larger alignment errors of that pair. The fraction of the HRMA image expected to be contained in a one arc second diameter aperture is shown for a number of energies in figure 25, and varies from about 80% at low energies to about 65% at the highest energies. The actual flight images also will be degraded slightly by detector resolution and aspect system performance.

Finally, it is instructive to compare the expected performance of AXAF with that of previous high resolution X-ray astronomy missions. Figure 26 shows a comparison of the effective areas of the Einstein, ROSAT, and AXAF mirrors as a function of energy. The larger effective area and wider bandwidth shown in the left hand panel will be valuable for the study of fainter and/or higher temperature sources. Figure 27 shows a comparison of the expected encircled energy distributions of the three missions. The better image distribution of the AXAF mirror will enhance sensitivity and enable the study of complex structures which have not been resolved by previous missions. Both Einstein and ROSAT were extremely successful missions and made valuable contributions to astronomy. The greatly improved performance we expect from AXAF should result in even more significant advances.

5 Some Future Developments and Conclusions

We can expect future improvements in both the design and the manufacture of X-ray optics. Mirrors can be designed for specific purposes. For example, the reflectivity can be enhanced over broad energy ranges using multilayers [19]; this is likely to be extremely important for observations at higher energies. The fabrication process also now is good enough so that we can consider increasing the number of optical elements [25] to improve collecting area or field of view.

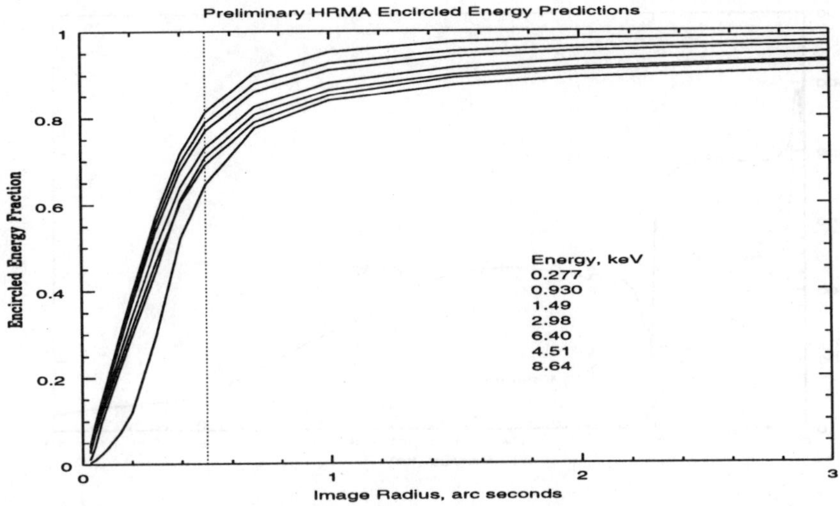

Figure 24: Predicted encircled energy fractions as a function of aperture diameter for the HRMA in flight. MP1, which is not quite as good as MP3 and MP4, contributes very little to the image at 6.4 keV; this results in slightly better performance at 6.4 than at 4.51 keV. The actual flight performance will be degraded slightly by detector resolution and aspect system performance.

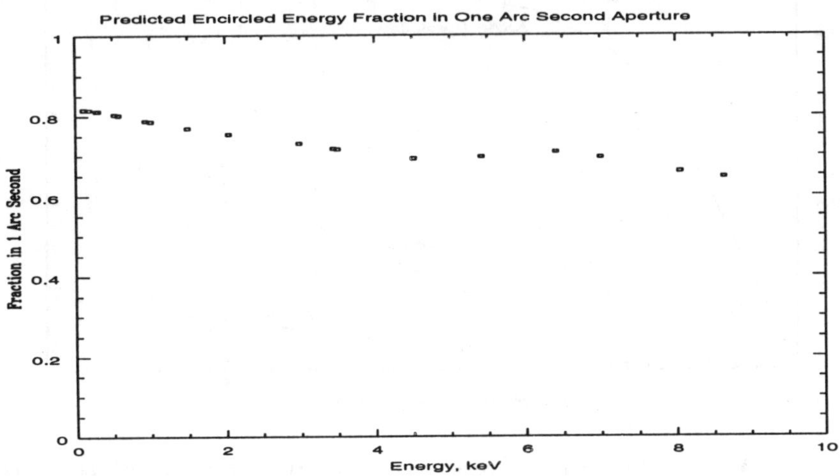

Figure 25: Fraction of the image energy expected to be contained in a one arc second diameter aperture. The actual flight performance will be degraded slightly by detector resolution and aspect system performance.

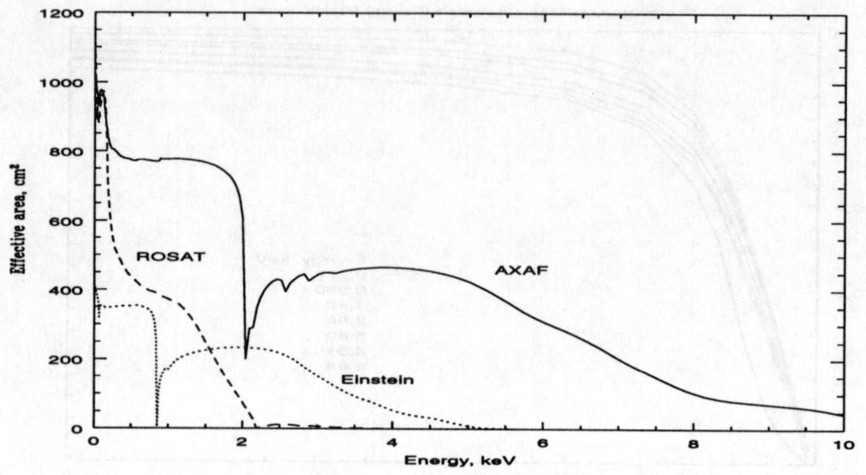

Figure 26: The expected effective area as a function of energy for three high resolution X-ray mirrors. The wider bandwidth of the AXAF mirror will be valuable for the study of higher temperature systems.

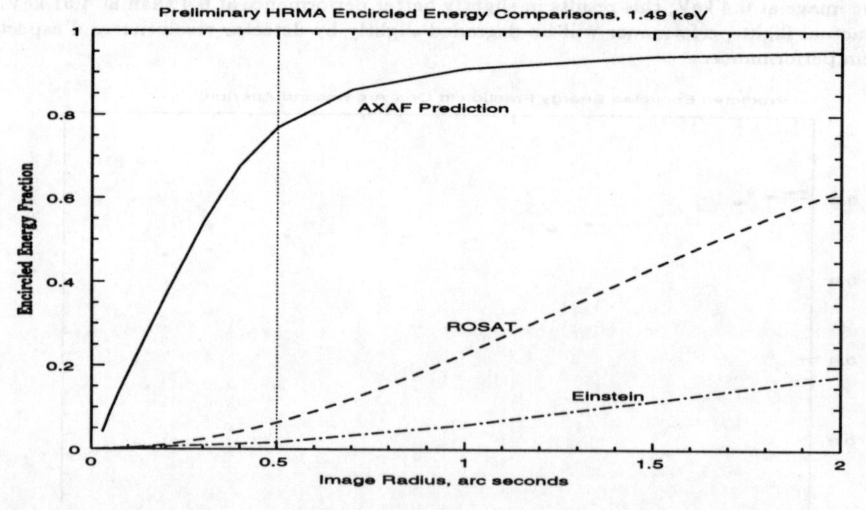

Figure 27: Comparison of the expected encircled energy performance of AXAF at 1.49 keV with previous high resolution X-ray astronomy missions; the scale has been selected to emphasize the improvement for one arc second images, which is smaller than the resolution specification for ROSAT. Both Einstein and ROSAT were extremely successful missions and made valuable contributions to astronomy; we also have high expectations for AXAF.

A more mature example of a design intended to be optimized for a specific purpose has been given by Burrows, Burg, and Giacconi[4] who designed a two reflection system similar to a Wolter type–I design, but with a polynomial correction calculated to optimized a specific figure of merit. These authors chose as a figure of merit the quantity:

$$\sqrt{\left(\int \theta\sigma^2 \, d\theta\right) / \left(\int \theta \, d\theta\right)}$$

where θ is the field angle and σ is the rms image spot size, and the integration takes place over the entire field of view. A slightly more general merit function is given by:

$$\bar{\sigma} = \left(\frac{\int_0^{\Theta} \theta \left(\frac{A}{A_0}\right)^a \sigma^p \, d\theta}{\int_0^{\Theta} \theta \, d\theta} \right)^{\frac{1}{p}}$$

where p, a, and Θ are input parameters, σ is the rms radius in angular units, A is the geometric area ($\sim$ the low energy effective area) of the telescope, and the normalizing area A_0 is the quantity A for on axis rays and the initial Wolter type–I design. Some results for the example $p = (-3/2)$, $a = (3/2)$[e], and various values of Θ are given in figure 28. There are substantial advantages of this type of design for survey experiments.

The relatively thick walls which have been required for high resolution mirrors result in an inefficient use of the frontal aperture and severely weight limited systems, and so existing X–ray mirrors have been designed for either good angular resolution or large area. However, future experiments will require mirrors which provide *both*, and hence it will be necessary to develop light, thin walled mirrors which still yield good resolution. The fabrication of such surfaces is almost possible with existing techniques because the preferred material removal processes are not sensitive to small distortions of the optical elements, and mirrors can be supported (with great care) to avoid distortions during the measurements of surface errors. However, the final alignment of such thin walled elements in an optical system may be more easily accomplished using mechanisms in a zero gravity environment.

[e]These values can be shown to maximize the number of sources detected from a population which is uniformly distributed in a Euclidean universe with an instrumentally dominated background.

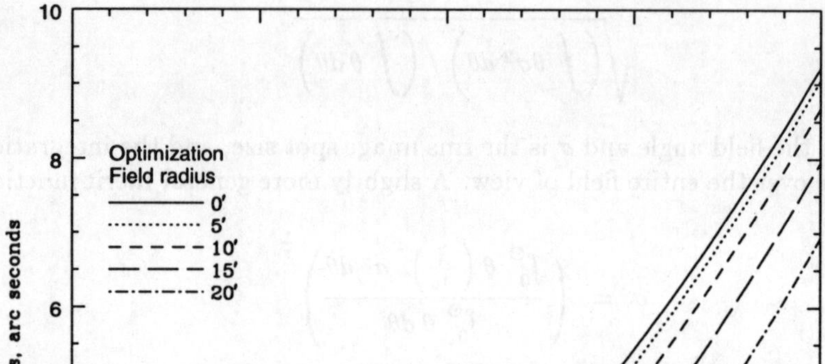

Figure 28: The design is optimized to maximize the number of sources detected within the optimization field radius for the case of uniformly distributed sources in a Euclidean universe with an instrumentally dominated background. The criterion is similar to that developed by Burrows, Burg, and Giacconi[4].

Acknowledgments

Many people contribute to large space projects; for example, hundreds of people at NASA, TRW, HDOS, EKC, OCLI, and SAO have made valuable contributions to the AXAF optics and to their calibration. Some of those who should be given extra credit are Paul Reid, Matt Magida, Tom Gordon, and John Patterson of HDOS, Bob Hahn and Gerry Johnson of OCLI, Charlie Atkinson and Gary Matthews of EKC, Paul Glenn of Bauer Associates, Steve Loer and Scott Texter of TRW, John Humphreys, Danny Johnston, Martin Weisskopf, and Steve O'Dell of MSFC, Harvey Tananbaum, Lester Cohen, Mark Freeman, Bill Podgorski, Dan Schwartz, and Ed Kellogg of SAO, and many, many more.

One person, Riccardo Giacconi, deserves much of the credit for the very existence of the field of X-ray astronomy, and also for the development of focusing X-ray optics, the essential technique for most of our current observations. His work has resulted in many of the dramatic improvements in our understanding of the universe which are discussed in other sections of this book.

References

1. B. Aschenbach, *Rep. Prog. Phys.* **48**, 579 (1985).
2. B. Aschenbach, *Appl. Optics* **27**, 1404 (1988).
3. P. Beckmann and A. Spizzichino, *The Scattering of Electromagnetic Waves from Rough Surfaces* (Pergamon Press, NY, 1963).
4. Christopher J. Burrows, Richard Burg, and Riccardo Giacconi, *ApJ* **392**, 760 (1992).
5. R. C. Chase and L. P. Van Speybroeck, *Appl. Optics* **12**, 1042 (1973).
6. Lester M. Cohen, *Proc. SPIE* **2515**, 375 (1995).
7. A. H. Compton, *Phil. Mag* **45**, 1121 (1923).
8. M. Freeman, J. Hughes, L. Van Speybroeck, J. Bilbro, and M. Weisskopf, *Proc. SPIE* **1742**, 136 (1992).
9. R. Giacconi, N. F. Harmon, R. F. Lacey, and Z. Szilagyi, *J. Opt. Soc. Am.* **55**, 345 (1965).
10. R. Giacconi, W. P. Reidy, G. S. Vaina, L. P. Van Speybroeck, and T. F. Zehnpfennig, *Space Sci. Rev.* **9**, 3 (1969).
11. R. Giacconi, W. P. Reidy, T. Zehnpfennig, J. C. Lindsay, and W. S. Muney, *ApJ* **142**, 1274 (1965).
12. R. Giacconi and B. Rossi, *J. Geophys. Res.* **65**, 773 (1960).
13. Paul Glenn, *Proc. SPIE* **2515**, 352 (1995).
14. Paul Glenn, Paul Reid, Albert Slomba, and Leon P. Van Speybroeck,

Appl. Optics **27**, 1539 (1988).
15. P.Gorenstein, *this volume*, (1997).
16. R. A. Harrison, E. C. Sawyer *et al*, Solar Phys. **162**, 233 (1995).
17. B. L. Henke, E. M. Gullikson, and J. C. Davis, *Atomic Data and Nuclear Data Tables* **54**, #2 (1993)(Revised September, 1995).
18. B. L. Henke, P. Lee, T. J. Tanaka, R. L. Shimabukuro, and B. K. Fujikawa, *Atomic Data and Nuclear Data Tables* **27**, 1 (1982).
19. K. D. Joensen, P. Høghøj, F. Christensen, P. Gorenstein, J. Susini, E. Ziegler, A. Freund, and J. Wood *Proc. SPIE* **2011**, 360 (1994).
20. P. Kirkpatrick and A. V. Baez, *J. Opt. Soc. Am.* **38**, 776 (1948).
21. D. Korsch, chapter 11.7 of *Reflective Optics* (Academic Press, Inc., San Diego, CA, 1991).
22. K.Pounds, *this volume*, (1997).
23. Paul B. Reid, *Proc. SPIE* **2515**, 361 (1995).
24. David L. Shealy and Timo T. Saha, *Appl. Optics* **29**, 2433 (1990).
25. A. Szentgyorgyi, M. Elvis, A. Roll and L. van Speybroeck, *Proc. SPIE* **2198**, 1048 (1994).
26. J.Trumper, *Adv. Space Res.* **2**, 241 (1983).
27. J.Trumper, *this volume*, (1997).
28. W.Tucker and H.Tananbaum, *this volume*, (1997).
29. J. H. Underwood and W. S. Muney, *Solar Phys.* **1**, 129 (1967).
30. J. H. Underwood and W. M. Neupert, Solar Phys. **35**, 241 (1974).
31. G. S. Vaiana, W. P. Reidy, T. Zehnpfennig, L. Van Speybroeck, and R. Giacconi, *Science* **161**, 564 (1968).
32. G. S. Vaiana, L. Van Speybroeck, M. V. Zombeck, A. S. Krieger, J. K. Silk, and A. Timothy, *Space Sci. Instr.* **3**, 19 (1977).
33. Leon P. Van Speybroeck, *Proc. SPIE* **184**, 2 (1979).
34. L. P. Van Speybroeck and R. C. Chase, *Appl. Optics* **11**, 440 (1972).
35. Leon. P. Van Speybroeck, D. Jerius, R. Edgar, T. Gaetz, P. Zhao, and P. Reid, *Proc. SPIE* **3113**, 89 (1997).
36. Leon. Van Speybroeck, Paul Reid, Daniel Schwartz, and James Bilbro, *Proc. SPIE* **1160**, 94 (1989).
37. Mark Waldman, *Proc. SPIE* **2515**, 330 (1995).
38. M. C. Weisskopf, S. L. O'Dell, R. F. Elsner and L. P. Van Speybroeck, *Proc. SPIE* **2515**, 312 (1995).
39. M. C. Weisskopf, S. L. O'Dell, and L. P. Van Speybroeck, *Proc. SPIE* **2805**, 2 (1996).
40. H. Wolter, *Ann. Physik* **10**, 94 (1952).
41. H. Wolter, *Ann. Physik* **10**, 286 (1952).

THE DIVERGENT EVOLUTION OF THE X-RAY TELESCOPE

PAUL GORENSTEIN
Harvard-Smithsonian Center for Astrophysics
Cambridge, MA 02138 USA
E-mail: goren@cfa.harvard.edu

The focusing telescope which preceded X-ray astronomy has become the principal instrument of an X-ray observatory and will be even more predominant in the future. We can expect the focusing telescope to evolve into several instruments with specific adaptations. The launch of XMM will initiate the era of high throughput telescopes which will be the keystone of the next generation X-ray observatories of the US and Europe. Focusing telescopes are also being developed for the 10 to 100 keV band. In addition the lobster eye geometry is receiving considerable attention on account of its very wide field of view capability which would permit new classes of instruments for broad sky monitoring and all sky surveys.

1 Introduction

The X-ray telescope predates X-ray astronomy. Two years before the discovery of cosmic X-rays a 1960 paper by Riccardo Giacconi and Bruno Rossi(1) proposed that a focusing X-ray telescope be employed to study cosmic X-rays. An idea that was introduced ahead of its time has matured into the primary tool of X-ray astronomy. The future of the focusing telescope is even more promising. It is at the heart of every major mission to be launched at the end of this century, namely AXAF, XMM, Astro-E, and Spectrum-X-Gamma. Indeed the focusing telescope is now so pervasive that it is virtually impossible to imagine a future X-ray observatory not based upon one. However, just as X-ray astronomy has evolved from an era of discovery and cataloging of sources to an era of specific studies such as AGNs, the evolution of structure of clusters and superclusters of galaxies, activity in stellar coronae, and temporal variations in all compact objects, we can expect the X-ray telescope itself to adapt to the special requirements of a particular study. Indeed this divergence has already begun as the focusing telescope begins evolving into a family of instruments with specific adaptations such as high angular resolution, high throughput, broad bandwidth, or wide field of view. The high angular resolution telescope has been the mainstay of X-ray astronomy, and will continue a play a major role. That subject will be addressed in another chapter by Leon Van Speybroeck. This chapter will discuss various X-ray telescopes optimized for other properties with the most emphasis upon high throughput.

This author has had the good fortune of witnessing much of and participating in some of the early development of the focusing X-ray telescope that1 took place at American Science and Engineering and the Smithsonian Astrophysical Observatory under the direction of Riccardo Giacconi. Consequently, with this experience I take the liberty of writing from a personal perspective, including a speculative section on a future high throughput telescope configuration. Reasonable speculation is perfectly consistent with the spirit of exploration and discovery that Riccardo Giacconi has consistently encouraged, among his colleagues and students and belongs in any anthology of work such as this which has benefitted so much from his influence.

2 High Throughput Telescopes

2.1 Introduction

The initial use of X-ray telescopes was high angular resolution imaging as exemplified by Einstein Observatory and ROSAT, the first cosmic X-ray satellites based exclusively on focusing optics. While their great success shows that high resolution X-ray imaging will always be an integral part of astronomy, we know that answering many fundamental questions in astrophysics and cosmology requires instruments with very large photon collecting power but not always with the highest angular resolution. High throughput focusing telescopes would combine the advantages of large area counters: large aperture, broad bandwidth, and wide field with the advantages of focusing: background reduction, confusion avoidance, and the ability to image the structure of a low surface brightness, low contrast object.

The need for high photon collecting power can be demonstrated by comparing the number of photons required for measuring the position of a source to the number required to obtain quantitative information from its spectrum. A high resolution focusing telescope can establish the existence of a point X-ray source and measure its position with merely ten photons. On the other hand, an X-ray line is often no more than ten per cent of the total spectral intensity. High statistical precision measurements of the line's centroid, width, and strength relative to another line require detecting more than a hundred photons in that line, or a thousand photons from the object. Furthermore, a very high resolution spectrometer based upon a dispersive element is inefficient. Hence, the total number of photons the telescope must concentrate for the highest resolution spectroscopy is higher by perhaps three orders of magnitude than merely positioning an object. Other studies requiring high throughput include future all sky surveys, imaging low surface brightness objects and timing studies. Therefore while much of the exploratory work, such as detecting, positioning and identifying sources may be accomplished by modest area high resolution telescopes, fundamental astrophysics and cosmology investigations will ultimately require focusing telescopes with much greater photon collecting power.

The principal attributes of a high throughput telescope compared to a high resolution telescope of equal aperture are significantly more effective area and bandwidth. It also has lower mass and cost per unit area allowing a mass or cost limited payload to have a larger aperture. It reflectors can be packed more densely because they are lighter and thinner. The angular resolution is invariably poorer because the optimum figure control processes, accurate machining and polishing of a rigid optical element, are not options. However, the angular resolution must not be compromised to the point where source confusion becomes the limitation. For very large area systems it should be no worse than a few arcseconds.

Four types of optics were proposed for high throughput X-ray telescopes. Two, epoxy replication and electroforming were first employed in the 1960's by Riccardo Giacconi's group at American Science and Engineering for use on rockets and an OSO satellite. Eventually both led to orbiting telescopes, the epoxy replicated Wolter 1 mirrors of EXOSAT and the electroformed mirrors of Beppo SAX. A third type, thin foil segmented telescopes, was developed at the Goddard Space Flight Center and is the basis of the ASCA mirrors. The fourth is a prototype mirror assembly consisting of two orthogonal stacks of paraboloidal reflectors. It is often referred to as a "Kirkpatrick-Baez" telescope.

The era of the high throughput telescope is rapidly approaching with the launches of XMM and Spectrum-X-Gamma currently scheduled for 1999. Those missions and even larger successors will dominate the future of X-ray astronomy. This chapter will describe each of these mirror types. It will include ideas on a very high throughput telescope system that could be built some two generations hence.

2.2 The Kirkpatrick-Baez Telescope and LAMAR

The first description of a high throughput focusing telescope system (at least known to this author) was made by Gorenstein in 1973 (2) in an unsolicited report to a working group organized by NASA to consider the future of high energy astronomy and uses of the "Space Laboratory", a concept that was a precursor to the International Space Station. The paper described a system of identical telescopes and detectors filling an aperture several tens of square meters in area. This configuration is totally equivalent to a single large telescope of equal aperture and angular resolution. However, it is much simpler to construct as the individual modules (telescope plus detector/spectrometer) are relatively light weight and do not have to be precisely coaligned. The idea was to develop a telescope module that could be copied over and over again at low cost. In this way an array could be built up incrementally on the Space Laboratory by delivering and attaching more modules. During all intermediate phases the system could be utilized in observations. This large area modular array of reflectors was called "LAMAR" in a subsequent paper.

My interest in high throughput modular arrays of focusing telescopes was influenced and strongly encouraged by Riccardo Giacconi. From the beginning, he advocated the position that the development of high resolution and high throughput telescopes proceed simultaneously parallel paths. Indeed, Riccardo Giacconi's initial concept for the HEAO observatory, known as "LOXT" included a high throughput as well as a high resolution telescope. The high throughput telescope was eliminated in the restructuring of the HEAO program.

We discussed the LAMAR experiment as a route for the development of a high throughput system. Indeed, an anthology of articles that describe the many contributions and positive influence of Riccardo Giacconi would be incomplete without a section devoted to high throughput focusing telescopes.

Two versions of the LAMAR experiment were selected for development. The first selection was for flights aboard the Space Shuttle. It occurred in 1979 and was a modest size system with 1000 cm^2 effective area at 1 keV. However, when the Challenger disaster in 1986 resulted in the loss of three years of flights many experiments including the LAMAR instrument were cancelled. The second proposed LAMAR instrument was larger (5000 cm^2 at 1 keV), included dispersive spectroscopy and was selected as an attached payload for the Space Station in 1989. Soon afterwards budgetary problems forced the Space Station program to undergo a complete restructuring that resulted in the cancellation of all attached payloads. Nevertheless, LAMAR'S concept of a modular system of relatively low cost multiple identical mirrors observing simultaneously, no doubt helped to inspire subsequent high throughput telescope systems which met with a more successful fate such as the European XMM.

The Kirkpatrick-Baez Telescope

The Kirkpatrick-Baez telescope consists of two orthogonal nested stacks of one dimensional parabolas. Each stack by itself forms a line image. Its shape is rectangular rather than the circular geometry of the Wolter types. A series of three telescopes of increasing size were built for and flown on sounding rockets by my group at AS&E in the late 60's, early 70s. Another was developed at Columbia by M. Weisskopf and co-workers. What made this device feasible and low cost was the availability of relatively thin, 1.8mm, commercial float glass with good flatness ($<$ 20 arcsec deviations after selection) and smoothness ($<$ 7 Å rms). Construction of a telescope consisted of setting the gold coated reflectors in place, adjusting the figure to the appropriate paraboloidal shape, and fixing them. In connection with the LAMAR Space Shuttle experiment we developed an automated figure formation system (3) that adjusted the positions of four points each along the top and bottom of the reflector as the 1D image was viewed simultaneously and deviations were fed back to the computer that controlled the system. The system was able to impart the correct the figure in a half hour of unattended operation. (However, it took considerably longer for the epoxy bonds that fixed the positions to set reliably.) The precision and reproduceability of the system was about 2 arcseconds but the resolution of the reflector was limited by figure errors along its length that are not correctable by a finite number of adjustment points and by elasticity effects resulting from bending a plate.

A prototype mirror assembly consisting of two orthogonal sections was built and tested at the MSFC long pipe facility. Its power width in each dimension was about 30 arcseconds (4). This is about what was expected from the uncorrectable figure errors.

There are no plans to include K-B X-ray mirrors on any mission currently scheduled or under discussion. However, there still may be an important role for this optic in the future. At the end of this section we describe a very high throughput, segmented mirror system that we believe is best implemented in the K-B geometry. As another example of a possible future K-B mirror, Cash has described a very high resolution mirror built in a four reflection configuration of spherical rather than paraboloidal reflectors (5).

2.3 *Foil Telescopes*

High throughput X-ray telescopes made of thin aluminum foils were first made at the Goddard Space Flight Center by P. Serlemitsos and co-workers for the BBXRT Space Shuttle experiment (6) which was flown in December, 1990 as part of the Astro-1 payload. Each aluminum foil, about 0.13 mm thick, was cut to its prescribed dimensions and then rolled into conical segments. The problem of the surface smoothness was solved in a unique way. The conical segments were dipped into an acrylic bath and withdrawn at the optimum rate in a controlled atmosphere. The acrylic liquid wet the material thoroughly and hardened to form a smooth surface. Gold was evaporated over the acrylic. The collection of conical segments was then mounted and fixed in a double cone geometry approximating a Wolter 1 mirror with a 0.4 m diameter and 3.8 m focal length Each complete mirror assembly had an angular resolution of about 3 arcmins half power diameter. The resolution was limited by larger scale irregularities in the foils that are not removed by the acrylic.

Four similar telescopes were made at GSFC for ASCA. These mirrors and their two CCD and two gas scintillating proportional counter detectors have provided the most important spectroscopic data of recent years.

Two larger foil mirrors, 0.6m diameter and 8 m focal length, were constructed at the Danish Space Research Institute for the SODART instrument (7) of the Spectrum-X-Gamma mission which is currently scheduled to be launched by Russia in 1999. The method of construction was similar to ASCA's but used larger and thicker (0.4 mm) foils.

Rolled foil mirrors are relatively simple to construct and have the advantage of being very light weight per unit effective area. Their limitation is poor angular resolution. This is likely to discourage the construction of any acrylic coated foil mirror systems larger than SODART's except possibly as a light bucket for the special purpose of measuring temporal variations.

Epoxy Replication

The first replication process to be employed was epoxy replication of two Wolter 1 mirrors for EXOSAT. This method consists of several phases. The first is constructing a substrate from a light weight material of approximately the correct final size and figure but slightly larger. The EXOSAT substrate material was beryllium. The mandrel is made in the negative of the desired figure.

The bulk material is aluminum, but its top surface is a layer of a polishable material like electroless nickel which is polished to as good a finish as possible. The mandrel is coated with gold which acts as both a separation agent and as the reflecting surface of the replica after separation. Epoxy of a low viscosity is injected into the space between the mandrel and substrate and allowed to set. The aluminum mandrel has a larger coefficient of thermal expansion than the substrate. Separation is achieved by cooling the mandrel with liquid nitrogen. This results in the mandrel contracting away from the substrate leaving the epoxy and gold coating on the replica. The new surface of the substrate has a figure and finish only slightly inferior to those of the mandrel. The initial surface of the substrate does have to be sufficiently smooth. Otherwise, an attenuated rough texture will "print through" to the active side when the epoxy contracts upon setting. Epoxy replication is a relatively expensive process because of the need to fabricate the substrate in approximately the correct shape. This could be accomplished another replication process, for example by chemical vapor deposition of silicon carbide upon a sacrificial carbon mandrel (8).

GSFC is combining epoxy replication with segmented foil substrates to produce four mirrors for the ASTRO-E payload to be launched by Japan in 1999 (9). Epoxy replication replaces the acrylic dipping process. It promises to improve the surface quality of the substrates which would improve the angular resolution. The segmented foils are mounted by the same method as before.

Electroforming

The process of electroforming integral cylindrical X-ray telescope shells from nickel was applied to satellite instruments by O. Citterio and his associates at the Brera Observatory. Production of flight units for three missions was carried out by the Italian company Media Lario. Electroforming requires a mandrel that is very similar to the mandrel of epoxy replication and separation is also done by thermal means. Nickel accumulates around the mandrel in an electrolytic bath. The process has been applied to three missions with mirror systems of increasingly larger size. The first space application was the construction of four sets of 30 nested mirror shells with diameters ranging from 5.5 to 15 cm and a focal length of 1.85 m for the Italian/Dutch Beppo SAX satellite which is currently operating, A photograph of that set of mirror shells is shown in **Fig. 1**. The angular resolution is about 1 arcmin HPD and is limited by the figure which is a double cone rather than the parabola-hyperbola of the Wolter 1 optic. The four SAX mirrors have performed as expected.

Table 1				
		SAX	JET-X	XMM
Effective Area (At 1.5 keV)	(cm^2)	80	160	1475
Angular Resolution HEW (at 1.5 keV	(ARCSEC)	60	15	15
Optical Design		Double Cones	Wolter I	Wolter I
Focal length	(mm)	1850	3500	7500
Diameter of the Shells (Min-Max)	(mm)	68-162	190-300	300-700
Length of the Shells	(mm)	300	600	600
Thickness of the Shells (Min-Max)	(mm)	0.2-0.4	0.65-1.1	0.47-1.1
Number of shells		30	12	58
Weight of a Module	(kg)	13	67	425

Principal characteristics of the optics of SAX, Jet-X and XMM(8).

The next application was the construction of two larger mirrors for the JET-X instrument for the Spectrum-X-Gamma mission. These mirrors have been completed and are ready to be integrated into the spacecraft. Three 700 mm diameter mirrors are nearing completing for XMM.

The properties of three generations of electroformed mirrors are summarized in **Table 1**. Electroforming is a very promising technology for future missions. However, it suffers from two limitations. The more significant is the high density of nickel. It may limit the size of the optics in light of the goal of new missions to reduce launch costs by minimizing the mass of payloads. Replicas can be thinned only until residual stress of electroforming or separation forces causes them to distort. It may not be possible to improve the resolution beyond 15 arcseconds.

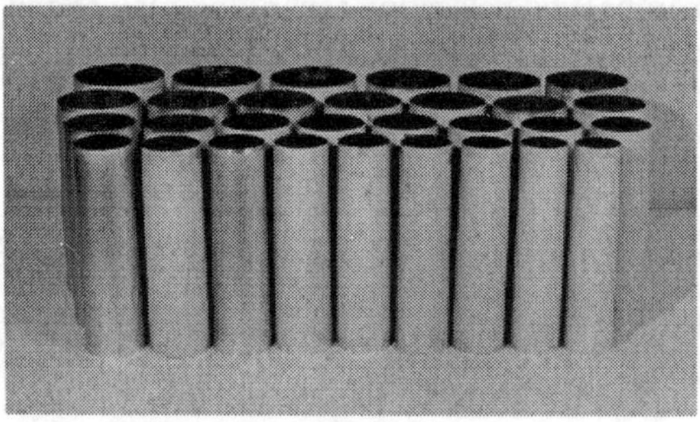

Figure 1: Electroformed double conical reflectors for one Beppo SAX Telescope.

Composite and Ceramic Substrates

Composite and ceramic integral cylindrical substrates are under investigation as a lighter weight alternative to electroformed nickel. They would be used in conjunction with epoxy replication as described above. A low viscosity epoxy is injected in the small space between a preformed substrate and a mandrel that is coated with a separation agent. Results obtained by Citterio and co-workers on silicon carbide substrates look quite promising (8). XMM studied carbon fiber reinforced epoxy (CFRP) substrates in conjunction with epoxy replication prior to adopting electroforming for their telescopes(10). CFRP proved to be too unstable, with respect to changes in the humidity environment. The search is on for more more stable materials and more progress is anticipated.

2.4 New High Throughput Missions

Introduction

As we prepare for the launches of the first true high throughput missions, XMM and Spectrum-X-Gamma, before the next century, the US, European, and Japanese X-ray astronomy communities are already planning future high throughput missions. Plans place a great deal of emphasis upon high resolution spectroscopy which is probably the important application of high throughput measurements. Thanks to the NASA New Mission Concept studies program initiated in 1995 the objectives and approach of the US program are rather well defined. The US astrophysics community has coalesced around the High Throughput X-ray Spectroscopy mission (HTXS, http://htxs.gsfc.nasa.gov). A European group has described the "Xeus" observatory. Xeus is a super XMM to be launched by an Ariane 5 rocket with possible participation by the International Space Station as an assembly or test platform. Everyone recognizes that a new or improved telescope technology is needed to provide a large factor increase in effective area with a much smaller factor increase in mass and with better angular resolution than XMM.

The High Throughput Spectroscopy Mission (HTXS)

The HTXS is a candidate for a new NASA mission to start in 2004. It will have a total effective area of at least 15,000 sq cm at 1 keV, 6,000 sq cm at 6.4 keV, and 1500 sq cm at 40 keV. It will contain two sets of coaligned telescopes to cover both the soft and hard band. High resolution spectroscopy will be achieved by the use of microcalorimeters for the 1 to 10 keV band and by dispersive reflection gratings for lower energies. The beam from the soft telescopes will be divided between the gratings and microcalorimeter similar to what is being done with gratings and the imaging CCDs of XMM. Hard X-ray telescopes will use moderate energy resolution detectors.

The innovative feature of HTXS is the use of six identical, relatively small satellites with launch intervals of 3-4 months. They will be in high Earth orbit or at the libration points for maximum viewing efficiency. The use of six identical smaller spacecraft as opposed to a single launch of a very large spacecraft will reduce cost and of course reduce the overall risk to the mission from a launch failure. The cost is lowered because the technology for the smaller spacecraft can be adapted from the new of generation of satellites and their launch vehicles that will be built in large numbers for global communications and television networks. This avoids the large development costs of a single, large custom spacecraft and the costs and risks of launching it with a very large, rarely used rocket.

The soft X-ray and hard X-ray telescopes of HTXS will have a focal length of 8.5 m and will utilize an extendible optical bench to permit a Delta class launch in a shorter configuration. A sketch of a single HTXS spacecraft is shown in **Fig. 2**.

2.5 The Ultimate High Throughput Telescope

The coming generation of high throughput telescopes, XMM and Spectrum-X-Gamma, has effective area of about 3000 sq cm. The next generation HTXS multiple satellite system and its European counterpart will provide nearly ten times as much effective area. Elvis and Fabbiano (11) have described a series of scientific objectives that require an additional order of magnitude larger effective area than HTXS, plus very good angular resolution. It would have to have a total aperture of 8 to 10 M to attain an effective area of 200,000 sq. cm. Although it is too early to begin developing such a telescope system one can still ask the question of how it could be constructed and launched.

We can say the following. One, since a 200,000 sq cm effective area telescope will, inevitably, be an expensive construction project, it should be a long lived observatory where detectors can be serviced or changed out as new technology becomes available. Two, as it would study very faint sources, its angular resolution will have to be better than a few arcseconds to avoid confusion and have the ability to image galaxies and clusters of galaxies in formation. (Fortunately, as a result of relativistic cosmology effects the decrease in angular size ultimately ceases as an object is removed to increasingly larger distance from the observer.) A single monolithic telescope is not a desirable option. The thickness of a reflector increases with size. Large radii reflectors become excessively massive as well as difficult to construct. Furthermore, it is not desirable to rely totally upon a single launch or even possible to launch a huge monolith with existing vehicles. The HTXS approach of multiple spacecraft is not extendible to another order of magnitude in size, i.e. the order of 60 spacecraft. Constructing and tracking such a network of spacecraft are very complex and expensive operations. Furthermore, it is not possible to service or update the detectors of such a far flung network. A multiple mirror array with many mirrors and detectors would also suffer from the inability to service or change detectors.

I describe an approach that appears to satisfy the constraints. It is a large diameter, (10 m diagonal) segmented single-focus telescope with a focal length of 50 to 100 m. Alternatively, it could be an array of a few telescopes of smaller size and shorter focal length. We dispense with an optical bench. Instead, we propose that the detectors and telescope be on separate spacecraft with their own power, communication and attitude control systems. The mirror array is situated on a larger spacecraft and is unfolded. The detector is situated on a smaller, SMEX like spacecraft. There can be several such spacecraft with different types of detectors. (For dispersive spectroscopy there is a third spacecraft in the alignment which accommodates a grating or crystal array.) The two (or three) spacecraft perform station keeping during an observation. The smaller spacecraft has a propulsion system and adjusts its orbital position to restore the detector to the focus when the telescope is pointed at another target. With a simple ranging system (and coarse positioning from the GPS) it

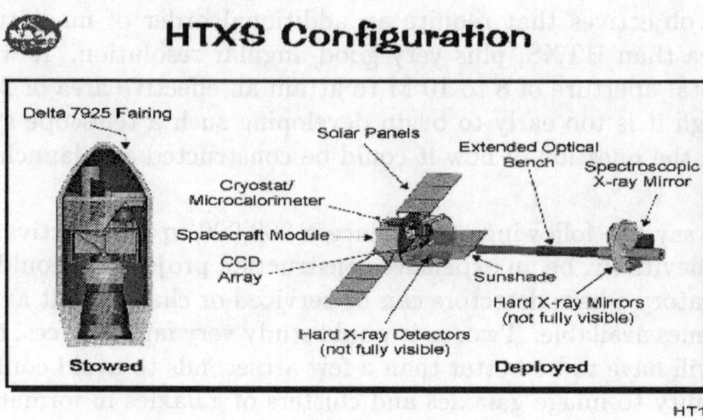

Figure 2: Sketch of one (of six) space craft with instruments for the proposed High Throughput X-ray Spectroscopy mission.

should be possible to place the detector within a few millimeters of the focus. This is sufficient accuracy for long focal lengths.

The Kirkpatrick-Baez type geometry seems well adapted to a large, segmented, unfoldable mirror system, perhaps better than the cylindrical Wolter geometry. **Fig. 3** shows a large monolithic K-B mirror together with a segmented mirror with the same total aperture. The positions and orientations of each segment are adjustable. Each of the telescope segments is secured at several corners by precise linear positioners with a limited range of movement in three dimensions. It also has a bright star sensor. This system of controls is able to position each segment into alignment with a bright star as often as needed. All segments are simple rectangular boxes with the same outer dimensions of about 50 cm cubed. Along the same column, the segments and the reflectors within are nearly identical and many are fully interchangeable with each other. All reflectors are identical flats before their figures are imparted. Within a segment, the reflectors themselves can be segmented along the direction of the optic axis. In fact, if an angular resolution of ten arcseconds is acceptable, each parabolic reflector can be approximated by a series of shorter flats oriented at slightly different angles along the optic axis direction. Imparting the correct parabolic curve or a close approximation of it to each segment will greatly improve the angular resolution. This is shown in **Fig. 4**.

The rear section of each module (hidden in **Fig. 3**), whose reflectors are orthogonal to the front section reflectors, can be either be in its own box

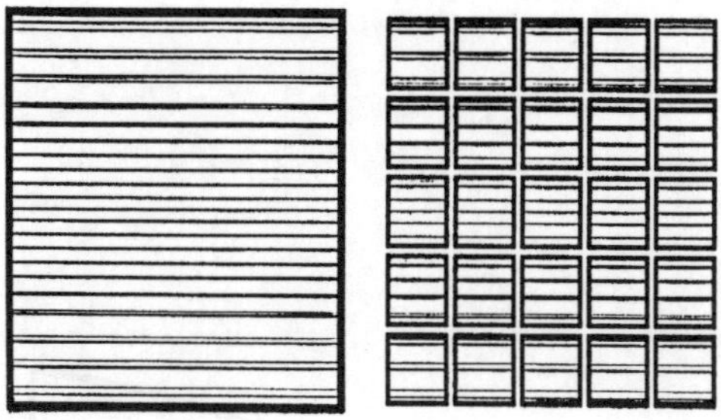

Figure 3: Monolithic and segmented Kirkpatrick-Baez mirrors.

with its own position adjustment system decoupled from the front section or be integrated with it in the same box and share a common, positioning system.

An orthogonal view (**Fig. 5**) illustrates how this mirror system could be folded into a more compact volume for launch and then deployed on orbit. An 8 m×8 m aperture can be folded into a 2 m× 2m × 8 m volume that fits into the Space Shuttle or a larger expendable rocket. It can be subdivided into four sections with each on a separate Delta rocket. The sections can either be joined on the Space Station and the array deployed in its own orbit or launched individually to form a few member multimirror array.

The intrinsic angular resolution advantages of an ideal Wolter mirror compared to a K-B mirror do not apply to a large, segmented single-focus telescope. The Wolter reflectors can no longer be complete cylinders of revolution. Therefore their superior rigidity is lost. The angular resolution of a K-B system with reflectors that are exact paraboloids is limited by the fact that the focus from the front section is rendered imperfect following reflection by the rear section. However, with a segmented mirror, this aberration can be corrected by a slightly varying the focal length of each front section segment. The focal length of the front section of each segment would vary approximately linearly with distance of its rear section from the optic axis. This results in a nearly perfect arcsecond focus. Perhaps most significant is the fact that the K-B reflectors are still nearly flats. On the other hand the Wolter reflectors are highly curved in the azimuthal direction and the curvature varies over a wide range. Production of many reflectors with dissimilar shapes will be a more complex process than a batch that are very similar to each other.

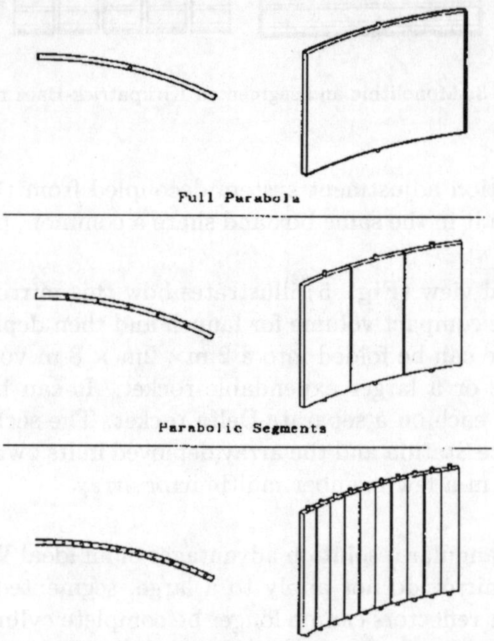

Figure 4: Parabolic reflector of K-B telescope and two methods of approximating it. One, a series of oriented flats which provides a resolution of 10 arcseconds and two, a series of oriented parabaloidal segments which is a better approximation.

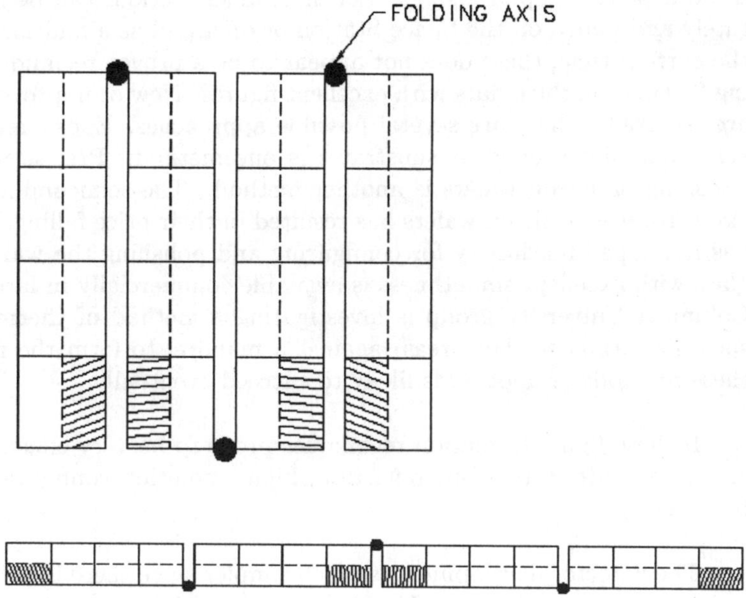

Figure 5: Folding a segmented K-B system into a more compact volume for launch (top) and reduced view of unfolded system (bottom).

The cost and feasibility of this system is determined ultimately by the cost and mass of the individual reflectors. A 10m telescope would require as many as fifteen thousand 50cm× 50cm reflectors. If the reflectors could be as thin as 200 microns the mass of the mirror system including an allowance for structure would be about 8 metric tons. This can easily be launched into low Earth orbit by the Space Shuttle or by larger expendable rockets such as the Ariane 5. To avoid depending upon a single launch, smaller sections can be launched separately and joined on the Space Station or operated as a multimirror array. At the current time, there does not appear to be a proven technique for fabricating 200 micron thick flats with excellent figures. New or improved technologies are required. There are several possible approaches. Epoxy replication upon very thin silicon or glass substrates is one method. Precise machining and polishing of silicon wafers is another method. The semiconductor industry's extensive use of silicon wafers has resulted in their price falling. This industry has developed machinery for configuring and polishing the wafers. Very thin glass with excellent smoothness is available commercially in large sheets. The Columbia University group is investigating a method of thermally slumping such glass (but smaller areas) against a mandrel to form the figure. One of these methods or another is likely to succeed eventually.

If these figure formation techniques prove to be very effective, the mirror could be made in the four reflection, high resolution configuration described by Cash(5).

The detector development is much simpler since there is only one or a few instruments for each type. The detector would be quite large if investigators seek to utilize the full field for a future survey. However, it would be in a single compact format and if placed aboard a small satellite it could be made retrievable if desired.

3 Hard X-Ray and Wide Field Telescopes

3.1 Hard X-Ray Telescopes

There are substantial reasons to endeavor to extend the band of focusing another order of magnitude from 10 to 100 keV. In this band or above it the typical AGNs emits most of its energy. Two decades of observing with soft X-ray telescopes, if not our intuition, has demonstrated how much more sensitive a focusing telescope is than an area detector or coded aperture with even ten times more effective area. Non-focusing instruments have detected relatively very few sources in the hard X-ray band. A future area detector or coded mask system with an order of magnitude more effective area than the OSSE instrument of CGRO can attain only a factor of three more sensitivity and at only a great cost in complexity and increased mass. High energy resolution germanium detectors exceeding a thousand sq cm are impractical. The only hope of being able to detect a significant number of extragalactic objects above 10 keV and study their spectra with high resolution detectors is through focusing. Fortunately, for distant objects, the redshift effectively extends the intrinsic bandwidth of the source that can be studied. For these reasons, we can expect there to be much effort devoted to developing focusing devices optimized for hard X-rays.

In all of the various techniques that have been proposed to extend the bandwidth, long focal lengths and small graze angles are beneficial. In fact, a 10 cm diameter optic with a gold coating can attain a 40 keV cutoff with a focal length of 10 m. The cutoff can be extended more than a factor of two by combining small graze angles with new approaches. The most promising of the new approaches are graded d-spacing multilayer coatings (13) and bundles of capillaries of the type developed by Kumakhov in Russia.

Capillaries are the grazing incidence analogue of fiber optics. In traversing a capillary successfully, an X-ray experiences many reflections that are well below the critical angle. The survival probability can be appreciable if the curvature is small. A capillary bundle can provide significant effective area up to 100 keV according to idealized models (14). The limitation of capillaries are a lack of true imaging (they are concentrators), the small field of view, and the fact that its optical design bears little resemblance to conventional X-ray telescopes. Constructing a capillary telescope is a niche technology that has been mastered at only a very few specialized centers and there is very little opportunity to build upon existing X-ray telescopes. Small capillary bundles are proving to be useful in the laboratory for concentrating and deflecting beams. It remains to be seen if this technology can be scaled up successfully to the large apertures required of cosmic telescopes.

In contrast, graded d-spaced multilayers are much easier to adopt because they are a type of coating that can be incorporated into nearly all existing optical designs by substituting them for gold or iridium. The coatings consist of alternating layers of a heavy and light material that can number up to several hundred bilayers. The bilayer period decrease continuously with depth into the coating. The key to the reflection is the sharp change in the index of refrac-

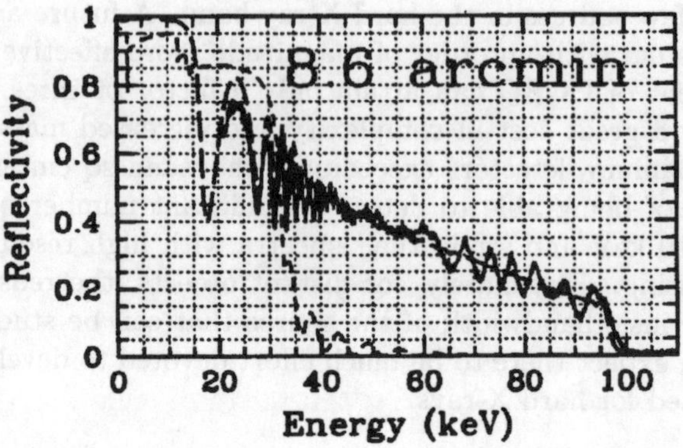

Figure 6: Measured hard X-ray reflectivity at fixed graze angle as a function of energy of Ni/C graded d-spacing multilayer deposited upon polished glass flat(14), (solid line) and theoretical reflectivity of gold (dashed line).

tion at the light-heavy interface and coherence among several adjacent layers of nearly equal period. This structure mimics the action of a Bragg crystal with a variable period or "d-space". An X-ray of a given energy and angle of incidence will penetrate the multilayer stack until it reaches the depth where the Bragg condition is satisfied locally. The X-ray is reflected. However, the efficiency is considerably less than 100%. X-rays are partially absorbed entering and exiting the multilayer stack. Also the transition between the heavy and light material is not perfectly sharp due to interdiffusion of the two materials and the roughness of the substrate which propagates to the next interface with the deposition of a new layer. **Fig. 6** illustrates the measured reflectivity of a Ni/C graded d-spacing multilayer coating and that of an ideal gold coating as a function of energy for an angle of incidence of 8 arcmin. Future programs in the US, Europe, and Japan all include plans for extending the bandwidth of telescopes by the use of multilayer coatings.

3.2 Wide Field Focusing Telescopes

The most familiar X-ray telescope geometry is conical, e.g. the Wolter 1 or the double cone. The field of view of conical telescopes is limited to the critical angle. Hence, the focusing telescope is not thought of as a wide field device particularly above a few keV. However, there is another geometry with essentially unlimited field, the "lobster eye" telescope. The lobster eye telescope indeed functions according the same principle as the eyes of crustaceans. It is the grazing incidence analogue of the spherical mirror. Two versions of the lobster eye design have been described in the literature (14, 15). One is based upon square channels and the other is a Kirkpatrick-Baez type (two orthogonal stacks of reflectors) design. Small scale devices based upon them have been built and tested (16, 17).

The orthogonal reflector stack geometry version of the lobster eye telescope and the focusing geometry are shown in **Fig. 7**. It is apparent from the symmetry that the field of view is large. It is also possible to construct a cylindrical one dimensional version of the lobster eye telescope with a coded mask consisting of circumferential slits pseudo-randomly distributed along the axis of the cylinder providing angular resolution in the other dimension (18). The theoretical angular resolution is equal to angular separation between adjacent reflectors. For the K-B geometry that would be several arcminutes with current substrates such as 0.5 mm thick glass. The Leicester University group has been studying square microchannel plates whose surface is shaped into a sphere as the focusing element. The theoretical angular resolution would be just a few arcseconds. However, thus far mechanical errors have limited the resolution to no better than a few arcminutes.

The lobster eye geometry has several disadvantages. One, in order to cover the full field the detector must be half as large in each dimension as the telescope aperture. Therefore, for a large area telescope the position sensitive proportional counter is the only practical device at present. Two, as indicated in **Fig. 8** the point response function is complex. In addition to the two dimensional image there are also two line images and a halo of unfocused radiation. The intensity of each component is equal on the average. On the other hand the point response function does not vary much across the field. Finally, the effective area at the center of the field is only about 30 to 40% that of a conical mirror with the same aperture. Nevertheless, the key factor is that the product of the two dimensional field and the effective area is one to two orders of magnitude larger for the lobster eye than for the conical mirror of equal aperture.

There are several potential applications of the lobster eye geometry. The most important one, is a wide field monitoring station for soft X-ray transients on all time scales. The broad field and good angular resolution result in continuous monitoring, high sensitivity, and excellent positioning. There is no problem of source confusion. Another potential application is a next genera-

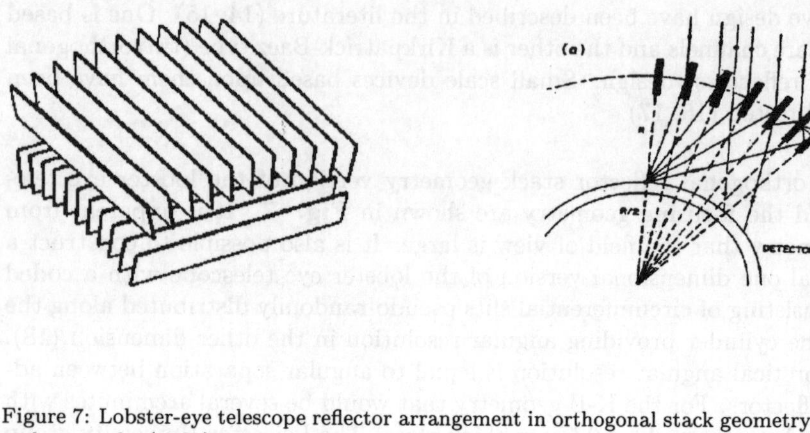

Figure 7: Lobster-eye telescope reflector arrangement in orthogonal stack geometry(15) (left) and ray paths (right).

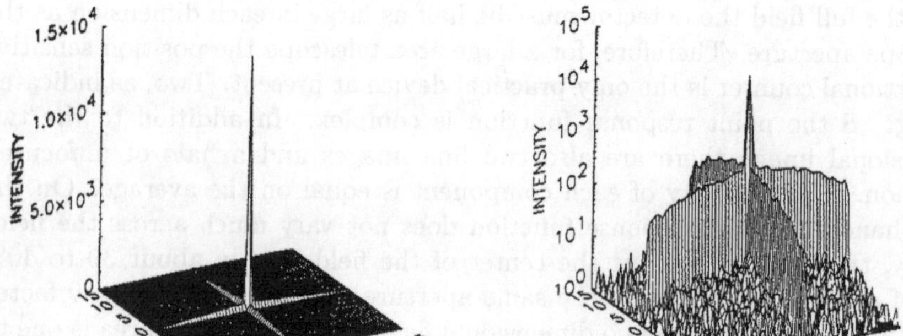

Figure 8: Point response function of lobster eye telescope on linear vertical scale (left) and logarithmic vertical scale (right).

tion all sky or broad sky ($>$ 5000 sq deg) survey with high photon statistics from each object. This survey could be performed with a device having a field of several hundred square degrees. The conical mirror, particularly the "wide field" design where the mean angular resolution is optimized over the one degree field (19) is much suited to deeper surveys of smaller regions but not well adapted to an all sky survey.

4 References

1. R. Giacconi, and B. Rossi, Journal of Geophysical Research, **65** 773, 1960
2. P. Gorenstein, *Modular Focusing Arrays for X-ray Observations of Very High Sensitivity from the Space Shuttle* Presentation to Working Group on High Energy Astrophysics, Woods Hole, MA July 1973.
3. D. Fabricant, M. Conroy, L. Cohen, and P. Gorenstein, Proc. SPIE, 640, 164, 1986.
4. D. Fabricant, L. Cohen, and P. Gorenstein, Applied Optics, 27, 1456, 1988.
5. W. Cash, in, The Next Generation of X-ray Observatories, (NGXO) M.J.L. Tuner and M.C. Watson, Eds. Leicester X-ray Astronomy Group Special Report XRA97/02, 147, 1997.
6. "BBXRT: A Preview to Astronomical X-ray Spectroscopy in the 90's" P.J. Serlemitsos and S.J. Shrader eds. Publication 93-73 of the Laboratory for High Energy Astrophysics, GSFC, 1993.
7. F.E. Christensen et al. SPIE Vol. 2279, 78, 1994.
8. O. Citterio, P. Conconi, M. Ghigo, F. Mazzolini, R. Buzzi, and G. Parodi, SPIE Vl. 3113, 334, 1997.
9. Y. Soong, L. Jalota, P. Serlemitsos, SPIE Vol. 2515, 64, 1995.
10. W. Egle, H. Balla, P. Kaufmann, B. Aschenbach, and H. Brauninger, SPIE Vol. 1160, 432, 1989.
11. M. Elvis and G. Fabbiano, in NGXO, Op. Cit. 33.
12. F.E. Christensen et al SPIE Vol., 1546, 60, 1992.
13. P. Gorenstein, SPIE 1546, 46, 1992.

14. W.K.H. Schmidt, Nuclear Instruments and Methods, 127, 285, 1975

15. J.R.P. Angel, Ap. J. 233, 64, 1979.

16. J.E. Lees, G.W. Fraser, A.N. Brunton, and R. Willingale, "Imaging in High Energy Astronomy" L. Bassani and G. Di Cocco Eds. Kluwer Academic Publishers, Dordrecht/Boston/London 1995.

17. P. Gorenstein, E. Whitbeck, G.K., Austin, A.T., Kenter, L. Pina, A. Inneman, R. Hudec, SPIE vol. 2805, 74, 1996.

18. P. Gorenstein, in "Variability of Galactic and Extragalactic X-ray Sources, A. Treves, Ed., 253 Associazione Per L'Avanzamento Dell'Astronomia, Milan-Bologna, 1986.

FROM STARFISH TO EXOSAT – MY VARIOUS INTERACTIONS WITH RICCARDO GIACCONI

K. A. POUNDS
Department of Physics & Astronomy, Leicester University, UK
E-mail: kap@star.le.ac.uk

1 The Early Years – Literary, Nuclear, and Personal Links

Many contributors to this book will be recalling Riccardo Giacconi's influence on their scientific career. Undoubtedly the dominant figure in the first two decades of X-ray Astronomy, his impact on astronomy in general was – and is – felt world-wide. As one of the first generation of X-ray astronomers, Riccardo's work affected my career from the initial, seminal discovery[1] of Sco X-1 in June 1962 At that time I was leading a small research group at Leicester University studying the Sun's X-ray emission. Our primary vehicle was the Skylark sounding rocket, which we had already launched successfully on several occasions at Woomera in South Australia. Looking back I admit to some initial doubts as to the validity of the American Science and Engineering group's discovery, primarily because the result was so unexpected. [That might have been due, in part, to my own mind-set having just previously had accepted a scientific proposal to NASA, with colleagues at University College London, for a cosmic X-ray instrument to be flown on the Orbiting Astronomical Observatory – eventually Copernicus – based on the search for stellar coronae.] However, confirmatory rocket-borne observations from both the AS&E group and Herbert Friedman's group at NRL [2,3] inspired my colleagues and me to divert some of our effort into the new field of Cosmic X-ray Astronomy. That effort led, in April 1967, to our first night-time rocket flight and to the successful detection of Sco X-1, Tau X-1 (the Crab Nebula), and a bright X-ray nova that we named Centaurus X-2, and – remarkably – was at the time 20% brighter than Sco X-1[4]. Although for the next 5 or 6 years we had a fair degree of success – and lots of fun – exploring the southern sky visible from Woomera, the scientific impact of our research in the evolving field of X-ray Astronomy remained well short of our US counterparts, with the AS&E and NRL groups being joined by other active research group at LRL and MIT. Then, of course, in December 1970, Dr Giacconi's group took a further major step forward, with the successful launch of the first satellite dedicated to X-ray Astronomy, Uhuru.

Before recounting my own brief involvement with Uhuru I should digress in order to recall a second influence that Riccardo had on my early scientific career, also back in the summer of 1962, and of which I was unaware until reading a more recent description of the famous discovery rocket flight by Herb Gursky[5]. Apparently the historic launch was left to Herb and a technician, while Riccardo and other "senior AS&E personnel" were attending a high-altitude nuclear test at Johnston Island in the Pacific Ocean. The nuclear device was detonated successfully at 0900 UT on 9 July at 400km altitude, creating a new radiation belt around the Earth, as well as spectacular auroral displays, magnetic fluctuations and ionospheric disturbances. Unfortunately for me, and colleagues in Leicester and UCL, the Ariel-1 satellite[6] launched a few weeks earlier, on 26 April, was subjected to the intense particle background. Although we were treated to some dramatically high count rates as Ariel-1 flew through the "Starfish debris"[7], my array of gas proportional counters, intended to monitor the solar X-ray spectrum, were rapidly degraded as the methane quench gas polymerised and collected on the anode wires. However, since more than 30 years passed by before I learned of Riccardo's complicity in this event, it is probably now too late to complain!

Moving forward again to 1967, that was the year I first got to know Riccardo personally, as my young family and I enjoyed six months in New England while I was working at AS&E. In fact most of my time was spent pursuing my initial research interest in solar physics, alongside Pippo Vaiana, as we both prepared for the launch later that year of solar X-ray instruments on NASA's OSO-4 satellite. However, I was able to share in the excitement of the X-ray Astronomy programmes at AS&E and MIT, nearby, while there was growing anticipation of the launch of SAS-A (later, Uhuru). The design of Uhuru was classic AS&E; simple, reliable and effective, and – from my perspective, at least – Uhuru was always destined to be a success.

In the event, that was so, with the discovery of X-ray binary stars and extended emission from clusters of galaxies being among the most significant in high energy astrophysics to this day. The Spring of 1971, a few months after the launch of Uhuru, saw my second extended visit to AS&E begin. During the subsequent summer months I was privileged to share in the unfolding drama, with discoveries being made on a weekly basis. That is a story best told by members of the Uhuru team; my own only recorded contribution was in being a member of the team that failed to identify the first X-ray cluster emission, preferring instead to explain the extended emission in Virgo as being due to the giant radio galaxy M87[8].

2 Ariel-5 – My Chance for Independence

Arriving back in England from that second stay at AS&E, I began serious work in preparing for our first X-ray Astronomy satellite, Ariel-5. The Leicester group had two experiments on the payload, the sky survey instrument (SSI) mounted to view out in the spin plane of the satellite, and a crystal spectrometer-cum-polarimeter viewing along the spin axis. Given the success of Uhuru, coupled with the fact that the SSI was only a little more sensitive, my own expectations centred more on the Bragg crystal instrument. In the event, following a successful launch in October 1974 from the same San Marco platform, near Malindi in Kenya, used by Uhuru, it was the SSI that proved much the more productive. Although both Leicester instruments worked well, until Ariel-5 ran out of control gas in March 1980, the brightest sources in the sky – the X-ray binaries – turned out not to exhibit either the narrow emission lines or the strong polarisation which the crystal spectrometer could have detected (e.g.[9]). On the other hand, the SSI showed just how fruitful it was simply to repeat the survey of a sky so rich and variable as that in the 2-10 keV band.

August 1975 presented those of us working on Ariel-5 with an ideal opportunity to expose our results to the world, as we hosted the first European Astronomy Meeting in Leicester. There was a strong attendance from MIT (who were successfully operating SAS-3 at that time), the new group at Goddard Space Flight Center (also involved in Ariel-5), and Japan, as well as X-ray and other astronomers from many European countries. Although unfortunately Riccardo was unable to be present I am sure he was in our thoughts as the drama of the week unfolded. I refer to the sighting in the SSI data, being received at Leicester and processed on our PDP-8, all within 24 hours of transmission to the Quito or Ascension Island ground stations, of a new transient source just before our visitors arrived in Leicester. Then, as the Conference got under way, the daily data dump showed the source to be brightening rapidly, exceeding the Crab Nebula after 2 days and Sco X-1 1 day later![10]. On that particular story, the rest is history – as they say. Mon X-1, or A0620-00 was – for several weeks – the brightest cosmic X-ray source ever seen, it was quickly identified with a 20th magnitude "star"[11], that apparently flares every 75 years or so, and the mass of the unseen, compact comparison turned out to be in the range that makes it a strong black hole candidate[12].

My next interaction with Riccardo, that I recall, occurred in Boston, where the 8th Relativistic Astrophysics Meeting was held, in December 1976. The meeting agenda had a strong X-ray Astronomy content, reflecting the new discoveries being made by that first generation of X-ray satellites (also includ-

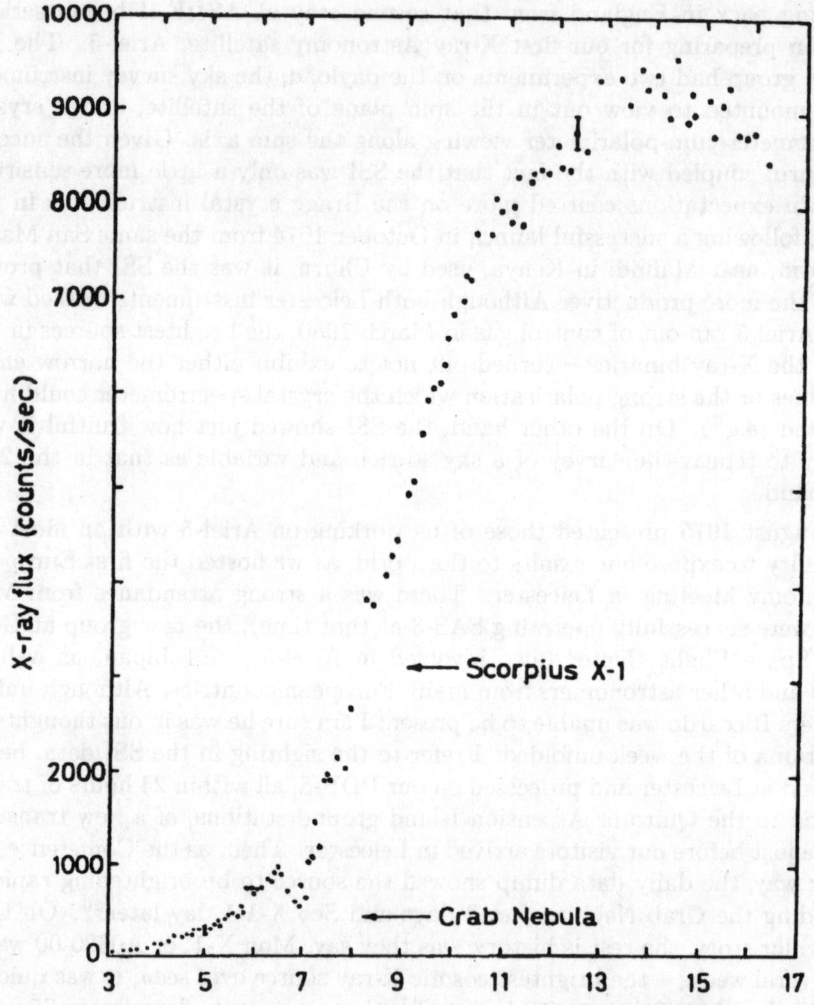

Figure 1: The 2–18 keV X-ray light-curve of A0620-00 as recorded by the Sky Survey Instrument on Ariel-5

ing ANS and SAS-3). There were major sessions on Close Binaries, X-ray Bursters, Neutron Stars and on Extragalactic X-ray sources. My own contribution was in the latter area, being a progress report on the Ariel-5 High Latitude Catalogue[13]. Based on the first two years in orbit, I was able to report the reliable detection of 107 sources above 10 degrees galactic latitude. 54 of these were confirmations of (mainly) Uhuru sources, and 53 were "new". At that time 31 other Uhuru (3U catalogue) sources were not confirmed, partly explained by the different sky coverage in these initial (2A and 3U) catalogues. Scientifically, the major content of my report was twofold. First, the number of identified galaxy cluster sources was doubled, from ≈20 to ≈40 (the uncertainty on each number being ≈10% at that time). Then, I was able to show that Seyfert galaxies represented a second major class of extragalactic X-ray source, with 11 reliable Ariel-5 identifications, compared with 2 based then on the 3U source list (NGC 4151 and NGC 1275). I recall the pleasure (and satisfaction) in being able to announce such an important result from the Ariel-5 sky survey, that had been missed in the Uhuru analysis – up to then. Having Riccardo as chairman of that session made the experience especially sweet for me, perhaps in the manner of a graduate student reporting an important result to his/her supervisor! (It may be, of course, that Riccardo didn't fully accept my claim that powerful X-ray emission was probably a common property of AGN – given that the quasar 3C 273 was also seen – until a similarly impressive list emerged somewhat later in the 4U catalogue, based on a second phase Uhuru analysis!)

One aspect of that talk which I had forgotten, until it was brought up again by Walter Lewin several years ago, was the demise of the UHGLS. For younger readers I should perhaps explain this acronym stood for "unidentified, high galactic latitude sources", a mystery surrounding some 43 sources (above 20 degrees latitude) in the 3U Catalog[16] without visible identification. In the finest traditions of astronomy the nature of these sources (conjectured to be a new form of "X-ray Galaxy") was – I suspect – used by X-ray astronomers at that time to stimulate support for future missions. In the event, the Ariel-5 survey data I reported in Boston included 15 former UHGLS that more precise positions showed were identified with clusters of galaxies and 9 with a variety of galactic objects or AGN. A majority of the remainder were not detected at flux levels below those reported in 3U, suggesting variability (AGN?) or a false detection. [In the event 12 of the remaining UHGLS were indeed not confirmed in the deeper 4U catalogue[15]].

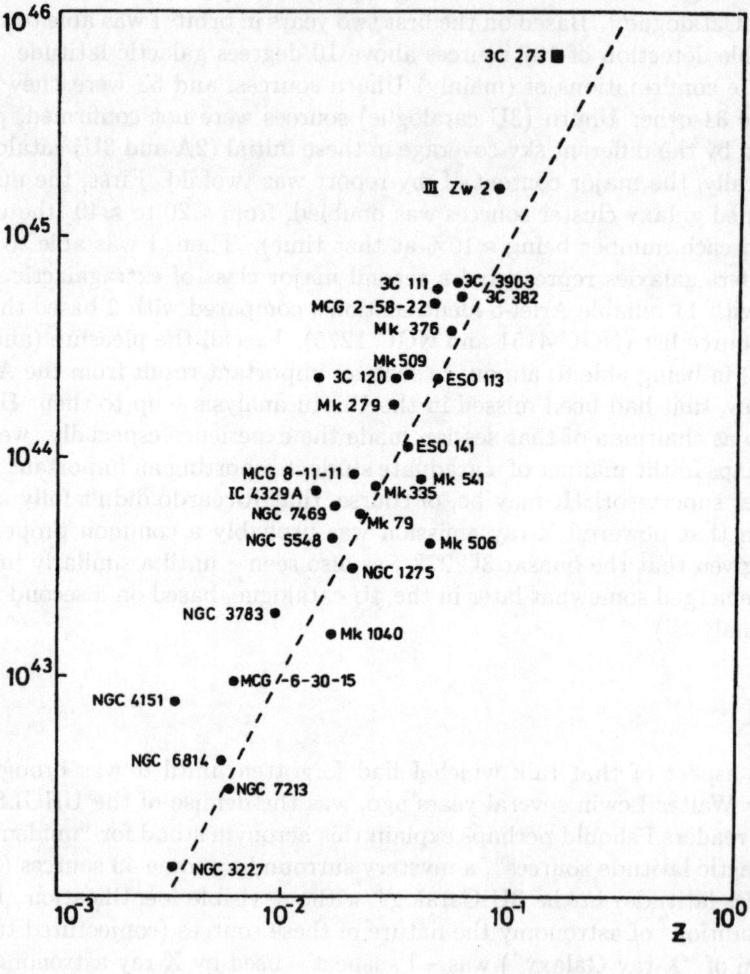

Figure 2: X-ray Seyferts identified by the late 1970s, primarily based on Ariel-5 and Uhuru survey data (from Pounds (1980))

3 Moving Apart – But Staying Close

Over the 30 years that I have known Riccardo personally I have enjoyed many stimulating discussions. They were almost always instructive (at least to me), though I rarely felt I had won many of the points debated. One of the few occasions where my viewpoint did survive the test of time - albeit somewhat fortuitously - related to the design and logic of ESRO/ESA's first X-ray mission, HELOS (later EXOSAT). I had the task, over the period 1969-72, along with several European colleagues (including Joachim Truemper), of defining an X-ray mission to meet the opportunity offered by ESRO as "compensation" for the Organisation having chosen COS-B ahead of the X-ray mission, COS-A, in 1968. Needing to find a mission complementary to HEAO-B (later, the Einstein Observatory) then under development by NASA, we went for a 3-axis pointed satellite, carrying both low- and medium-energy detectors, into a deep space orbit. The main reason for the choice of that – at the time – unique orbit (for an astronomy mission) was to allow accurate location of medium energy sources by the lunar occultation technique, over a broad band about the ecliptic plane. Riccardo made a point whenever we met about that time of telling me we were wasting payload mass by that choice of orbit. In the event, the high orbit, and resulting long, uninterrupted observations of galactic and extragalactic sources proved to be a highlight, so far unique to EXOSAT. While the high orbit benefited EXOSAT via the secondary reason for its choice, it was, nevertheless, a source of quiet satisfaction to find – a decade later – that NASA had revised the orbit for AXAF from LEO to HEO.

Another product of a discussion I had in RG's office, by then at CfA, when the case for AXAF was being prepared, is the rather famous graph reproduced in figure 4. With hindsight, the idea for that graphic way of putting NASA's Great Observatories in context was probably one of my most enduring contributions as an original member of the AXAF Science Working Group (which Riccardo chaired). Even then I should acknowledge that Martin Elvis – who had recently moved to CfA from Leicester – did the research that eventually brought the data together on one plot.

Since Riccardo moved to the STScI more than 10 years ago our paths have crossed less often. Coincidentally, the end of operations of the Einstein Observatory in 1981 began a period of 15 years, ended only recently with the launch of XTE, when US leadership in X-ray Astronomy passed to Europe and Japan. It may well be that that change in the "balance of power" would have happened anyway, but I can't help thinking NASA would not have been allowed to get away with a 15-year gap between US-led X-ray Astronomy missions had Riccardo not been called away to run the HST project!

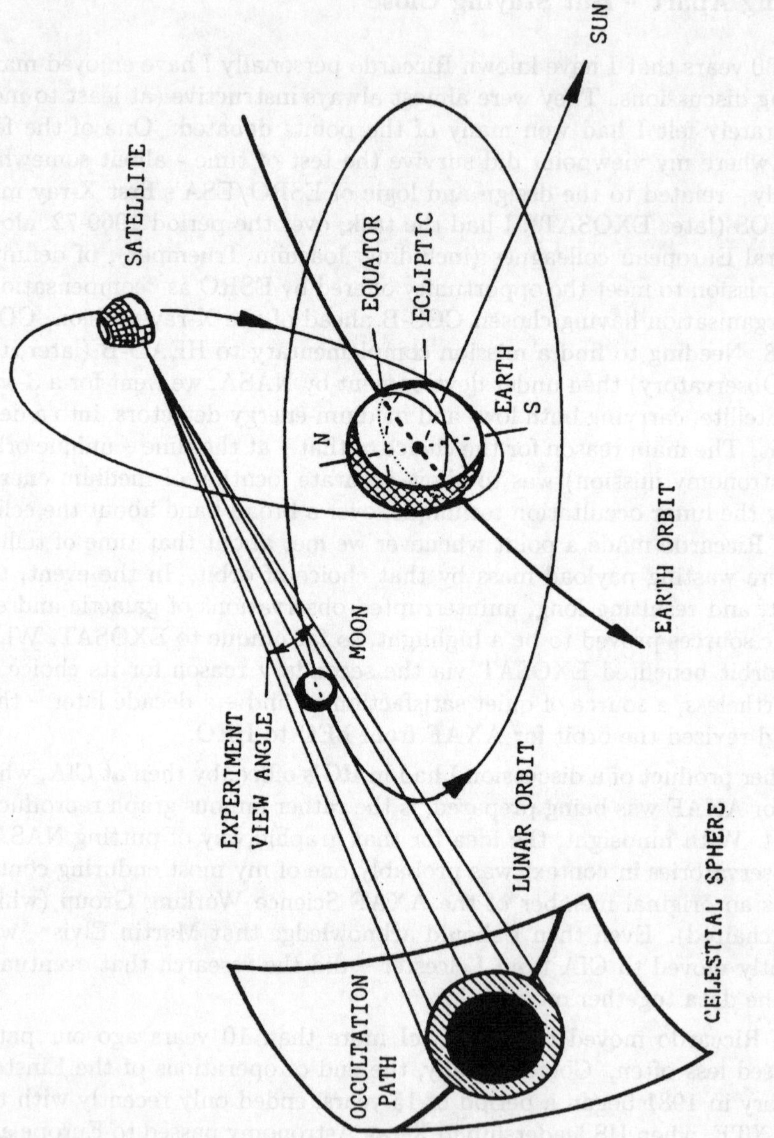

Figure 3: The high apogee orbit chosen for the HELOS spacecraft, which eventually yielded the extended observations for which EXOSAT is now famous

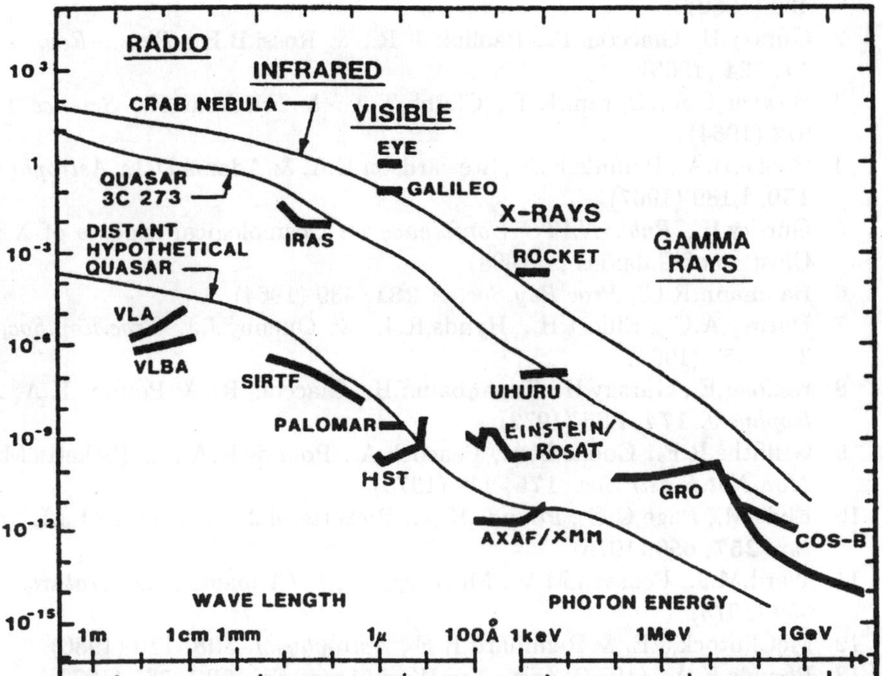

Figure 4: The relative point-source sensitivities of historical and current astronomical telescopes, initially prepared to support the AXAF mission

Looking back now, over 35 years since I first became aware of a man called Riccardo Giacconi, I am happy to count him as an important long-term colleague and good friend.

References

1. Giacconi,R., Gursky,H., Paolini,F.R., & Rossi,B.B., *Phys. Rev. Lett.* **9**, 439 (1962)
2. Gursky,H, Giacconi,R., Paolini, F.R., & Rossi,B.B., *Phys. Rev. Lett.* **11**, 524 (1963)
3. Bowyer,C.S., Byram,E.T., Chubb,T.A., & Friedman,H., *Science* **146**, 912 (1964)
4. Cooke,B.A., Pounds,K.A., Stewardson,E.A. & Adams,D.J., *Astrophys.J.* **150**, L189 (1967)
5. Gursky,H., *Publ. NATO Conference on* Cosmological aspects of X-ray Clusters of Galaxies , (1994)
6. Baumann,R.C., *Proc.Roy.Soc. A* **281**, 439 (1964)
7. Durney,A.C., Elliott,H., Hynds,R.J., & Quenby,J.J. *Proc.Roy.Soc. A* **281**, 553 (1964)
8. Kellogg,E., Gursky,H., Tananbaum,H., Giacconi,R., & Pounds,K.A. *Astrophys.J.* **174**, L65 (1972)
9. Griffiths,R.E., Cooke,B.A., Peacock,A., Pounds,K.A., & Ricketts,M.J., *Mon.Not.R.astr.Soc.* **175**, 449 (1976)
10. Elvis,M., Page,C.G., Pounds,K.A., Ricketts,M.J., & Turner,M.J.L., *Nature* **257**, 656 (1976)
11. Ward,M.J., Penston,M.V., Murray,C.A., & Clements,E.D., *Nature* **257**, 659 (1975)
12. McClintock,J.E. & Remillard,R.E., *Astrophys.J.* **308**, 110 (1986)
13. Pounds,K.A., (1977) *Ann. New York Acad. Sci.* **302**, 361 (1977)
14. Pounds,K.A., in *X-ray Astronomy* Eds. Giacconi, R, & Setti, G., D.Reidel Publ. Co., 273, (1980).
15. Forman,W., Jones,C., Cominsky,L., Julien,P., Murray,S., Peters,G., Tananbaum,H., & Giacconi,R., *Astrophys.J.Supp.* **38**, 357 (1978)
16. Giacconi,R., Murray, S., Gursky, H., Kellogg, E., Schreier, E., Matilsky, T, Koch, D., & Tananbaum, H. *Astrophys.J.Supp.* **27**, 37 (1974)

X-RAY ASTRONOMICAL SPECTROSCOPY

STEPHEN S. HOLT
Code 600, NASA/GSFC, Greenbelt MD, 20771, USA
E-mail: steve.holt@gsfc.nasa.gov

Spectroscopy is traditionally characterized by the sacrifice of quantum efficiency for high spectral resolution. Since X-ray astronomy is a photon-limited discipline, the choice between high resolution for very few sources versus much lower resolution for many more has not always been an easy one. The development of new thermal detectors with both high resolving power and high efficiency offers the opportunity to avoid such compromises in the future.

1 Introduction

As measured from the first discovery of extra-solar X-rays by Riccardo Giacconi and his colleagues in 1962, X-ray astronomy will soon be entering its fifth decade.

For the first twenty years, the major advances in our understanding of X-ray sources arose primarily from temporal measurements. This is not to say that source identifications and early spectroscopic measurements were not important, but the discoveries of periodicities (and their rates of change) in strong galactic X-ray sources led directly to our understanding of their nature as neutron stars that are accreting mass from their companions in binary stellar systems. Non-periodic temporal measurements also produced important results during these early years, including the discoveries of both X-ray and gamma-ray burst sources, unambiguous source identification via simultaneous X-ray and radio/optical variations, and the "shot noise" character of the X-ray emission from the first black hole candidates. Giacconi and his colleagues were very much at the forefront of most of these discoveries.

Similarly, as a pioneer in the development of grazing incidence optics and as the Principal Investigator for the Einstein Observatory, Giacconi played the major role in providing X-ray astronomy with imaging as its primary observational tool for its second twenty years. I had the privilege of being the NASA Project Scientist for that mission, as well as being the PI for its primary spectroscopic instrument, the Solid State Spectrometer. I distinctly remember Giacconi's alarm upon first learning that, if poorly designed, the spectrometer's cryostat could build up sufficient pressure to become hazardous to the mission. It is no secret that Giacconi was much less interested in spectroscopy than he was in imaging, and it took us quite a while to demonstrate that the spectroscopic "bomb" was worth its potential risk.

With the emergence of the High Thoughput X-ray Spectroscopy mission as the next major observatory after AXAF, it would appear that spectroscopy will be the primary tool of the third twenty years of X-ray astronomy research. This paper will not review past accomplishments. Instead, it will concentrate on the manner in which spectroscopic instruments can be best designed for future research applications.

2 Characteristics of an "Ideal" Spectrometer

The most important attributes of a spectrometer depend upon the specific scientific objectives of a particular investigation, but there are two which are so generally useful that they can be safely assumed to be characteristic of an ideal spectrometer. These two attributes are energy resolution sufficient to address the most important scientific objectives for all classes of sources, and sensitivity sufficient to collect enough photons to take advantage of the energy resolution.

Traditionally, the highest resolving powers are accomplished with "dispersive" spectrometers, which spatially separate photons by wavelength prior to detection. Such instruments have increasingly higher separation for longer wavelengths, so that their resolving powers get better with decreasing photon energy.

Non-dispersive instruments, which generally attempt to measure total energy deposition in an absorbing medium, can achieve simultaneous sensitivity over a wide bandpass with high efficiency. Because they have energy resolution which is dominated by either the statistics of energy deposition in the detection medium or by amplification noise (see next section), they tend to have resolving powers which get better with increasing energy.

3 Non-dispersive Spectrometers

The primary characteristic of any spectrometer is its energy resolution, or equivalently, its resolving power. There may be a number of variances $(\sigma_i)^2$ which contribute to the total variance σ^2 in the extraction X-ray energy information:

$$\sigma^2 = \sum(\sigma_i)^2. \tag{1}$$

The FWHM resolution at energy E is defined as:

$$\delta E(E) = 2.36\sigma(E), \tag{2}$$

and the resolving power at energy E is defined:

$$R(E) = E/[\delta E(E)], \tag{3}$$

i.e. the X-ray energy measured in units of the resolution. In the following, we shall omit the (E) for brevity.

For a traditional detector that creates charge, i.e. a number $N = E/w$ electrons, where w is the average energy to create ion pair, the variance should depend only on the statistics of electron production. We should have

$$\sigma/E = \sigma_N/N \tag{4}$$

if there are no additional sources of noise, so that

$$R^{-1} = \delta E/E = 2.36\sigma/E = 2.36\sigma_N/N \tag{5}$$

Since $(\sigma_N)^2 = FN$, where F is the "Fano factor," we have

$$R^{-1} = 2.36\sqrt{F/N} = 2.36\sqrt{wF/E}. \tag{6}$$

For proportional counters with gas factor G, we can define

$$f = (\sigma_G/G)^2 \tag{7}$$

so that

$$(\sigma/E)^2 = (\sigma_N/N)^2 + [(\sigma_G/G)^2]/N = (F+f)/N \tag{8}$$

and

$$R^{-1} = 2.36\sqrt{w(F+f)/E}. \tag{9}$$

Typical numbers for gas counters are: $w \approx 25$ eV/ion pair, and $F \approx 0.2$, so that for $f = 0$ we could achieve $N_{6\text{keV}} \approx 240$ and $R_{6\text{keV}} \approx 15$. Since $f \approx 0.6$ for a standard proportional counter, the best that we can obtain is $R_{6\text{keV}} \approx 7$ or, as a function of energy,

$$R \approx 3\sqrt{E_{\text{keV}}}. \tag{10}$$

We can actually approach $f = 0$ for gas scintillation proportional counters, for which we can then achieve $R \approx 6\sqrt{E_{\text{keV}}}$.

We can do better with solid state detectors, where with typical parameters $w \approx 3$ eV/ion pair, and $F \approx 0.1$, we could expect to achieve $R_{6\text{keV}} \approx 60$, or $R \approx 25\sqrt{E_{\text{keV}}}$, if $\sigma = \sigma_N$ (i.e., if there are no additional noise sources). Without the advantage of internal multiplication G, the extraction of low levels of charge makes the preamplifier noise non-trivial, so that we cannot ignore the noise component that is approximately linear with the capacitance on the front end of the preamplifier, which can be represented by σ_A (equivalent to N electrons). We then have

$$\delta E = 2.36\sqrt{wFE + w^2N^2} = 2.36\sqrt{[wFW + (15 \text{ eV})^2]} \tag{11}$$

for $N = 5$, in which case $\delta E_{min} = 35$ eV and $R_{max} = E/\delta E_{min}$ so that

$$R \leq 30E_{\text{keV}}. \tag{12}$$

4 Microcalorimeters

An X-ray calorimeter works by converting the energy of a single X-ray photon entirely into heat, and then allowing the precise measurement of the attendant temperature rise. For a calorimeter to function as a spectrometer for individual X-rays, the effective noise in the temperature measurement must be small compared to that temperature rise.

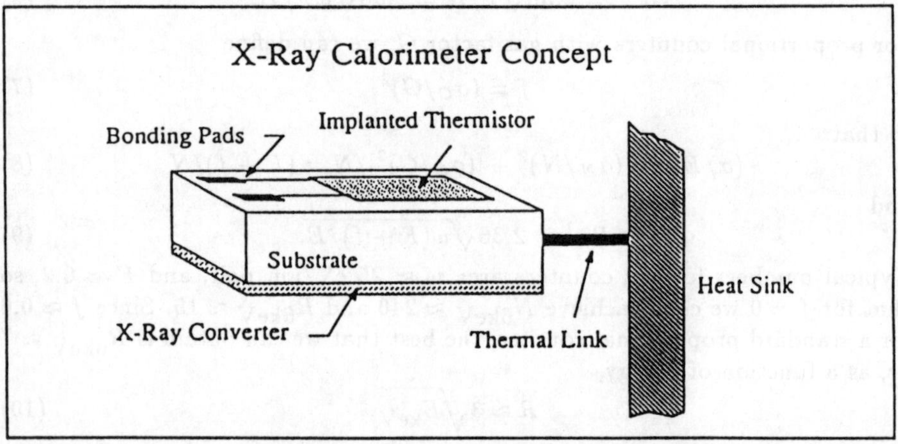

Figure 1: Schematic of the X-ray calorimeter concept.

As displayed in Figure 1, the functional elements of such a calorimeter are an absorber and a thermometer mounted on a substrate that has small heat capacity C_0 at the heat sink temperature T_0, connected by a weak thermal link of conductance G_0 to the heat sink. When an X-ray of energy E is absorbed, the temperature of the substrate is increased by an amount

$$\Delta T = T - T_0 = E/C_0 \tag{13}$$

in a time short compared to the time constant

$$\tau_0 = C_0/G_0 \tag{14}$$

associated with the recovery of the substrate back to T_0. The random transfer of energy between the substrate and heat sink produces fluctuations in the energy content of the substrate, which can be thought of as $\sqrt{N}$ variations in the number of phonons contained in the substrate that represent the background noise level against which the increase in energy due to an absorbed X-ray event must be measured. Formally,

$$R = E/(\delta E) = \Delta T/(\delta T_0) \approx N_{tot}/(\delta N_{tot}), \tag{15}$$

so that if we have

$$N_{tot} \approx (\sigma_{Ntot})^2 \approx C_0/k \tag{16}$$

and

$$E_N \approx kT_0, \tag{17}$$

then

$$\sigma^2 = (E_N)^2(\sigma_{Ntot})^2 = (kT_0)^2 C_0/k = kT_0^2 C_0, \tag{18}$$

where k is Boltzmann's constant. Note that this is independent of G_0 and any details of the thermal link. The advantages of low temperature operation are obvious, particularly when consideration is given to the fact that C_0 scales as T_0^3 with the proper choice of materials. As an example, a piece of silicon 0.5 mm on a side and 25 mm thick with heat capacity 4×10^{-15} J/K at 100 mK would have a limiting fluctuation noise of about $\delta E = 0.5$ eV (FWHM).

The most important additional noise sources are those which might arise from the conversion of X-ray energy to heat (e. g., if some energy is trapped in states that are long-lived compared to the rise time of the heat pulse), and from the measurement of the temperature. Conversion noise may (in principle) be made arbitrarily small, but the Johnson noise in ion-implanted thermistors provides an unavoidable increase in the minimum system noise achievable for this thermometry technique, as demonstrated in the seminal work of Moseley, Mather and McCammon[1] on this subject, based on analyses of noise sources in infra-red bolometers. These authors demonstrate that in the case where only Johnson noise is an important addition, the expression for achievable noise can be approximated with a multiplicative factor ξ, so that the best that we can expect to do is

$$\delta E = 2.36\xi(kT_0^2 C_0)^{1/2} \tag{19}$$

The key parameter characterizing the sensitivity of the thermometry is the logarithmic derivative of the resistance of the thermistor with temperature:

$$\alpha = d(\log R)/d(\log T) \tag{20}$$

where ξ scales as $1/\sqrt{\alpha}$ for $\alpha > 3$. For doped semiconductors, $3 < \alpha < 6$ and $\xi \approx 2$.

The baseline system for Astro-E has an array of 32 such individual calorimeters in the telescope focal plane, covering a total field-of-view of a few square arc minutes. It is necessary to utilize an array in order to simultaneously cover this large a field-of-view while keeping the heat capacity in each element low, and we obtain the added benefit of introducing some imaging into the calorimeter system.

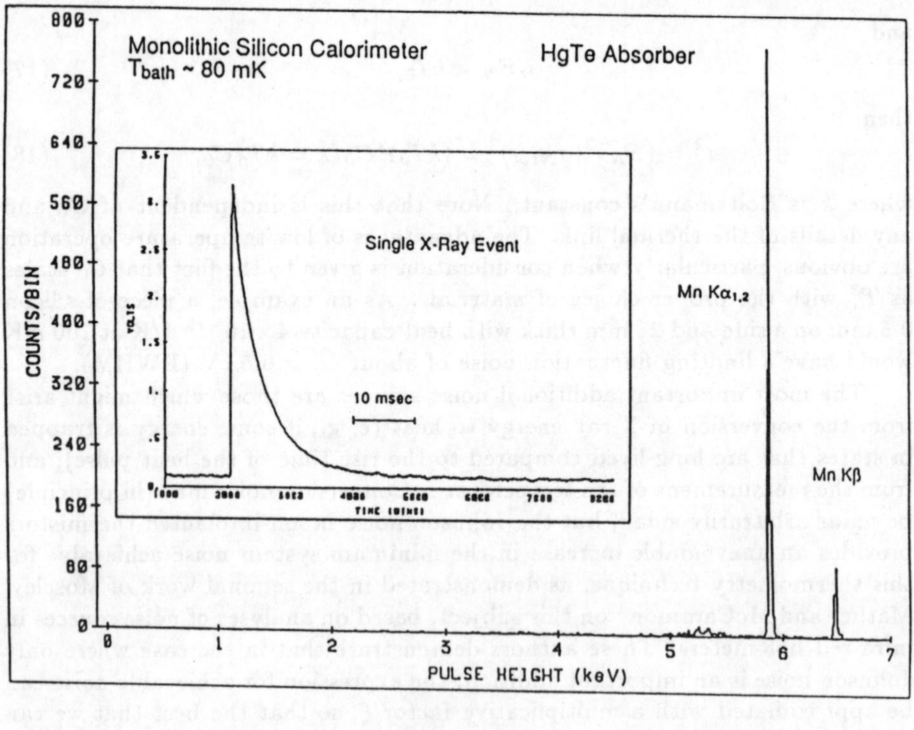

Figure 2: Response of a prototype AXAF X-ray calorimeter to X-rays from Fe^{55} (Mn K). The inset demonstrates the temporal response to a single X-ray. The pulse height spectrum demonstrates the clear separation of the Kα and Kβ peaks, and can be fit with a system resolution of 11.8 eV FWHM for these lines. The small broad peak near 5.2 keV is due to X-rays that were absorbed in the silicon substrate (which was not completely covered by the HgTe absorber for this test). In more recent tests, we have achieved resolution as good as 7.3 eV for Fe^{55}.

The detailed solid-state nature of silicon prevents it from satisfying the criterion of allowing the complete degradation of all the deposited energy in a time small compared with τ_0, so that it is not an ideal material for the absorption and thermalization of X-rays. It does have a low heat capacity and it is well-suited for accommodating implanted thermometers, however, so that a good compromise for Astro-E is an X-ray absorber mounted on a silicon substrate, as indicated in Figure 1. HgTe has been baselined as this absorber, but some superconductors might also be usable. A discussion of such design issues for the version of this calorimeter array which was originally part of the focal plane instrumentation chosen for AXAF is given in McCammon *et al.*[2] Figure 2 demonstrates the achievement of approximately 10 eV (FWHM) for 6 keV X-rays from a radioactive Fe^{55} source with a system close to the baseline Astro-E geometry. For a detector about twice as good (about the best that we could hope to do with this design) the limiting resolving power would be

$$R = E/\delta E = 200 E_{\mathrm{keV}}. \tag{21}$$

5 How Much Resolving Power Is Enough?

Clearly, higher resolving power is "better," but it is worth examining whether or not a case can be made for a practical upper bound to the resolving power based upon purely astrophysical considerations.

A good starting place for the determination of the required resolving power derives from an examination of standard coronal equilibrium spectra in the temperature range 10^{6-8} K. Figure 3 displays such spectra for solar abundances as viewed by a detector at the focus of an X-ray telescope with flat response over the bandpass 0.1- 10 keV. The three traces in each panel are for $\delta E = 1$, 10 and 100 eV.

In general, the identification and interpretation of features that are superposed on the continuum requires increasingly better spectral resolution as the features become weaker, or as they become confused with other features from which they cannot be resolved. At temperatures near 10^8 K, an equilibrium thermal spectrum is dominated by bremsstrahlung continuum, and exhibits strong isolated Fe K-emission at energies > 6.6 keV, as well as Fe L-lines near 1.2 keV. Note, however, that the "completely ionized" lower-Z elements exhibit strong recombination lines: H-like Si XIV (at 2.0 keV) and S XVI (at 2.6 keV), each with about 30 eV equivalent continuum widths, are barely discernible with $\delta E = 100$ eV, but $\delta E = 10$ eV can reveal O VIII at 0.65 keV and even separate Ne X and Mg XII from the Fe L-band emission. At temperatures near 10^7 K, where the power emitted in the lines and the continuum are comparable, $\delta E = 100$ eV can separate the K-emission complexes of the abundant elements (here dominated by He-like rather than H-like analogs for $Z > 10$), while $\delta E = 10$ eV uncovers most of the detail that $\delta E = 1$ eV can fully resolve. At 10^6 K, the equilibrium spectra are line-dominated; here O and Ne may be recognized with $\delta E = 100$ eV, but $\delta E = 10$ eV is necessary to extract quantitative information.

Table 1: Line Separations.

Line pairs	Approximate energy separation (eV)
Lyα - Resonance	10 Z
Resonance - Intercombination	Z
Resonance - Forbidden	2 Z
Resonance - Neutral	10 Z

Atomic lines provide diagnostics for electron temperature, ionization temperature, density, mass motion and elemental abundances. The model-dependent interpretation of these diagnostics can also be used to distinguish among thermalization, shock heating or photoionization, as well as the extent to which the plasma is in the ionizing or recombining phase, so that even its history may be gleaned from its current status. The required minimum sensitivities depend, of course, on the details of the scientific problems which the observations are intended to address. The utilization of these diagnostics can be found in many references [3,4].

For L-shell to K-shell transitions, which dominate the $E > 1.3$ keV spectrum of hot plasmas, the most important lines are those from:

- the fluorescence of cold (neutral) material,
- the analog of Lyα from hydrogen-like material, and
- the resonance (r), forbidden (f) and intercombination (i) transitions of helium-like material.

Fluorescence and Lyα-analog recombination are clearly the low and high temperature limiting cases, and the helium-analog lines are important because there is a wide range of plasma conditions and temperatures for which there is substantial population of this ionization state. Here the ratio $(i + f)/r$ can

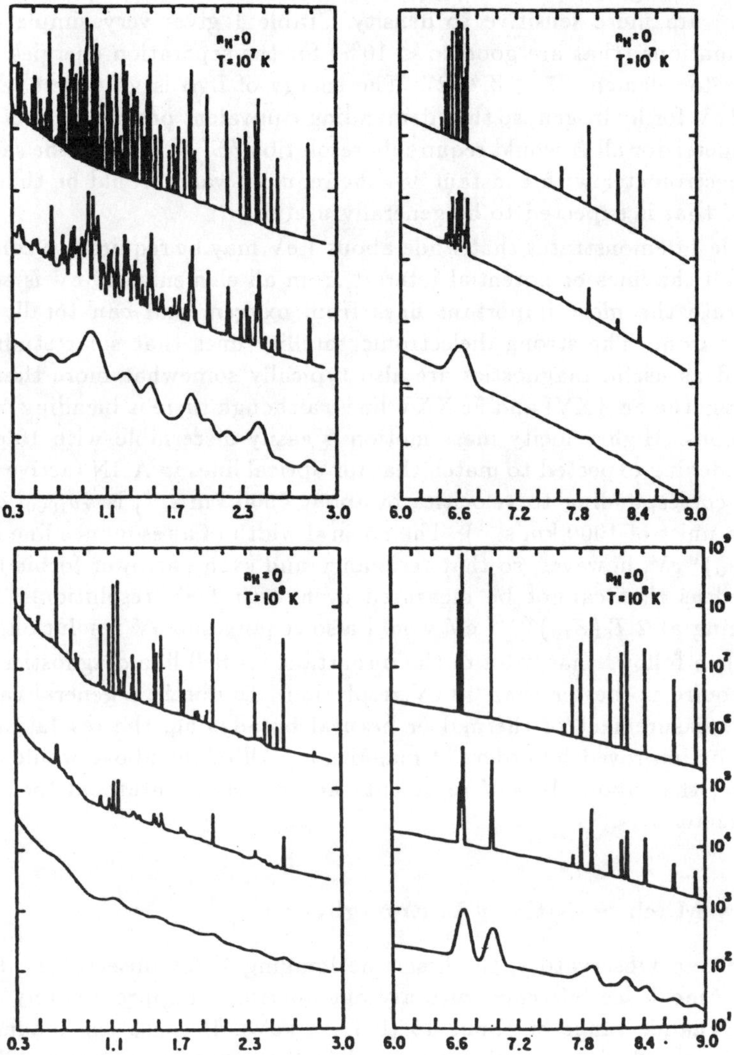

Figure 3: Coronal equilibrium spectra (arbitrary logarithmic intensity) of plasmas with solar abundances, in the energy bands (keV) containing Fe L- and Fe K-emission, as viewed with detectors having FWHM resolutions of 1, 10 and 100 eV (upper, middle and lower trace of each panel).

be a useful diagnostic for electron temperature, for example, while f/i and $f/(i+r)$ are more sensitive to density. Table 1 gives very simple analytic approximations (that are good to $< 10\%$) for the separation energies of these five lines for elements $7 < Z < 27$. The energy of Lyα is, of course, Z^2 times the 10.2 eV for hydrogen, so that demanding equivalent performance of a single spectrometer for all Z would require its resolution $\delta E(E)$ to scale the same way. For a spectrometer with constant δE, the required value would be that for the lowest Z that is expected to be generally useful.

Table 1 demonstrates that while about 1 eV may be required to completely resolve all the lines of potential interest from all elements, 10 eV is sufficient to separate the most important lines from oxygen, and can totally resolve them for iron. The strong dielectronic satellite lines that solar studies have identified as useful diagnostics are also typically somewhat more than 10 eV apart from the Fe XXVI and Fe XXV lines, although there is blending of fainter companions. High velocity mass motion is easily discernible with 10 eV, e.g., the broadening expected to match that for optical lines in AGN (active galactic nuclei), corresponding to velocities of about 5000 km s^{-1}, is $4v_{1000}(Z_{10})^2$ eV (v_{1000} in units of 1000 km s^{-1}). The natural width of a resonance line is about $10^{-2}(Z_{10})^4$ eV, however, so that resonance and even narrower forbidden lines have widths that cannot be measured even with 1 eV resolution. Thermal broadening at $2[T_8(Z_{10})^3]^{1/2}$ eV would also require sub-eV resolution.

It then follows that most of the important K-shell line diagnostics from O to Fe require no better than 10 eV resolution. To obtain a general capability for the measurement of thermal or natural broadening, the resolution would have to be improved by orders of magnitude. All of the above would suggest, therefore, that about 10 keV represents an important threshold for an X-ray spectrometer.

6 How Much Sensitivity Is Enough?

The Einstein Observatory, the first true imaging X-ray observatory, featured a grazing incidence telescope with average aperture of approximately 100 cm^2 over a dynamic range of about 10 (0.3 - 3 keV). Because the magnification of telescopes allows the utilization of detectors that are much smaller than the telescope apertures, source detection limits are determined from the CXB (Cosmic X-Ray Background, i. e., the combined effect of those sources that are below the detection limit) rather than the internal background of the detector. The coming generation of X-ray telescopes (those on Astro-E, AXAF and XMM) will feature apertures and dynamic ranges which are both an order of magnitude larger than those of the Einstein observatory. With such telescopes, it is easy to show [5] from the distribution of extragalactic sources with logN-logS slope -3/2 that there are $> 10^3$ sources which will provide > 1 count/second for a focal plane detector with unit quantum efficiency over the whole telescope bandpass.

In contrast to the quantum efficiency $\epsilon(E)$ at a particular energy E, a "grasp" parameter Φ can be defined as the ratio of the photons actually detected to those incident at the detector (or at the aperture of the telescope in front of the detector) over its entire bandpass, using a standard (e. g., typical Seyfert galaxy) spectrum. The combination of the two parameters $R(E)$ and Φ is then a fairer measure of the utility of any spectrometer for use with a specific telescope than is the resolving power $R(E)$ alone.

The generalization of Φ to an absolute parameter Ψ (rather than one related to a specific telescope) might be accomplished with a multiplicative factor that reflects the bandpass over which the instrument can be sensitive, so that instruments that take a bigger "bite" out of the spectrum get an appropriately higher factor. I have arbitrarily chosen this normalizing factor to be the logarithm of its dynamic range, i. e. a detector that could measure all incident photons with perfect efficiency over an energy range of exactly two orders of magnitude would have $\Psi = 2$. Because the sensitivity to a narrow line feature (i. e. narrower than the FWHM resolution $\delta E(E)$) will be proportional to the square root of most observational parameters (the quantum efficiency $\epsilon(E)$, the exposure, and $R(E)$, for example) I should include an effective background parameter upon which the line sensitivity similarly depends, for completeness. Such a parameter that appropriately reduces to Ψ for zero background is $\Psi/[1 + B(E)/S(E)]$, where $B(E)$ are the extra counts (whatever the cause) measured with the line counts $S(E)$ within $\delta(E)$.

A convenient measure of the sensitivity for line detection is EW (equivalent width — the energy range centered about the candidate line over which the continuum contains the same number of photons as does the line). For the microcalorimeter instrument that will be flown on Astro-E, the 5-σ sensitivity is approximately given by:

$$\mathrm{EW} = E_{\mathrm{keV}}(t_4 I_{\mathrm{mC}})^{-1/2}\mathrm{eV}, \tag{22}$$

where E_{keV} is the energy measured in keV, t_4 is the observation time in units of 10^4 seconds, and I_{mC} is the intensity of the source in units very peculiar to X-ray astronomy — mC is a "milliCrab" or 10^{-3} the intensity of one of the brightest (and non-variable) X-ray sources, the Crab nebula.

7 What next?

There are still lots of ways to increase the overall capabilities of the Astro-E detectors. Perhaps the most obvious is with respect to spectroscopic resolution. Recently, transition edge thermometers have been implemented to microcalorimeters [6]. These have the advantage of very high values of α (up to ≈ 1000 has already been achieved) so that we can achieve values of $\xi << 1$. This means that we "violate" what had appeared to be the thermodynamic limit of the simple calculation associated with phonon exchange in the initial description of thermal detectors given above. Detectors that have the capability of achieving sub-eV resolution will allow us to measure line shapes at smaller and smaller velocities, and ultimately their natural widths.

There are lots of other advantages offered by new alternatives to implanted thermistors, not the least of which is some relief from the natural recovery time of the thermal link τ_0. The higher level of electrothermal feedback could decrease the pulse response time by a factor that can approach α, so that the next generation of microcalorimeters can be orders of magnitude faster than those that will be flown aboard Astro-E. There should also be a variety of ways to increase the array spatial resolution from the current 32 pixels of Astro-E to 1000 or more.

8 Summary

The current state-of-the-art in non-dispersive spectrometers for X-ray astronomy is the Astro-E calorimeter. Such detectors have Ψ well in excess of unity because they can operate over a dynamic range of more than 100 with absolute quantum efficiency $\epsilon(E)$ approaching unity. Since non-dispersive detectors tend to have noise-dominated (approximately constant with energy) resolution, their resolving powers get better with increasing energy. The currently verified resolution of about 10 eV (FWHM) then corresponds to resolving powers of about 100 for Fe L lines near 1 keV, and about 700 for Fe K lines near 7 keV. It is relatively straightforward to provide simulations of calorimeter data based upon conventional wisdom or popular models. The coronal equilibrium spectra exhibited in Figure 3 provide a few such examples.

As impressive as the sensitivities discussed in previous sections appear to be, they still cannot convey the new *qualitative* spectroscopic capability of the calorimeter. The generality of an instrument that *simultaneously* samples the entire telescope bandpass with high spectral resolution *and* high sensitivity cannot be overemphasized. The following are examples of important astrophysical measurements that can be made *uniquely* by such an instrument:

- The detection of X-ray line emission from the hot intercloud medium of AGN (i. e., 6.7 keV He-like Fe emission in the presence of much stronger 6.4 keV fluorescent Fe emission)

- The determination of absolute elemental and ionization state column densities via absorption edges in the broadline clouds of AGN and the ISM

- The measurement of resonant absorption lines (if present) in systems such as AGN and OB stellar winds

- The independent determination of H_0 and q_0 from quasar emission lines and absorption lines from the gas in clusters of galaxies in the line of sight [7]

- The tracing of plasma diagnostics and elemental abundances in low surface brightness areas of SNR

- The determination of the sites at which Fe line emission originates in all compact binary X-ray sources (i. e., in each one separately)

- The detailed history of cooling flows in clusters of galaxies and in giant elliptical galaxies

- Sensitivity high enough to survey large samples of each source category

References

1. S.H. Moseley, J.C. Mather, and D. McCammon, *J. Appl. Phys.* **56**, 1257 (1984).
2. D. McCammon, W. Cui, M. Juda, J. Morgentahler, J. Zhong, R.L. Kelley, S.S. Holt, G.M. Madejski, S.H. Moseley, and A.E. Szymkowiak, *Nucl. Instrum. Methods* **A326**, 157 (1993).
3. J.N. Bahcall and C.L. Sarazin, *Ap. J.* **219**, 781 (1978).
4. A.K. Pradhan and J.M. Shull, *Ap. J.* **263**, 477 (1981).
5. T. Maccacaro, E. D. Feigelson, M. Fener, R. Giacconi, I.M. Gioia, R. E. Griffiths, S. S. Murray, G. Zamorani, J. Stocke, and J. Liebert, *Ap. J.* **253**, 504 (1982).
6. K.D. Irwin, G.C. Hilton, D. A. Wollman, and J. M. Martinis, *Appl. Phys. Lett.* **69**, 1945 (1996).
7. J.H. Krolik and J.C. Raymond, *Ap. J. Lett.* **335**, L39 (1988).

THE X–RAY BACKGROUND AND THE ROSAT DEEP SURVEYS

G. HASINGER
Astrophysikalisches Institut Potsdam, An der Sternwarte 16,
14482 Potsdam, Germany
E-mail : ghasinger@aip.de

G. ZAMORANI
Osservatorio Astronomico, Via Zamboni 33,
Bologna, 40126, Italy
E-mail: zamorani@astbo3.bo.astro.it

In this article we review the measurements and understanding of the X-ray background (XRB), discovered by Giacconi and collaborators 35 years ago. We start from the early history and the debate whether the XRB is due to a single, homogeneous physical process or to the summed emission of discrete sources, which was finally settled by COBE and ROSAT. We then describe in detail the progress from ROSAT deep surveys and optical identifications of the faint X-ray source population. In particular we discuss the role of active galactic nuclei (AGNs) as dominant contributors for the XRB, and argue that so far there is no need to postulate a hypothesized new population of X-ray sources. The recent advances in the understanding of X-ray spectra of AGN is reviewed and a population synthesis model, based on the unified AGN schemes, is presented. This model is so far the most promising to explain all observational constraints. Future sensitive X-ray surveys in the harder X-ray band will be able to unambiguously test this picture.

1 Introduction

It is a heavy responsibility for us, as it would be for everybody else, to write a paper on the X–ray Deep Surveys and the X–ray background (XRB) for this book, in honour of Riccardo Giacconi. Everybody recognizes the importance of Riccardo's contribution in both these fields. The discovery of the XRB using proportional counters in a rocket flight, the UHURU satellite, the first X–ray telescope with focusing optics "Einstein" are milestones in X–ray astronomy and Riccardo played a leading role in these experiments. More recently he gave a fundamental contribution to the planning, execution and analysis of the ROSAT Deep Surveys. As a consequence of this, a significant fraction of the results which we will describe in this paper either are his own results or are based on experiments which he conceived and led to success.

In Section 2 we give a brief historical overview of the XRB problem, from its discovery up to the results obtained in the eighties with the HEAO–1 and Einstein missions. During these years the origin of the XRB has been discussed mainly in terms of two alternative interpretations: the diffuse hypothesis (e.g. hot intergalactic gas) and the discrete source hypothesis. The existence of these radically alternative hypotheses has not been "neutral" with respect to devising experiments which wanted to study the XRB. In fact, if the XRB were mainly due to discrete sources, experiments aimed at studying the single sources responsible for it would obviously need high angular resolution in order to detect and resolve the large number of expected faint sources. Vice versa, if the XRB were mainly diffuse, source confusion would not be a problem and one could safely abandon the high angular resolution option. In this case the crucial experiment would be a measurement as accurate as possible of the spectrum in order to reveal the dominant physical production processes. The two working hypotheses led various groups of scientists to design very different sets of experiments[1]. In Section 3 we present and discuss some recent results from deep surveys with ROSAT. These surveys have already resolved into discrete sources $\sim 70\%$ of the measured XRB in the 1–2 keV band. The available optical identifications, still in progress, suggest that AGNs are the dominant population at these faint X–ray fluxes. Finally, in Section 4 we discuss some models which, taking into account the existing observational data on AGNs, are able to produce reasonably good fits to the XRB spectrum up to ~ 100 keV. The discussion is summarized in section 5.

2 The Early History of the X–ray Background

The existence of a diffuse XRB was discovered more than 35 years ago[2]. The aim of the experiment was to measure the X–ray emission from the Moon, but the data showed something unexpected: a strong X–ray source about 30 degrees away from the Moon (Sco X–1) and a diffuse emission approximately constant from all the directions observed during the flight. After this discovery, the first real improvement in our knowledge of the XRB has been made with the first all–sky surveys (UHURU and ARIEL V) at the beginning of the seventies. The high degree of isotropy revealed by these surveys led immediately to realize that the origin of the XRB has to be mainly extragalactic. Moreover, under the discrete source hypothesis, the number of sources contributing to the XRB has to be very large[3] ($N > 10^6\, sr^{-1}$).

In the same years a number of experiments were set up to measure the spectrum of the XRB over a large range of energy. It was found that over the energy range 3–1000 keV the XRB spectrum is reasonably well fit with two power laws with energy indices $\alpha_1 \sim 0.4$ for $E \leq 25$ keV and $\alpha_2 \sim 1.4$ for $E > 25$ keV (see Figure 1 in[4]).

At the beginning of the eighties two different sets of measurements led additional fire to the debate between supporters of the discrete source and diffuse hypotheses. On the one hand, the excellent HEAO–1 data showed that in the energy range 3–50 keV the shape of the XRB is very well fit by an isothermal bremsstrahlung model corresponding to an optically thin, hot plasma with kT of the order of 40 keV[5]. Moreover, it was shown that essentially all Seyfert 1 galaxies with reliable 2–20 keV spectra ($\sim$ 30 objects, mostly from HEAO–1 data) were well fit by a single power law with an average spectral index of the order of 0.65, significantly different from the slope of the XRB in the same energy range [6]. Since Active Galactic Nuclei (AGNs) were already considered the most likely candidates for the production of the XRB under the discrete source hypothesis [7], these two observational facts were taken as clear "evidence" in favour of the diffuse thermal hypothesis. On the other hand, the Einstein observations were suggesting a different scenario:

a) Pointed observations of previously known objects very soon showed that AGNs, as a class, are luminous X–ray emitters [8].

b) When a larger AGN sample became available, it was confirmed that AGNs could contribute most of the diffuse soft XRB [9]. Actually, in order to avoid a contribution from AGNs larger than the observed background, it was concluded [9] that the optical counts of AGNs had to flatten at magnitudes slightly fainter than the limit of the optically selected samples existing at that time. Such a flattening was later seen in deeper optical surveys.

c) Deep Einstein surveys showed that about 20% of the soft XRB (1–3 keV) is resolved into discrete sources [10,11,12,13] at fluxes of the order of a few $\times\, 10^{-14} erg\, cm^{-2} s^{-1}$. A large fraction of these faint X–ray sources were identified with AGNs.

Because of the difference between the spectra of the XRB and those of the few bright AGNs with good spectral data, the supporters of the diffuse, hot plasma hypothesis had to play down as much as possible the contribution of AGNs to the XRB to a limit which was close to be in conflict with an even mild extrapolation of the observed log(N)–log(S). Actually, a number of papers were published in which it was "demonstrated" that even in the soft X–ray band AGNs could not contribute much more than what had already been detected at the Einstein limit.

At that time we personally think that there was already evidence (for those who wanted to see it...) that the diffuse thermal emission as main contributor to the background was not tenable (see, for example, [14]). Very simple arguments in this direction were given in [15]. On the basis of reasonable extrapolations of the X–ray properties and the optical counts of known extragalactic X–ray sources (mainly AGNs and galaxies), Giacconi and Zamorani concluded that it is unlikely that their contribution to the soft X–ray background is smaller than 50%. Given this constraint, they then discussed two possibilities:

i) either faint AGNs have the so–called (at that time) "canonical" spectrum observed for brighter AGNs. In this case the residual XRB (i.e. the spectrum resulting after subtraction of the contribution from known sources) would not be fitted anymore by optically thin bremsstrahlung;

ii) or spectral evolution for AGNs is allowed. In this case, in order not to destroy the excellent thermal fit in the 3–50 keV data, diffuse emission could still be accommodated only if discrete sources have essentially the same spectrum as the XRB. On this basis, they concluded that "since in this scenario we would already require that the average spectrum of faint sources yielding 50% of the soft XRB is essentially the same as the observed XRB, there is nothing that prevents us from concluding that the entire background may well be due to the same class of discrete sources, at even fainter fluxes".

In other words, reversing the usual line of thought, the excellent thermal fit of the 3–50 keV XRB spectrum was shown by these arguments to be a point in favour of the discrete source hypothesis, rather than of the hot gas hypothesis! These conclusions, however, were not well received in a large fraction of the X–ray community; probably, they had the defect of being too simple and direct...

Thus, the debate between the supporters of the two hypotheses continued, until the final resolution of the controversy came from the incredibly neat results obtained with the FIRAS instrument on board COBE: the absence of any detectable deviation from a pure black body of the cosmic microwave background set an upper limit to the comptonization parameter $y < 10^{-3}$, more than ten times smaller than the value required by the hot intergalactic gas model [16]. The most recent upper limit for the comptonization parameter [17] is now $y < 2.5 \times 10^{-5}$. Discussing these data, in [18] it is concluded that a uniform, hot intergalactic gas produces at most 10^{-4} of the observed XRB!

3 The ROSAT Deep Surveys

3.1 Instrumental considerations

The detailed preparation for ROSAT Deep Survey observations started in 1984, well in advance of the actual ROSAT mission. At this time the engineering model of the ROSAT Proportional Counter PSPC [19] had already been calibrated in the laboratory and the task was at hand to understand and try to correct for the various electronic and geometric distortions and gain non–linearities, which in general plague imaging proportional counters. Non–linearities in the performance of the Einstein IPC have e.g. set the final sensitivity threshold for Einstein Deep Surveys [20,21]. The ROSAT PSPC, even at the raw coordinate level, shows a higher degree of uniformity than

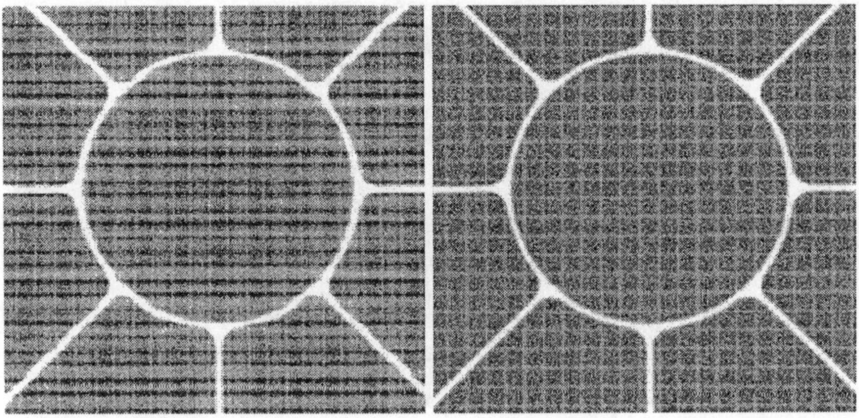

Figure 1: *The ROSAT PSPC flat-field in detector coordinates. Left: uncorrected; right: corrected for geometric distortions (the web-like shadow is due to the mechanical support structure of the PSPC window).*

the IPC. It was therefore possible to largely correct the significant distortions present in the PSPC data[22]. Figure 1 shows the comparison of a PSPC ground calibration flat field before and after correction of the distortion effects.

These images immediately indicate another geometrical problem. While it is obviously possible to remove the distortions created by the detector itself, the shadows cast by the complicated wire-mesh in front of the PSPC cannot be corrected for. The mere fact that the 100μ wire shadow can be detected, demonstrates the relatively good angular resolution of the PSPC. In the early days we were still hoping that the satellite pointing instability would wash out any residual flat-field inhomogeneities.

Roughly at the same time we started prototyping the ROSAT scientific analysis software. Early ideas about local and map-detect algorithms, background estimation etc. were taken over from the Einstein analysis system. However, substantial improvements were incorporated, the most important of which was probably the rigorous application of Poisson statistics and maximum likelihood estimators in all statistical computations.

In order to test and understand the analysis algorithms we developed a science simulator system, which turned out to be one of our most powerful tools in the preparation of the science mission. During the course of time the simulation models became more and more realistic, including extended and point sources, time variability, different spectral models and realistic number counts for the X–ray sources as well as cosmic and solar scattering backgrounds.

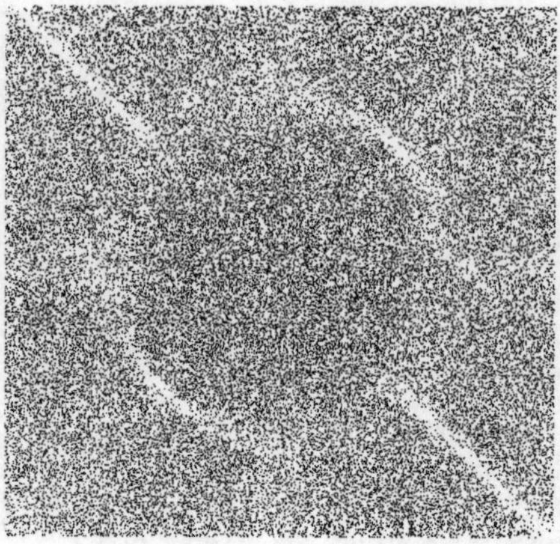

Figure 2: *The ROSAT PSPC flat-field in sky coordinates after the application of the wobble mode.*

Orbital variation of the exposure time and background components due to the radiation belts were included as well. Every individual photon was traced through a realistic model of the X-ray mirror system and the detector [23]. Comparing input and output information, the detection algorithms could be tested, calibrated and, if necessary, improved.

In 1985 the first data about the performance of the ROSAT attitude control system became available from dynamical hardware-in-the-loop tests. It turned out that the attitude control was much more stable than specified and anticipated. Simulations including realistic attitude data indicated that sources in the center of the PSPC field could easily get lost behind shadows of the PSPC support grid. Less than a year from the originally planned launch (1987) we were able to convince the funding agency and industry to introduce a "wobble mode" into the attitude control system, which was to become the standard ROSAT pointing mode. Instead of wandering around at random, the spacecraft pointing direction is guided smoothly into a linear periodic zigzag motion with a period of 402 seconds and an amplitude of ±3 arcmin (±1 arcmin for the HRI) diagonal with respect to the detector coordinates. The wobble-mode, while introducing artificial periodic power into the measured light curves of X-ray sources, turned out to be very valuable for the study of the X-ray background or other diffuse sources. Figure 2 shows the PSPC flat-field expected after the wobble motion has been applied to it. The rms variation of the wobbled PSPC flat field is about 3 percent [24].

3.2 Science preparation

In 1985 we assembled an international team of astronomers interested in deep X–ray surveys. Riccardo Giacconi, together with Richard Burg, at that time both at STScI, brought the Einstein experience into the team and originally suggested the deep survey in the Lockman Hole as our prime study area. This hole had just been discovered as the direction with an absolute minimum of interstellar hydrogen column density [25]. Gianni Zamorani brought the experience and the data of the deep optical survey in the Marano field [26], which we defined as a second study area in the southern hemisphere. Maarten Schmidt, who had previously modeled [27] the contribution of various classes of sources to the X–ray background, was planning the optical identification work in the North, first using the Palomar 200" and later the Keck telescopes. Joachim Trümper, Gisela Hartner and Günther Hasinger were responsible for securing the ROSAT data, i.e. calibration, simulation and finally observation and analysis. Since then the group met regularly at 1/2-1 year intervals. We recall these meetings as extremely productive but also quite exhaustive and often with violent disputes. In particular we all appreciated Riccardo's rigorous and constructive criticism.

The Lockman Hole is actually several degrees across. At first we chose a $\sim 0.3\ deg^2$ sky region close to the absolute minimum of N_H, but away from bright stars. Well in advance of the actual X–ray observations we performed optical UBVRI and radio observations in this field. In 1988 a mosaique in U, B and R was taken with the University of Hawaii 2.2m telescope and in 1989 drift scans in V and I were performed with the 4-shooter camera at the Palomar 200"-telescope. A 16 hr mosaique of observations [28] at 20 cm in the VLA D-configuration was taken in 1990. In the meantime other groups have picked up our fields as deep study areas. Both the Lockman Hole and the Marano field are covered by deep ISOCAM FIR imaging studies (Cesarsky et al.), the Lockman Hole is e.g. targeted in the Heidelberg CADIS NIR survey, in Luppino's wide field weak lensing survey and in ASCA deep survey observations.

3.3 ROSAT observations

Deep X–ray survey observations in the Lockman Hole commenced in the ROSAT AO-1 (spring of 1991) with a 100 ksec PSPC pointing exposure. The PSPC exposure was aimed from the beginning to reach the ultimate instrumental limits, while an HRI raster scan was planned to improve the PSPC positions. Since the spatial homogeneity of ROSAT observations is very good (see above), their ultimate sensitivity limit is set by confusion. In order to fight confusion, we had to obtain a very good understanding of the PSPC point-spread function [29], develop a completely new algorithm for X–ray crowded field analysis, and run large numbers of simulations in order to understand the systematic subtleties and the limitations of the observations. A substantial fraction of this work was already available before the actual observations. The knowledge about instrumental limitations has led us to the final choice of the exposure time and to a conservative selection of flux limits and off-axis ranges for the complete samples to be analyzed with PSPC data.

Because of the expected confusion in the PSPC, it was also clear from the beginning that HRI data would be necessary to augment the PSPC identification process. Because of the HRI smaller field-of-view, lower quantum efficiency and higher intrinsic background, we figured that it was necessary to invest about a factor four more HRI time than PSPC time to cover the same field with the same sensitivity, based on pre-launch knowledge. We therefore started a raster scan in AO-1, with ~ 100 pointings of 2 ksec each across the survey region. The remaining 200 ksec of HRI raster observations were approved in AO-2. When, however, the first HRI in-flight performance figures became available [30], we realized that the anticipated sensitivity would not be reached with the HRI raster scan. Due to the increased quantum efficiency, compared to the Einstein HRI, the ROSAT HRI is also more susceptible to background induced by particles in orbit. An increased halo of the HRI point-spread function as well as irreproducible attitude errors of about 5" are responsible for a further loss of sensitivity. Knowing this, we were able to trade the 200 ksec HRI time remaining for AO-2 into an extra PSPC observation of 100 ksec, which was performed in spring of 1992.

Two years later, after the PSPC observations of the Marano field had been successfully completed and the PSPC had run out of gas, we started to apply for an HRI ultradeep survey aimed for a total observing time of 1 Msec in a single pointing direction. This survey was planned to push the unconfused sensitivity limit deeper than the PSPC exposure in a substantial fraction of the PSPC field. In order to allow an X–ray "shift and add" procedure, correcting for the erratic ROSAT pointing errors, we selected a pointing direction for the ultradeep HRI exposure which is inside the PSPC field of view, but shifted about 8 arcmin to the North–East of the PSPC center, this way covering a region containing about 10 relatively bright X–ray sources known from the PSPC and the HRI raster scan. The 1 Msec HRI survey was approved for ROSAT AO-6 and AO-7 and is still going on. So far an exposure time of 880 ksec has been accumulated.

The observations in the *Lockman Hole* represent the deepest X–ray survey ever performed. The total observing time invested is quite comparable to that of other major astronomical projects, like e.g. the Hubble Deep Field [31]. Figure

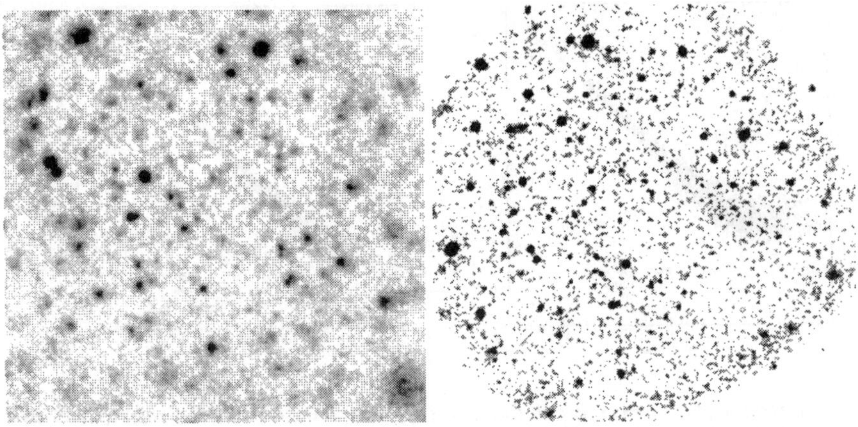

Figure 3: *ROSAT images of the Lockman Hole. Left: 207 ksec PSPC observation, right: 1 Msec HRI observation. The image size is* 34 × 34 *arcmin.*

3 shows the 200 ksec ROSAT PSPC image and the 1 Msec HRI image of the Lockman Hole. The details of the X–ray observations and data analysis, as well as a full catalogue of X–ray sources are being published [32].

3.4 Fighting confusion

The ROSAT deep survey exposures are probing the limits of observational and data analysis procedures. In order to obtain a reliable and quantitative characterization and calibration of the source detection procedure, detailed simulations of large numbers of artificial fields, analysed through exactly the same detection and parameter estimation procedure as the real data, are required. Early simulations [24] had already demonstrated that source confusion sets the ultimate limit in ROSAT deep survey work with the PSPC. The crowded–field multi–ML detection algorithm used in [32] was specifically designed to better cope with source confusion. However, we felt it necessary to calibrate its efficiency and verify its accuracy through new simulations. We have simulated sets of PSPC and HRI observations with 200 ksec and 1 Msec exposure, respectively, approximating our current observation times. The final results are relatively independent of the actual log(N)–log(S) parameters chosen for the artificial fields [24]. In the simulations point sources are placed at random within the field of view, with fluxes drawn at random from the log(N)–log(S) function down to a minimum source flux, where all the X–ray background is resolved for the assumed source counts. For each source the expected number of counts

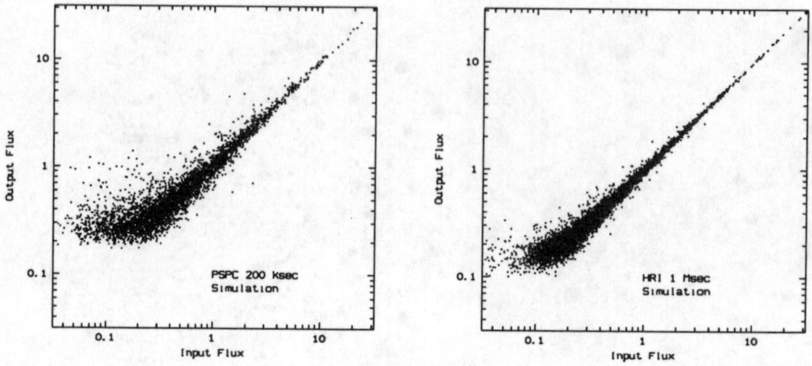

Figure 4: *Detected flux versus input flux for (a) 66 simulated PSPC fields of 200 ksec exposure, and (b) 100 simulated HRI fields of 1 Msec exposure each. The PSPC data is for off–axis angles smaller than 12.5 arcmin, for the HRI the limit is 10 arcmin. Fluxes are in units of* 10^{-14} $erg\ cm^{-2}\ s^{-1}$.

are computed taking into account the ROSAT vignetting correction, the appropriate energy to counts conversion factor and the exposure time of the simulated image. The actual counts of each source are drawn from a Poissonian distribution and folded through the point spread function. The realistic multi–component point spread function model is taken from [29] for the PSPC and from [30] for the HRI. Finally, all events missing in the field, i.e. particle background and non–source diffuse background are added as a smooth distribution to the image.

For each detected source the process of "source identification" has been approximated by a simple positional coincidence check. A detected source was identified with the counterpart from the input list, which appeared closest to the X–ray position within a radius of 30 arcsec. The faintest sources detected in the PSPC at small off–axis angles have a flux of 2×10^{-15} $erg\ cm^{-2}\ s^{-1}$. At larger off–axis angles the sensitivity is reduced correspondingly. The faintest HRI sources reach down to fluxes of 10^{-15} $erg\ cm^{-2}\ s^{-1}$. Again, at larger angles the sensitivity is reduced.

The detected source catalogues are affected by biases and selection effects present in the source–detection procedure. The most famous of those is the Eddington bias, which produces a net gain of the number of sources detected above a given flux limit as a consequence of statistical errors in the measured flux (see discussion in [24]). Another selection effect, most important in the deep fields considered here, is source confusion. The net effect of source confusion is difficult to quantify analytically, because it can affect the derived source catalogue relation in different ways. Two sub–threshold sources could be present in the same resolution element and thus mimic a single detected source. This leads to a net gain in the number of sources, similar to the Eddington bias. Two sources above the threshold could merge into a single brighter source. In this case one source is lost and one is detected at a higher flux. Whether the total flux is conserved or not depends on the distance between the two sources and on the details of the source detection algorithm. The detection algorithm cannot discriminate close sources with very different brightness, which results in a net loss of fainter sources.

The effects of confusion become immediately obvious in figure 4 where for each detected source the detected flux is compared to the flux of the nearest input source within 30 arcsec. While for bright sources there is an almost perfect match, there is a significant systematic deviation for fainter X–ray sources, where most detected objects appear at fluxes significantly brighter than their input counterparts. This is a direct indication of confusion because every detected source is only associated with one input source while its flux may be contributed from several sources. Source confusion effects are much less pronounced in the HRI data.

While it is obviously possible to correct for confusion effects on the source counts in a statistical way, the optical identification process relies on the position of individual sources. Confusion effects like those described here can cause the position of the detected sources to be significantly offset with respect to the true source position. This happens more frequently near the detection limits and therefore complicates the identification of complete samples of faint sources based on PSPC data alone.

3.5 The log(N)–log(S) relation

The first ROSAT deep survey observation, from a ~ 50 *ksec* PSPC pointing during the verification-phase in the direction of the North-ecliptic pole (NEP), was presented in 1991 in [33]. The log(N)–log(S) function, which reached flux levels about a factor of four fainter than the Einstein deep survey limit, clearly showed a flattening of the source counts below the Euclidean slope, which was anticipated previously [21]. The detailed shape below the break was, however, quite uncertain because of large statistical errors. In the following years source counts were presented from a number of different ROSAT deep surveys, which basically agreed with the early findings [24,34,35]. The best constraints on the faint-end slope of the source counts is obtained by fluctuation analyses of PSPC

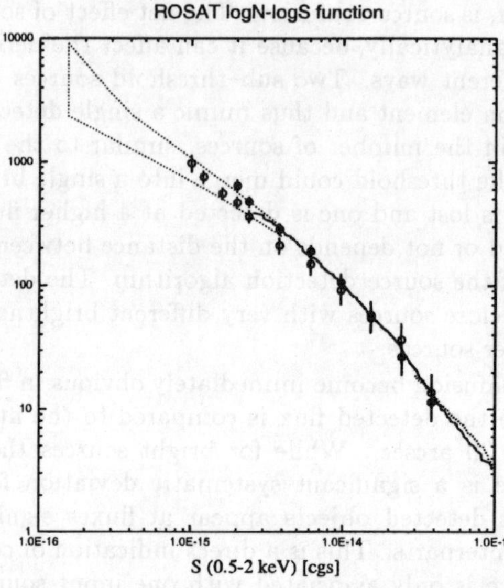

Figure 5: *Measured ROSAT log(N)–log(S) function in the Lockman Hole. Filled circles give the source counts from the 207 ksec PSPC observation. Open circles are from the ultradeep HRI observation (880 ksec). The data are plotted on top of the source counts (solid line) and fluctuation limits (dotted area) from* [24]*. The dotted line at bright fluxes refers to the total source counts in the RIXOS survey* [37]*.*

fields with exposure times of 100-150 ksec [24,36].

The final limiting sensitivity for the detection of discrete sources in the Lockman Hole is about 2×10^{-15} *erg* cm^{-2} s^{-1} for the PSPC and about a factor of two fainter with the HRI. Figure 5 shows the new PSPC and HRI log(N)–log(S) function [32], which is in good agreement with data published previously. The HRI source counts reaches a surface density of 970 ± 150 deg^{-2}, about a factor of two higher than any previous X–ray determination. About 70-80% of the X–ray background measured in the 1-2 keV band has been resolved into discrete sources now.

3.6 Optical identifications

Shanks et al. [38] have carried out a program of optical spectroscopy of sources detected in a 30 ksec PSPC verification phase observation in one of the AAT deep optical QSO fields and could quickly identify an impressive fraction of faint X–ray sources as classical broad-line AGNs (mainly QSOs). These studies have been continued by the same group on additional fields and deeper X–ray data in the following years[39], until finally a large enough sample of AGNs was available to calculate an X–ray luminosity function (XLF). Using data from medium-deep ROSAT fields combined with the Einstein Medium Sensitivity Survey[40], Boyle et al.[41] could improve significantly on the derivation of the AGN XLF and its cosmological evolution. Their data is consistent with pure luminosity evolution proportional to $(1 + z)^{2.7}$ up to a redshift $z_{max} \approx 1.5$, similar to what was found previously in the optical range. This result has been confirmed and improved later on by more extensive or deeper studies of the AGN XLF, e.g. the RIXOS project[42] or the UK deep survey project[43].

All these studies agree that at most half of the faint X–ray source counts and, correspondingly, half of the soft X–ray background can be explained by classical broad-line AGN based on the luminosity evolution models. Hasinger et al.[24] discussed the possibility that either the XLF models have to be modified considerably or a new population of X–ray sources has to contribute to the X–ray background. There was, indeed, mounting evidence that a new class of sources might start to contribute to the XRB at faint X–ray fluxes. The faintest X–ray sources in ROSAT deep surveys on average show a harder spectrum than the identified QSOs[24]. In medium-deep pointings a number of optically "innocent" narrow-emission line galaxies (NELGs) at moderate redshifts (z<0.4) were identified as X–ray sources, which was in excess of those expected from spurious identifications with field galaxies[44,39,45]. Roche et al.[46] have found a significant correlation of X–ray fluctuations with optically faint galaxies. Finally, in an attempt to push optical identifications to the so far faintest X–ray fluxes, McHardy et al.[47] claimed that broad-line AGN practically cease to exist at fluxes below 5×10^{-15} $erg\ cm^{-2}\ s^{-1}$, while the NELG number counts still keep increasing, so that they would dominate below a flux of 10^{-15} $erg\ cm^{-2}\ s^{-1}$.

While this is obviously an interesting possibility, it is useful to remind that all these findings are based on identifications near the limit of deep PSPC surveys, at fluxes where our simulations suggest that the PSPC data start to be severely confused. Because moderate-redshift field galaxies in general show emission lines, there is the possibility to either misidentify a field galaxy as the counterpart of an X–ray source which in reality is associated to a different optical object (e.g. a fainter AGN) or to misclassify an intrinsically faint AGN hidden in a NELG-type spectrum as a new class of X–ray sources (see also[48]). On the contrary, the X–ray positions in our Lockman Hole survey are largely determined by the HRI raster scan and ultradeep pointing. Instead of confused PSPC error boxes of (realistically) 15-20 arcsec radius, we therefore have error box radii of 2-5 arcsec. Using optical spectroscopy from the Keck telescopes we have recently been able to complete the optical identifications in the Lockman Hole down to a flux limit of 5.5×10^{-15} $erg\ cm^{-2}\ s^{-1}$. Out of the 50 X–ray sources in the complete X–ray sample, we identified 38 AGNs, 4 groups of galaxies, 1 normal galaxy and 3 galactic stars. Four X–ray sources remain unidentified so far, all of which are likely to be groups of galaxies[49]. In this survey, which has one of the highest rates of optical identifications among existing deep X–ray surveys and has a high degree of reliability, we see no evidence for the emergence of a new source population at low X–ray fluxes. We are currently working on Keck LRIS multi-slit spectroscopy to complete our identifications down to a flux of 1.5×10^{-15} $erg\ cm^{-2}\ s^{-1}$ and on the basis of the data acquired so far we do not see a need to alter the above conclusions.

3.7 Constraints from the angular correlation function

Since, as discussed above, the X–ray background is made up largely from discrete sources one would expect some variance in the spatial distribution of the background due to these sources. A signal in the angular correlation function (ACF) can in principle give strong constraints on the clustering properties of the sources contributing to the X–ray background. However, the XRB is remarkably smooth. Until recently no signal could be detected in the XRB ACF neither at soft X–ray nor at hard X–ray energies[50]. A significant signal at small angular separations (1-3 arcmin) could be found in a 1-2 keV analysis of 50 deep ROSAT pointed observations in about 10% of the fields[51]. However, this signal was clearly associated with a few extended, very-low X–ray surface brightness clusters or groups of galaxies at moderate redshift. The most prominent one of these, the NEP blotch[33], has been in the meantime identified with low surface brightness cluster or supercluster emission at moderate redshift[52,53]. In trying to obtain an upper limit on the spatial structure in the background due to the clustering of sources below the detection limit, the fields with significant cluster emission have been excluded from the following analysis. Indeed, once those fields were excluded, only upper limits for the ACF could be obtained[51]. However, those limits strongly constrain the nature and clustering properties of the sources which, being below the detection limit, contribute to the "residual" X–ray background. According to this analysis, less than 35% of this residual background can be due to objects with clustering properties similar to those of bright quasars. The objects which make up the residual background, i.e. the soft XRB not resolved in the ROSAT deep surveys, must have a clustering length smaller than normal galaxies and/or show very strong cosmologic evolution of their clustering[51]. The smoothness of the XRB therefore provides an important additional clue to unravel its composition.

4 AGN Spectra and Fits to the XRB spectrum

The recent significant advances in X–ray spectroscopy made possible with GINGA, ASCA and BeppoSAX have changed substantially our views on the spectral characteristics of AGNs and have shown that the X–ray spectra of AGNs are much more complex than what was thought only a few years ago. Detailed observations of an increasing number of relatively bright AGNs have detected a number of spectral features in addition to the power law continuum, such as the Compton reflection "hump" due to reprocessed emission, absorption from cold material probably in the torus, the warm absorbers, the Fe lines, the soft excess at low energies (see [54] for a recent review).

Already GINGA data had shown convincingly [55,56] that the typical spectrum of Seyfert 1 galaxies shows a flattening at $\sim$ 10 keV, due to the reflected component, with respect to the observed power law slope in the range 2–10 keV. These observations showed that the average spectrum for these objects is very similar to the shape of the spectrum hypothesized in [57], where it was shown that such a spectrum, integrated through redshift with reasonable assumptions on the cosmological evolution, could provide an adequate fit to the shape of the observed XRB above 3 keV.

The Ginga data have immediately led a number of groups to construct models for fitting the XRB spectrum with various combinations of AGN spectra and evolution (see, for example, [58,59,60,61]). Most of these models require an AGN population whose hard X–ray spectrum is dominated by the reflected component. Actually, the main difficulty of these models is the extremely large required contribution of such component. Moreover, although qualitatively in agreement with the overall shape of the XRB in the energy range 3–100 keV, it has been shown in [62] that these models are not able to fit satisfactorily the position and the width of the peak of the XRB spectrum.

Alternatively, other studies [63,64] have explored in detail the possibility that the dominant contribution to the XRB is due to the combination of the emission from a population of unabsorbed and absorbed AGNs, distributed according to the unified AGN scheme [65], as originally proposed in [66]. This class of models appears to be highly successful not only in fitting the spectrum of the XRB but also in reproducing a number of other observational constraints.

As an example, we show in Figure 6 the results of a fit to the XRB spectrum obtained by Comastri *et al.* [64]. The main ingredients of this model, which takes into account the observed spectral properties of different classes of AGNs over a broad energy range, are the following:

a) The X–ray primary continuum spectrum of AGNs is described by two power laws with spectral indices α = 1.3 below 1.5 keV and α = 0.9 above

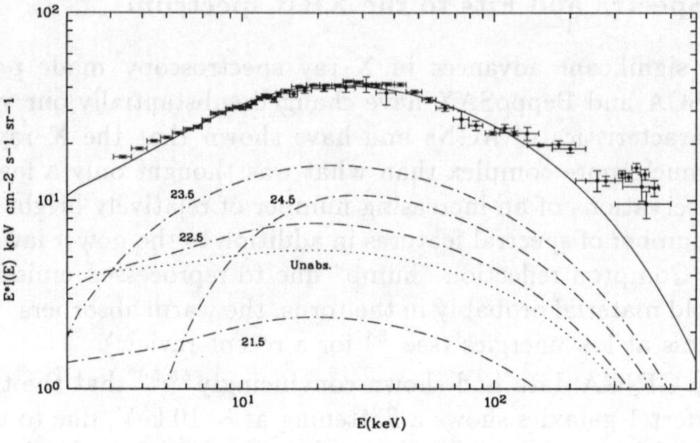

Figure 6: *The XRB spectrum above 3 keV: comparison between model (continuous line) and data. The data above 3 keV are taken from a compilation of the best experimental results*[67]. *The curves, labelled with values of log N_H, show the expected contribution from AGNs with different amount of intrinsic absorption.*

this energy. A Compton reflection component (50% of the primary spectrum) has been added in the continuum of the low luminosity AGNs [55] (i.e. Seyfert 1 galaxies), while no such component has been considered in high luminosity AGNs [68]. In agreement with the OSSE data [69], the power law spectrum is assumed to be exponentially cut–off at an e–folding energy of about 320 keV.

b) As required by the adopted unified scheme, the intrinsic X–ray spectrum of type 2 AGNs is assumed to be the same as the spectrum of type 1 AGNs, but modified by absorption effects [70]. A distribution of N_H, i.e. the column density of the absorbing material, in the range $10^{21} - 10^{25}$ atoms cm^{-2}, consistent with the observed distribution [71] is derived.

With these assumptions, the fit shown in Figure 6 has been obtained by integrating the local X–ray luminosity function of AGNs and assuming a luminosity evolution up to z $\sim$ 2. At higher redshift the X–ray luminosity function has been assumed to remain constant. This model, as well as other models based on similar assumptions (see, for example, [72]), fits the XRB spectrum well up to $\sim$ 100 keV, while it underpredicts the data at $\sim$ 500 keV by a factor of a few. It has been shown in [73] that at this energy the contribution from a different AGN population, namely flat spectrum radio quasars and "MeV blazars", is already substantial. These objects are likely to contribute most of the observed background in the energy range $1 - 10^3$ MeV.

As shown in the figure, the model predicts that in the hard X–ray band

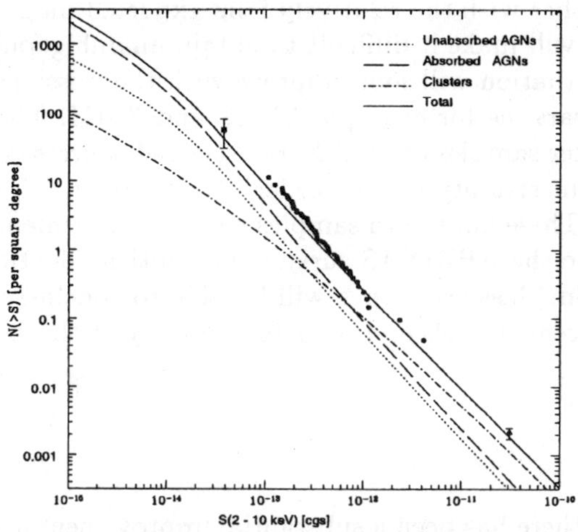

Figure 7: *The 2–10 keV log(N)–log(S). The circles represent the counts derived from the 60 ASCA sources discussed in*[74]*; the square shows the preliminary results of a different ASCA survey in a more limited sky area*[75]*; the triangle shows the extragalactic surface density from the HEAO-1 survey*[76]*. The curves show the predictions for various classes of objects from the model discussed in* [64]*.*

most of the contribution to the XRB comes from significantly absorbed objects, which are almost absent in the soft band, even at the faintest ROSAT limit. As a consequence, a significant test for this model would be the comparison of its predictions with the results of optical identifications of a complete sample of X–ray sources selected at low fluxes in the hard X–ray band.

Recently, ASCA data have been used [74] to derive the 2–10 keV log(N)–log(S) down to fluxes slightly below $10^{-13} erg\ cm^{-2}\ s^{-1}$. This data, based on 60 X–ray sources detected in ASCA images, is shown in Figure 7 together with the predictions for various classes of objects from the Comastri *et al.* model. As discussed in Cagnoni *et al.* [74], if one uses the ROSAT log(N)–log(S) and the average spectral properties of the ROSAT sources to predict the ASCA log(N)–log(S), the prediction falls a factor ~ 2 below the data at a flux of the order of $\sim 10^{-13} erg\ cm^{-2}\ s^{-1}$. On the other hand, as shown in the figure, the data are in good agreement with the predictions of the Comastri *et al.* model. The dashed line in Figure 7 shows that a significant fraction of the AGNs in this ASCA survey, and an even larger fraction at fainter fluxes, should be substantially absorbed and therefore their optical counterparts are expected to have optical spectra typical of Seyfert 2 galaxies. A program to optically identify these sources has already started, but the large positional uncertainty

($\sim$ 2 arcmin) together with the relatively faint expected magnitudes of the optical counterparts will make it difficult to obtain unambiguous identifications. Fortunately, the situation will soon improve with the missions to be launched in the next few years, as for example AXAF and XMM. These missions will provide much fainter samples of hard X–ray selected sources, with significantly better positional uncertainty (of the order of 1–2 arcsec for AXAF and 5–10 arcsec for XMM). These faint hard samples will be complemented by the bright all-sky sample from the ABRIXAS survey. The optical identifications of these sources with the 8m class telescopes will be able to conclusively test the predictions of our current models and, hopefully, finally settle the question of the X-ray background.

5 Conclusions

In the last decade there has been a substantial improvement in our understanding of both the X-ray background and the spectra of active galactic nuclei. To first order we can regard the problem of the X-ray background as solved. For the first time since its discovery we can explain the bulk of the X-ray background over a wide range of energies (0.1-100 keV) as due to a known class of sources, namely unabsorbed and absorbed active galactic nuclei, whose statistical properties like average spectra, absorption distribution, luminosity function etc. are consistent with those measured locally and which are strongly evolving with cosmic time.

In the soft X-ray band, mainly based on ROSAT deep surveys, the majority of identified sources down to the faintest fluxes are indeed AGNs and the background models can account for all observable constraints like log(N)-log(S) function, redshift distribution, average spectral index etc. Observations at harder X-ray energies, e.g. by ASCA, are consistent with this picture. However, only deeper survey observations with the next generation of sensitive X-ray observatories with good angular resolution in the harder X-ray band (i.e. AXAF and XMM) and optical follow-up spectroscopy from 8-10m telescopes will be able to unambiguously confirm the AGN XRB model and to tighten the constraints on some of the still relatively uncertain model parameters.

Acknowledgements

G.H. acknowledges the grant 50 OR 9403 5 by the Deutsche Agentur für Raumfahrtangelegenheiten (DARA) and G.Z. the Italian Spqce Agency (ASI) contract ASI 95-RS-152. We also warmly thank all our collaborators in the ROSAT Deep Survey project and A. Comastri for fruitful discussions.

References

1. Giacconi R. and Burg R. 1992, in *The X–Ray Background*, X. Barcons and A.C. Fabian eds., (Cambridge: Cambridge Univ. Press), 3.
2. Giacconi R., Gursky H., Paolini F.R. and Rossi B.B. 1962, *Phys.Rev.Letters*, **9**, 439.
3. Schwartz D.A. 1980, *Phys.Scripta*, **21**, 644.
4. Tanaka Y. 1992, in *X–Ray Emission from Active Galactic Nuclei and the Cosmic X–Ray Background*, W. Brinkmann and J. Trümper eds., MPE Report **235**, 303.
5. Marshall F.E. *et al.* 1980, *Ap.J.*, **235**, 4.
6. Mushotzky R.F. 1984, *Advances in Space Research*, Vol. **3**, no. 10–12, p.157.
7. Setti G. and Woltjer, L. 1973, in IAU Symposium No. 55, *X- and Gamma–Ray Astronomy*, H. Bradt and R. Giacconi eds., (Reidel, Dordrecht), p. 187
8. Tananbaum R. *et al.* 1979, *Ap.J.Letters*, **234**, L9.
9. Zamorani G. *et al.* 1981, *Ap.J.*, **245**, 357.
10. Giacconi R. *et al.* 1979, *Ap.J.Letters*, **234**, L1.
11. Griffiths R.E. *et al.* 1983, *Ap.J.*, **269**, 375.
12. Primini F.A. *et al.* 1991, *Ap.J.*, **374**, 440.
13. Hamilton T.T., Helfand D.J. and Wu X. 1991, *Ap.J.*, **379**, 576.
14. Setti G. 1985, in *Non–Thermal and Very High Temperature Phenomena in X–Ray Astronomy*, G.C. Perola and M. Salvati eds., p. 159.
15. Giacconi R. and Zamorani G. 1987, *Ap.J.*, **313**, 20.
16. Mather J.C. *et al.* 1990, *Ap.J.Letters*, **354**, L37.
17. Mather J.C. *et al.* 1994, *Ap.J.*, **420**, 439.
18. Wright *et al.* 1994, *Ap.J.*, **420**, 450.
19. Pfeffermann E. *et al.* 1986, *SPIE*, **733**, 519.
20. Soltan A.M. 1991, *MNRAS*, **250**, 241.
21. Hamilton T.T. 1992, in *The X–Ray Background*, X. Barcons and A.C. Fabian eds., (Cambridge: Cambridge Univ. Press), 138.
22. Briel U.G. *et al.* 1988, *SPIE*, **982**, 401.
23. Hartner G. 1986, Diplomarbeit TU München
24. Hasinger G. *et al.* 1993, *A&A* **275**, 1.
25. Lockman F.J., Jahoda K. and McCammon D. 1986, *Ap.J.*, **302**, 432.
26. Marano B., Zamorani G. and Zitelli V. 1988, *MNRAS*, **212**, 111.
27. Schmidt M. and Green R.P. 1983, *Ap.J.*, **269**, 352.
28. de Ruiter H. *et al.* 1997, *A&A*, **319**, 7.
29. Hasinger G. *et al.* 1994, *Legacy*, **4**, 40; MPE/OGIP Calibration Memo.
30. David F.R. *et al.* 1996, in *The ROSAT Users Handbook*
31. Williams R.E. *et al.* 1996. *A.J.*, **112**, 1335.
32. Hasinger *et al.* 1997, *A&A*, submitted.
33. Hasinger G., Schmidt M. and Trümper J. 1991, *A&A*, **246**, L2.
34. Branduardi-Raymont G. *et al.* 1994, *MNRAS*, **270**, 947.
35. Vikhlinin A. *et al.* 1995, *Ap.J.*, **451**, 553.

36. Barcons X. *et al.* 1994, *MNRAS*, **268**, 833.
37. Mason, K. *et al.* 1996, priv. comm.
38. Shanks T. *et al.* 1991, *Nature*, **353**, 315.
39. Georgantopoulos I. *et al.* 1996, *MNRAS*, **280**, 276.
40. Gioia I.M. *et al.* 1990, *Ap.J.Suppl.*, **72**, 567.
41. Boyle B.J. *et al.* 1994, *MNRAS*, **260**, 49.
42. Page M.J. *et al.* 1996, *MNRAS*, **281**, 579.
43. Jones L.R. *et al.* 1996, *MNRAS*, submitted, (astro-ph/9610124).
44. Boyle B.J. *et al.* 1995, *MNRAS*, **272**, 462.
45. Griffiths R.E. *et al.* 1996, *MNRAS*, **281**, 71.
46. Roche N. *et al.* 1995, *MNRAS*, **273**, L15.
47. McHardy I.M. *et al.* 1997, *MNRAS*, submitted, (astro-ph/9703163)
48. Hasinger G. 1996, *A&A Suppl.*, **120**, C607.
49. Schmidt M. *et al.* 1997, *A&A*, submitted.
50. Fabian A.C. and Barcons X. 1992, *ARA&A*, **30**, 429.
51. Soltan A. and Hasinger G. 1994, *A&A*, 288, 77.
52. Burg R. *et al.* 1992, *A&A*, **259**, L9.
53. Ashby M.N.L. *et al.* 1996, *Ap.J.*, **456**, 428.
54. Mushotzky R.F. 1997, in *Mass Ejection from AGN*, in press, (astro-ph/9705004).
55. Pounds K.A. *et al.* 1990, *Nature*, **344**, 132.
56. Nandra K. 1991, *Ph.D. Thesis*, Leicester University.
57. Schwartz D.A. and Tucker W.H. 1988, *Ap.J.*, **332**, 157.
58. Morisawa K., Matsuoka M., Takahara F. and Piro L. 1990, *A&A*, **263**, 299.
59. Fabian A.C., George I.M., Miyoshi S. and Rees M.J. 1990, *MNRAS*, **242**, 14P.
60. Terasawa N. 1991, *Ap.J.Letters*, **378**, L11.
61. Rogers R.D. and Field G.B. 1991, *Ap.J.Letters*, **370**, L57.
62. Zdziarski A.A., Zycki P.T., Svensson R. and Boldt E. 1993, *Ap.J.*, **405**, 125.
63. Madau P., Ghisellini G. and Fabian A.C. 1994, *MNRAS*, **270**, 117.
64. Comastri A., Setti G., Zamorani G. and Hasinger G. 1995, *A&A*, **296**, 1.
65. Antonucci R.J. 1995, *ARA&A*, **31**, 473.
66. Setti G. and Woltjer L. 1989, *A&A*, **224**, L21.
67. Gruber D.E. 1992, in *The X-Ray Background*, X. Barcons and A.C. Fabian eds., (Cambridge: Cambridge Univ. Press), 44.
68. Williams O.R. *et al.* 1992, *Ap.J.*, **389**, 157.
69. Zdziarski A.A. *et al.* 1995, *Ap.J.Letters*, **438**, L63.
70. Awaki H., Koyama K., Inoue H. and Halpern J.P. 1991, *PASJ*, **43**, 195.
71. Schartel N. *et al.* 1997, *A&A* **320**, 696.
72. Celotti A., Fabian A.C., Ghisellini G. and Madau P. 1996, *MNRAS*, **277**, 1169.
73. Comastri A., Di Girolamo T. and Setti G. 1996, *A&A Suppl.*, **120**, C627.
74. Cagnoni I., Della Ceca R. and Maccacaro T. 1997, *Ap.J.*, submitted.
75. Ogasaka Y. 1997, in *X-ray Imaging and Spectroscopy of Cosmic Hot Plasmas*, F. Makino and K. Mitsuda eds., 25.
76. Piccinotti G. *et al.* 1982, *Ap.J.*, **253**, 485.

MUSINGS ABOUT X-RAY LUMINOSITY FUNCTIONS AND SOURCE COUNTS

Maarten Schmidt
California Institute of Technology, Pasadena, CA 91125, USA
E-mail: mxs@deimos.caltech.edu

We discuss the role of the luminosity function in the interpretation of source counts of extragalactic X-ray sources. Given that the luminosity function derived from a given complete sample is a linear combination of the contributions of each of the sample sources, we illustrate various effects on the basis of source counts contributed by a single source. In particular, we show the effect on the source counts of density evolution, luminosity evolution, spectral index, and absorption.

1 Introduction

The luminosity function of extragalactic X-ray sources plays a central role in the interpretation of counts and redshifts of X-ray sources. As soon as a luminosity function is needed, most investigators make an initial assumption about its shape (usually a broken power law). This can be avoided, since the luminosity function can be directly determined from the observed distribution $n(S, z)$ of flux S and redshift z of sources in a sample complete to limiting flux S_{lim}. Moreover, since the luminosity function is a linear combination of contributions from the individual objects, we can evaluate the effect of each object in an observed sample on the luminosity function and the source counts. This brief contribution to Riccardo Giacconi's Festschrift is intended to illustrate the effect of one object on the source counts for various assumptions about its spectral energy distribution and its cosmological evolution. In so reducing the problem to its ultimate simplicity, we hope that it will give rise to discussions with Giacconi that are always challenging and enjoyable.

2 Role of the Luminosity Function

The luminosity function of a given type of astronomical sources is the distribution of their luminosities, L, per unit volume of space. Generally, the luminosity function may depend on other source properties, and in particular on distance or redshift z, representing cosmological evolution.

The luminosity function is an observer-independent distribution of absolute luminosities, and as such represents a physical property of the sources that in principle can be studied in terms of their formation and evolution. In practice, the shape of the luminosity function of extragalactic X-ray sources at any cosmic epoch is not understood.

The luminosity function plays a central role in setting the distribution of fluxes S and redshifts of a sample of sources that is complete above a given flux S_{lim}. For each of the sample sources, we can derive the total volume of space V_{max} over which it would be observable within the limits of the complete sample. This derivation involves q_o and H_o characterizing a cosmological model, the X-ray spectral energy distribution characterized by the spectral index and the absorption, and assumptions about the cosmological evolution. The contribution to the luminosity function by one of the sample sources is a delta function of amplitude $1/V_{max}$ at its luminosity L, and the total luminosity function is the sum of the delta functions of the sample sources. Details of the derivation are given in Schmidt and Green (1986).

Given this luminosity function, it is straightforward to derive the expected source counts in a different sample that is complete above, say, flux S_o. If we again consider the one source discussed above, then if V_o is the total volume in which it is observable, we expect V_o/V_{max} such sources in the new sample (cf. Schmidt and Green 1986). The total source count is the sum over all sources in the original sample.

3 Source Counts Corresponding to One Sample Source

We can consider the source counts generated by a single source in the original sample as a building block in constructing the overall source counts. We will illustrate what happens to this building block when we adopt different spectral energy distributions or different evolution parameters.

We use a single source observed in the ROSAT Bright Survey (RBS), which for simplicity we assume to be complete above a flux limit of $2.9\text{x}10^{-12}$ cgs in the $0.5 - 2.0$ keV band over 20,320 deg^2. We derive source counts, also in the $0.5 - 2.0$ keV band, along the lines discussed in the previous section. The cosmological model is characterized by $H_o = 50$ km s^{-1} Mpc^{-1} and $q_o = 0.5$. We assume that any cosmological evolution ends at redshift 1.65 (Hewett et al. 1993), and that for redshifts larger than 2.7 the co-moving density of sources declines by a factor of 2.7 per unit redshift (Schmidt et al. 1995).

Since at small redshifts space is approximately euclidean and evolution is assumed to be negligable, the source counts $N(> S)$ at high fluxes will be proportional to $S^{-1.5}$. We plot the *reduced counts*, $N(> S)/S^{-1.5}$ normalized to unity at $S = 2.9\text{x}10^{-12}$, to allow an easy comparison between different illustrations.

3.1 Effect of Density Evolution

In the case of density evolution of the luminosity function, we multiply the co-moving volume element by a density factor that is a function of redshift or cosmic time, and which is unity at $z = 0$. In choosing an example for illustration, it is useful to recall that the slope $d \log N / d \log S$ of X-ray AGN counts down to 10^{-14} cgs is close to the euclidean value of -1.5. Therefore we concentrate on parameters that produce reduced counts $N(> S)/S^{-1.5}$ that are of the order of unity.

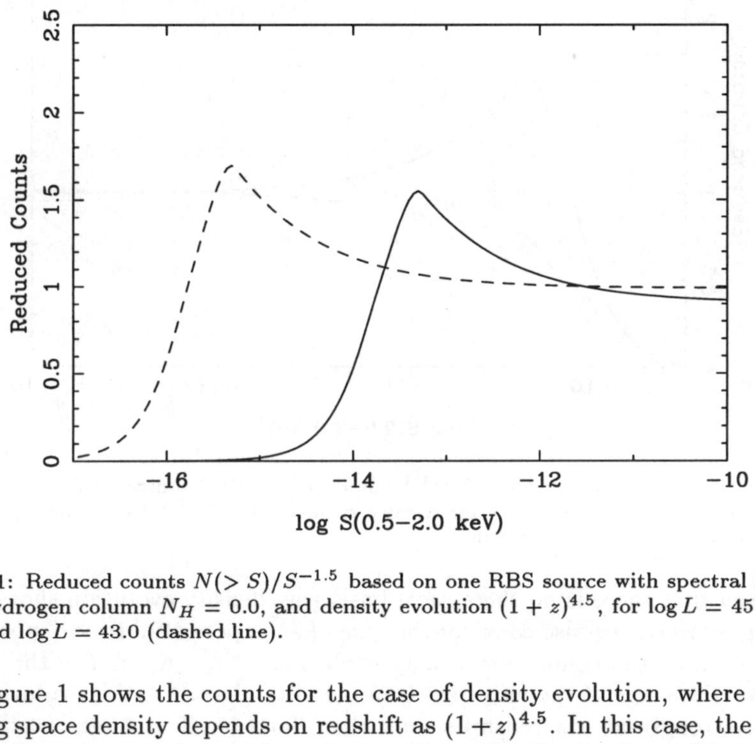

Figure 1: Reduced counts $N(> S)/S^{-1.5}$ based on one RBS source with spectral index = -1.0, hydrogen column $N_H = 0.0$, and density evolution $(1 + z)^{4.5}$, for $\log L = 45.0$ (solid line) and $\log L = 43.0$ (dashed line).

Figure 1 shows the counts for the case of density evolution, where the co-moving space density depends on redshift as $(1+z)^{4.5}$. In this case, the counts at large flux are very close to those for a uniform distribution in euclidean space. The sharp turn-around of the source counts is caused by the termination of the density evolution at $z = 1.65$. As a consequence, the turn-around flux scales with the luminosity as shown by the counts for $\log L = 43.0$ and 45.0. The total source counts, derived from a complete sample of sources will be the sum of their individual counts, each reflecting their luminosity L.

3.2 Various Cases of Density Evolution

In further examples, we primarily consider sources with $\log L = 43.0$. Figure 2 illustrates the difference between density evolution of $(1+z)^{3.5}$ and $(1+z)^{4.5}$. Clearly, density evolution of $(1+z)^{3.5}$ or less cannot explain the approximately euclidean slope of the source counts.

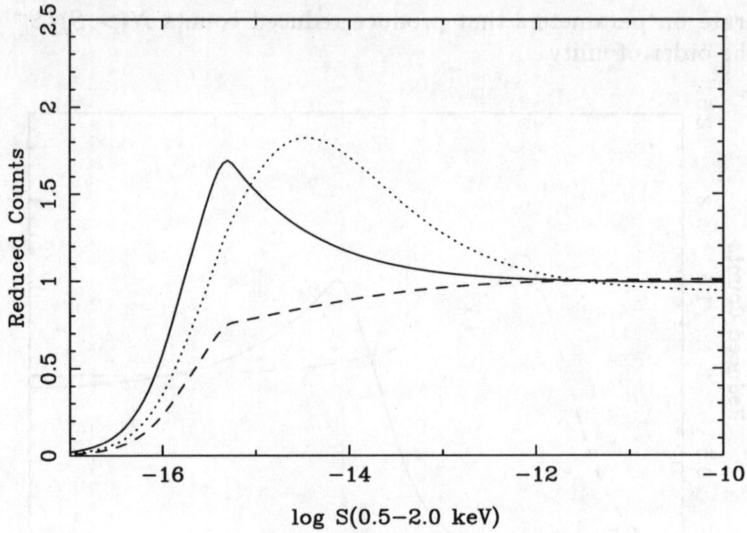

Figure 2: Reduced counts $N(>S)/S^{-1.5}$ based on one RBS source with $\log L = 43.0$, spectral index = -1.0, $N_H = 0.0$, for density evolution $(1+z)^{4.5}$ (solid line), $(1+z)^{3.5}$ (dashed line), and $e^{-5t/T}$ (dotted line).

Given that there is no theoretical basis why density evolution should be a $(1+z)$ power law, we also consider the case of an exponential time dependence. Figure 2 shows the counts for density evolution $e^{-5t/T}$ where t is the cosmic time and T the age of the universe. Compared to $(1+z)^{4.5}$, this evolution sets in more strongly at low redshifts, and then flattens off at large redshifts, as reflected in the counts.

3.3 Effect of Spectral Index

Determinations of the energy spectral index of AGNs range from around −0.7 to −1.3. We show in Figure 3 the effect of spectral index on the counts.

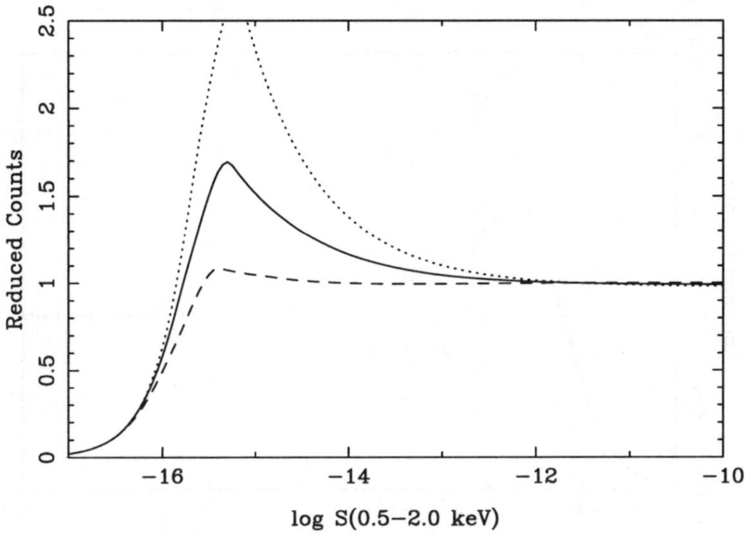

Figure 3: Reduced counts $N(>S)/S^{-1.5}$ based on one RBS source with $\log L = 43.0$, $N_H = 0.0$, density evolution $(1+z)^{4.5}$, for spectral index = –1.0 (solid line), –1.3 (dashed line), and –0.7 (dotted line).

Sources with a flatter spectrum generate larger source counts. This is an effect of the larger redshift of the weaker sources, which causes the observed bandpass of $0.5 - 2.0$ keV to correspond to higher rest-frame energies.

If the sources in a complete sample exhibit a range of spectral indices, the single-object approach can handle the situation without difficulty: the total source counts are still generated through adding the single-source counts, each now derived for its individual spectral index. If the spectral index of the sources is correlated with another property, say luminosity, this procedure produces the correct total source counts. This happens even if the investigator is not explicitly aware of the correlation!

3.4 Effect of Absorption

The first order effect of absorption will be to harden the spectrum, which will increase the slope of the source counts. Figure 4 shows the effect of a modest hydrogen column $N_H = 1.0\text{x}10^{21}$ cm^{-2} on the counts.

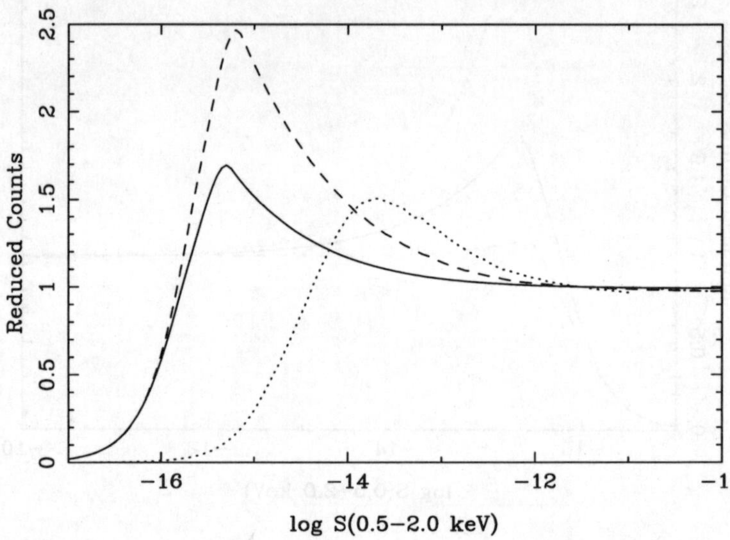

Figure 4: Reduced counts $N(>S)/S^{-1.5}$ based on one RBS source with $\log L = 43.0$, spectral index = −1.0, density evolution $(1+z)^{4.5}$, for $N_H = 0.0$ (solid line) and $N_H = 1.0\text{x}10^{21}$ cm^{-2} (dashed line); as well as for no evolution and $N_H = 2.0\text{x}10^{22}$ cm^{-2} (dotted line).

For larger hydrogen columns, the absorption in the bandpass of $0.5 - 2.0$ keV will be very strong and the effect on the source counts can be dramatic. As we move the source out to larger redshifts, the $0.5-2.0$ keV bandpass measures the source at higher energies, where the absorption declines steeply. The effect is similar to that of positive luminosity evolution, discussed in the next section. As shown in Figure 4, for a hydrogen column as large as $N_H = 2.0\text{x}10^{22}$ cm^{-2} we can produce approximately euclidean counts for zero density evolution!

If the absorption of each individual source in the complete sample is known and used in the procedure, the projected source counts will reflect the absorption distribution properly (cf. Sec. 3.3). In fact, the count contributions from each individual source can be tagged not only with its luminosity, but also its spectral index and absorption, allowing a detailed prediction of source properties as a function of flux S.

3.5 Effect of Luminosity Evolution

In the case of luminosity evolution, we increase the luminosity of the source by a factor $(1+z)^c$ as we move it out to larger distances. As a consequence, at a given flux the source is visible to a larger redshift, i.e. over a larger volume of space, leading to an increase in source counts.

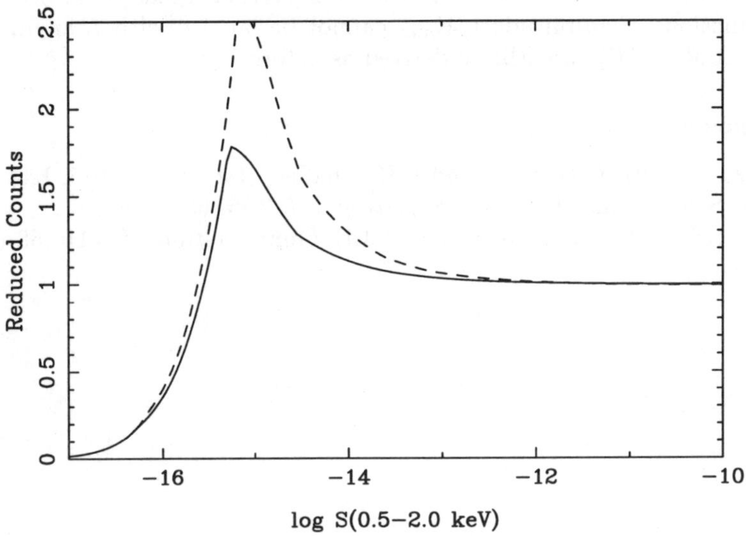

Figure 5: Reduced counts $N(>S)/S^{-1.5}$ based on one RBS source with zero-redshift luminosity $\log L = 42.0$, spectral index = −1.0, luminosity evolution $(1+z)^{2.5}$, for $N_H = 0.0$ (solid line) and $N_H = 1.0\text{x}10^{21}$ cm^{-2} (dotted line).

We illustrate in Figure 5 the case of luminosity evolution $(1+z)^{2.5}$ for an unevolved (zero-redshift) luminosity $\log L = 42.0$. For higher luminosities, the turn-down in the source counts will occur at a proportionally higher flux.

4 Multi–Parameter Detection

In the examples we discussed above, detection merely required that the flux of the point source exceeded a limiting flux, $S > S_{lim}$. For extended sources, such as clusters of galaxies, the point-source concepts of S and S_{lim} strictly do not apply. A more complex detection criterion will be required that probably involves the measured flux within one or more angular apertures. The single-object method of deriving the luminosity function and source counts is particularly suitable for detection algorithms involving more than one parameter.

For extended sources, the complex detection criterion $\mathcal{D}$ effectively takes the place of S_{lim} for point-source detection, and the source counts $N(>S)$ are now replaced by the counts $N(\mathcal{D})$ of objects that fulfill the detection criterion $\mathcal{D}$. Using our single-object analysis, the derivation of the luminosity function and the source counts is essentially the same as that discussed in Sec. 2. The delta functions of amplitude $1/V_{max}$ cannot be labeled with L anymore, but source counts $N(\mathcal{D})$ can still be derived as before.

References

1. P.C. Hewett, C.B. Foltz and F.H. Chaffee, *Astroph. J.* **406**, L43 (1993).
2. M. Schmidt and R.F. Green, *Astroph. J.* **305**, 68 (1986).
3. M. Schmidt, D.P. Schneider and J.E. Gunn, *Astron. J.* **110**, 68 (1995).

THE BRIGHTEST EXTRAGALACTIC RADIO SOURCES

Malcolm S. Longair
Cavendish Laboratory, Madingley Road, Cambridge CB3 0HE

The astrophysics of extragalactic radio sources intersects with many of Riccardo Giacconi's central interests — X-ray observations of the hot intergalactic gas in galaxies and clusters of galaxies, studies of active galaxies with the Hubble Space telescope, the cosmological evolution of active galaxies, and so on. A brief history of the contribution of studies of extragalactic radio sources to astrophysical cosmology is presented. New HST and infrared observations of a complete sample of 3CR radio galaxies are described. These are shown to be among the most massive galaxies at redshifts $z > 1$ and provide important clues to the formation of these massive galaxies and the nature of active galactic nuclei. The evolutionary history of these objects with cosmic epoch can now be related to the star-formation history of galaxies as a whole, thanks to observations with the Hubble Space Telescope.

1 Encounters with Riccardo Giacconi

Riccardo Giacconi has been at the centre of so many different aspects of the great flowering of modern astrophysics that it is difficult to do justice to his achievement in a short space. Those astronomers who, like me, began their PhDs in about 1963 belong to that lucky generation which has lived through the great revolutions of modern astrophysics and cosmology. In 1963, radio astronomy was reaching a state of maturity when the surveys were becoming much more reliable and the discovery of quasars had shaken many of the cherished beliefs of astrophysicists. The radio astronomers were about 17 years ahead of the X-ray astronomers and we watched with a combination of amusement and fascination as the first reports of the discoveries of the first rocket-borne X-ray experiments were announced, many of them associated with Riccardo's name.

To outsiders, the X-ray story appeared confused until 1972 and the remarkable discoveries of the UHURU satellite, which are fully celebrated elsewhere in this volume. For the extragalactic radio astronomer, the most intriguing discovery was the X-ray emission from the hot intergalactic gas in clusters of galaxies, which provided key information which we lacked until then. At last, it became possible to define the gaseous environments of some of the luminous radio sources in some detail and this was needed in order to account for the confinement of their radio structures. One of my bad luck stories was the attempt which Peter Willmore and I made to determine the structure of the X-ray emission from Cygnus A with the tiny X-ray detector on board the Copernicus satellite. We had detected the X-ray emission from Cygnus A in our first set of observations and were about to attempt the resolve its structure when the X-ray detector on Copernicus became non-functional (Longair and Willmore 1974).

A further encounter with X-ray astronomy occurred with the launch of the Einstein X-ray Observatory and the invitation by Riccardo to present a review of the relation between the X-ray and radio properties of extragalactic radio sources (Longair 1981). This presentation resulted in a collaboration with Pepi Fabbiano and Lance Miller in which we used the Einstein Observatory to study a complete sample of the brightest extragalactic radio sources in the northern sky (Fabbiano *et al.* 1984). This was an intriguing, but slightly frustrating, set of observations. As radio astronomers, Lance and I were desperate to discover the X-ray emission from the hot gas which must be present to confine the radio lobes, but it proved very elusive (Miller *et al.* 1985) — instead, we found a splendid relation between the radio and X-ray properties of these active galactic nuclei.

During this period, as a member of the Institute Visiting Committee and then the Space Telescope Institute Council, I had the pleasure of working with Riccardo during the early days of the build-up of the Space Telescope Science Institute in preparation for the launch of the HST. This was an extraordinarily demanding time for Riccardo, but I attempted to lighten his burden slightly when he invited me to present the first Hubble Public Lecture in 1984 on *Alice and the Space Telescope* (Longair 1989). Nobody believes me when say that this is the deepest piece of astronomical writing I have done — I believe that I wrote more home truths about the subject than in anything else I have written about astronomy.

The tragedy and triumph of the HST is now history and we all owe Riccardo an enormous debt of gratitude for his determination to 'do it right'. What I want to do in this paper is to show how many of these apparently diverse interactions with Riccardo have come together in our most recent observations with the Hubble Space Telescope.

2 The 3CR Radio Galaxies - A Personal History

I have recently written a brief history of the development of radio astronomy as part of a larger history of 20th century astrophysics and cosmology and more details of the early parts of that story can be found in that article (Longair 1995). As I have already stated, my own involvement began in 1963 when I joined the Mullard Radio Astronomy Observatory as a research student and was immediately told by Martin Ryle, 'If you are going to solve the theory of radio sources and cosmology, start by doing something useful.' That involved the optical identification of the radio sources in the Revised Third Cambridge

Catalogue (3CR), which lists the brightest 328 radio sources in the northern sky. The accuracy of the radio centroids was only about 1 - 2 arcmin and so statistically it was not feasible to attempt to identify radio sources with objects fainter than about 18th magnitude on the Palomar Sky Survey prints. The redshift of 3C273 had been measured in 1962 by Maarten Schmidt and the astronomical community was still coming to terms with the remarkable properties of the quasi-stellar radio sources, or quasars. Fortunately, as improved interferometric measurements of the 3CR radio source positions were made, it became remarkably easy to find the quasars in the 3CR sample on positional grounds alone. My particular favourite was 3C9, which was one of my first quasar identifications (Ryle and Sandage 1964, Longair 1965), and which is the largest redshift quasar in the 3CR complete sample at $z = 2.016$. For many years, it was the largest redshift quasar.

The construction of the Cambridge One-Mile Telescope and subsequently the 5-km array, now known as the Ryle Telescope, revolutionised the identification of the 3CR radio sources. The structures of the radio sources, as well as their positions, became known with an accuracy of the order of a few arcseconds rather than arcminutes. This resulted in the identification of essentially all the 3CR radio quasars rather quickly since they are all brighter than about 20th magnitude and quasar studies flourished. The radio galaxies proved very much harder. Only a few optical astronomers were interested in pursuing these difficult observations, but I was very lucky to collaborate with Jim Gunn who dedicated some of his precious Palomar 200-inch telescope time to obtaining deep optical images of the 3CR radio source fields. We battled on through the early 1970s with image tubes, but the real breakthrough came in 1978 when the first Texas Instruments CCD chips became available to Jim as part of the development programme for the Wide Field-Planetary Camera for the Hubble Space Telescope. Jim and his colleagues built a simple prototype CCD camera known as PHUEI (standing for Palomar-Hale Universal Extragalactic Instrument) and the results were remarkable. We observed all the unidentified fields in a complete 3CR sample and found very faint galaxies close to the centroids of most of the radio galaxies (Gunn *et al.* 1980). Incidentally, I was able to use these first CCD images as part of the case for the construction of the 4.2-metre William Herschel Telescope, a particularly fortunate piece of timing.

The next task was to obtain the redshifts and spectra for the radio galaxies and once again the radio astronomers had remarkable luck. Obtaining the redshifts of galaxies fainter than 20th magnitude was very challenging at that time, but fortunately the faint radio galaxies had strong narrow emission lines which enabled Hyron Spinrad and his colleagues to obtain spectra and redshifts for the majority of the 3CR radio galaxies over the period 1975 to 1980. The resulting redshift distributions and $V/V_{\rm max}$ statistics for the radio galaxies and radio quasars were very similar, confirming the inference from their source counts alone that they display similar evolutionary trends with cosmic epoch (see, for example, Longair 1997).

At this point, two separate but related developments took place. The first was that, in the light of the success of the 3CR programme, my colleagues and I at the MRAO decided to make a major onslaught on the problem of defining with much better statistics the evolutionary properties of the radio source population. Samples of sources were selected at low and high radio frequencies and at different limiting flux densities. The 3CR sample was re-examined for its completeness (Laing, Riley and Longair 1983) and a new deeper survey, essentially a 4C sample, of about 60 sources was analysed by Jeremy Allington-Smith (1982) (see also Allington-Smith *et al.* 1982). At 2.7 GHz, John Peacock and Jasper Wall defined the high frequency counterpart of the 3CR sample (Peacock and Wall 1985) and a further deeper sample at 100 mJy (Downes *et al.* 1986); in both cases, the samples consisted of about 200-300 sources. The key decision was to study samples for which there was a reasonable chance of winning sufficient telescope time to carry out aperture synthesis mapping of the radio sources, make the optical identifications of the radio sources and obtain optical spectra for the candidate identifications. At the same time, John Peacock began studying the optimum way of defining the evolutionary properties of the source populations in a free-form way (Peacock and Gull 1981, Peacock 1985). The culmination of all these endeavours was the paper by Dunlop and Peacock (1990) in which they summarised the results of the enormous effort needed to make substantial inroads into the definition of the evolutionary history of the complete radio source population with cosmic epoch. Dunlop (1997) has described the natural continuation of this programme to define as precisely as possible the decrease of the comoving space density of radio sources at large redshifts. Notice the enormous effort needed to make significant advances in this programme — the greatest problem was to obtain the observing time to study these relatively modest samples of radio sources.

The second realisation was the importance of making infrared observations of the radio galaxies. It only became feasible to make single element observations of faint galaxies in the 1 to 2.2 μm waveband in the late 1970s using InSb detectors. Simon Lilly and I began a systematic programme of infrared photometry of the 3CR narrow-lined radio galaxies in 1980 and by 1984 had derived the K$-z$ relation for a complete sample of the narrow-lined FR2 radio galaxies. What was very striking was the narrow dispersion about the mean K$--z$ relation which we interpreted in terms of the passive evolution of the old stellar populations of these massive galaxies with cosmic epoch (Lilly and Longair 1984). As a result, it became essential to obtain infrared photometry for all the radio galaxies in the complete samples as well. These have proved to be essential in interpreting the evolutionary behaviour of the overall radio source populations.

This state of innocence lasted until 1987, when Chambers, McCarthy and their colleagues discovered that the optical images of the faint radio galaxies were aligned with their radio structures. This really threw the cat among the pigeons since it indicated that the radio properties of the radio galaxies influenced their optical properties and so all bets about the usefulness of the radio galaxies as cosmological probes were off. Fortunately, however, we had an ideal way of tackling the problem — we were awaiting HST observations of these very sources.

3 Hubble Space Telescope Observations of 3CR Radio Galaxies

The proposal to study the 3CR radio galaxies with the Hubble Space Telescope was accepted by the project in 1977 when I joined the Space Telescope Science Working Group as an interdisciplinary scientist — the title of my proposal was 'Optical Identifications and Studies of the Physical Properties of Extragalactic Radio Sources to the Faintest Optical Magnitudes using the Space Telescope'. This was submitted *before* all the 3CR galaxy identifications were made and long before the alignment effect was discovered. By great good fortune, the planned observations were ideal for studying the alignment effect. Incidentally, I included in the proposal the remark, 'At present, I have not identified my team of co-investigators'. They turned out to be Philip Best, who was only four years old when the proposal was submitted, and Huub Röttgering, who came to Cambridge as a post-doctoral fellow. The first observations were received in the summer of 1994, 17 years after the proposal was submitted.

I invested all my guaranteed observing time in one programme — the study of an almost complete sample of 28 3CR radio galaxies in the redshift interval 0.6 to 1.8. To summarise briefly what we have found, strong alignments are observed in the HST images for most of the radio galaxies with redshifts $z > 1$. A sample of the data, consisting of all 8 radio galaxies in the redshift interval $1 < z < 1.3$, is shown in Figures 1 and 2 along with infrared images taken with the IRCAM3 infrared camera of UKIRT. It can be seen that the infrared images at 2.2 μm look exactly like the classic picture of a double radio source associated with a giant elliptical galaxy. In the HST images, however, a wide variety of high surface brightness structures is observed aligned with the radio structures and these bear little resemblance to the images of giant elliptical galaxies.

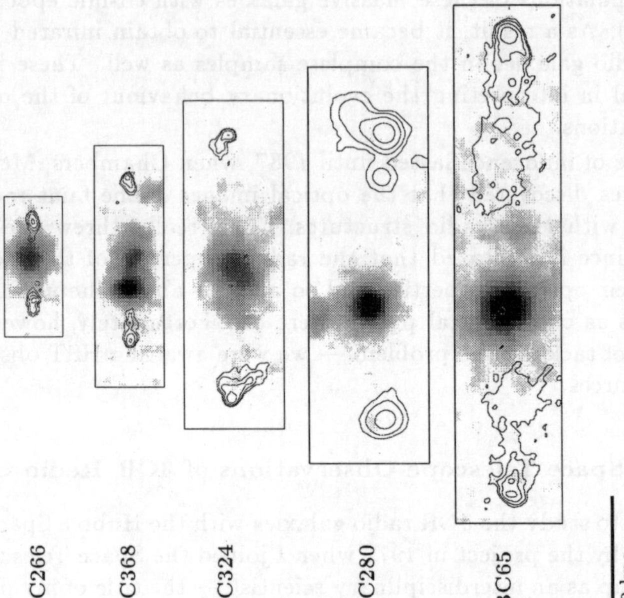

Figure 1: HST and UKIRT images of the radio galaxies 3C 266, 368, 324, 280 and 65 with the VLA radio contours superimposed. The images are drawn on the same physical scale (Best, Longair and Röttgering 1996).

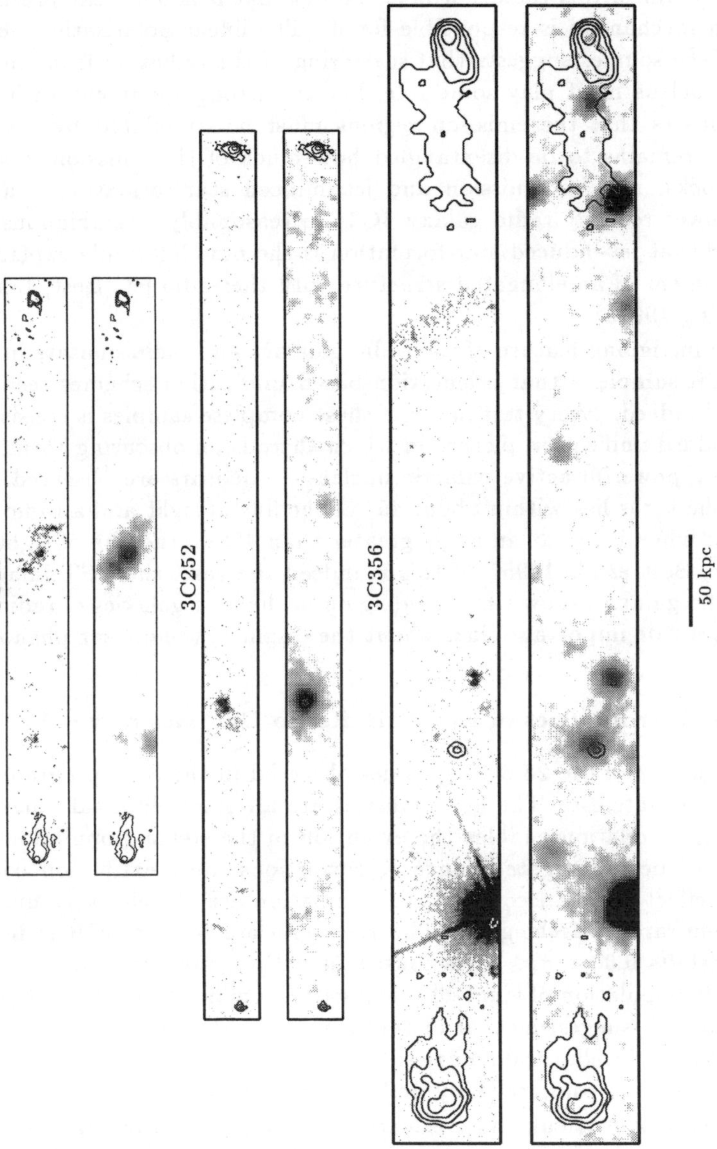

Figure 2: HST and UKIRT images of the radio galaxies 3C 267, 252 and 356 with the VLA radio contours superimposed. The images are drawn on the same physical scale (Best, Longair and Röttgering 1996).

The spectrum of this radiation must be flat, but it is not clear precisely what emission mechanism is responsible for it. The linear polarisation observed in some of the sources suggests that scattering of the radiation from an obscured quasar nucleus must play some role, but the strong alignment with the radio jets indicates that the emission regions must be stimulated by the radio jet itself. It remains to be disentangled how much of the emission is associated with shocks, nebular emission and jet-induced star formation. In the case of the lower redshift radio galaxy 3C34, a reasonably unambiguous case can be made that jet-induced star-formation is the most plausible explanation for the alignment of an elongated structure with the radio jet (Best, Longair and Röttgering 1997a).

One intriguing feature of the radio galaxies and radio quasars in the complete 3CR sample is that orientation-based unification schemes seem to work very well indeed. Every test made of these complete samples is consistent with the standard unification picture in which there is an obscuring torus about an extremely powerful active galactic nucleus — quasars are observed when the axis of the torus lies within about 45° of the line of sight and a radio galaxy is observed when it lies at an angle greater than 45° to the line of sight (see, for example, Best *et al.* 1995). If this is indeed the case, the HST images of the 3CR radio galaxies show what is going on in the host galaxies of radio quasars, and so provide important clues about the origin of the quasar phenomenon.

4 The Astrophysics of the 3CR Radio Galaxies at $z \sim 1$

The properties of the 28 3CR galaxies observed in the HST programme are of the greatest interest. The host galaxies are all giant elliptical galaxies; their Hubble $K-z$ relation enables the evolution of the stellar populations of these galaxies to be investigated, since K–corrections, dust extinction corrections, and the effect of any secondary star formation are all relatively unimportant at near–infrared wavelengths. The $K-z$ relation for the 3CR radio galaxies with redshifts $0.03 < z < 1.6$ is consistent with passive, 'closed–box' evolution of a single population of giant elliptical galaxies which formed at large redshifts (see Figure 3), suggesting that distant elliptical galaxies are precisely the same objects as nearby elliptical galaxies, but observed at an epoch when their stellar populations were younger (Lilly and Longair 1984).

What is of particular interest are the galactic environments in which the 3CR radio galaxies are found at large and small redshifts. The amplitudes of the spatial cross–correlation functions for galaxies in the vicinity of nearby 3CR FR II radio sources are similar to those for normal elliptical galaxies (Prestage and Peacock 1988), whilst the host radio galaxies themselves have

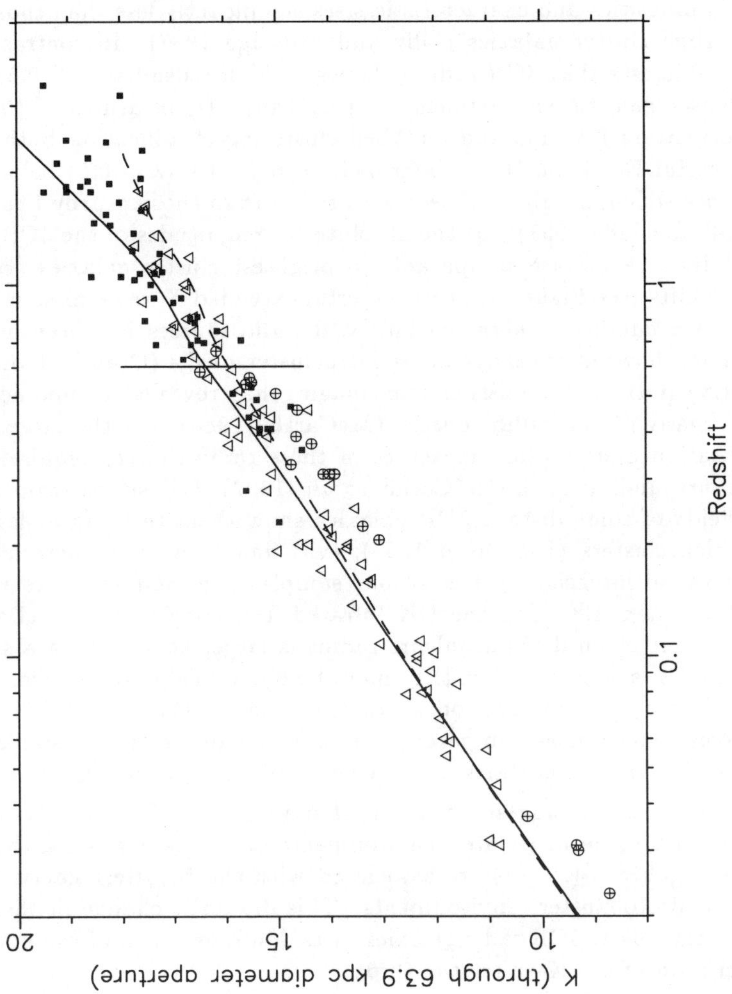

Figure 3: The K$-z$ relation for 3CR radio galaxies (open triangles), 6C radio galaxies (filled squares) and BCGs (crossed circles). The data for the 6C galaxies are from Eales *et al.* (1997), the 3CR galaxies from Lilly and Longair (1984) and Best *et al.* (1997b), and the BCGs from the sample of Aragòn-Salamanca (1993), converted to a 63.9 kpc metric aperture using a standard galaxy light curve. The solid line shows the relation expected for a non-evolving stellar population, whilst the dashed line shows that for a passively evolving population which formed at redshift $z = 5$. These relations have been normalised to match the magnitudes of the 3CR galaxies at low redshifts. The vertical line at redshift $z \sim 0.6$ provides a useful division between the high and low redshift populations.

optical luminosities and characteristic sizes significantly less than those of first ranked Abell cluster galaxies (Lilly and Prestage 1987). In contrast, recent evidence suggests that 3CR radio galaxies at high redshifts ($z \geq 0.5$) tend to lie in denser galactic environments. Specifically: (i) measures of the galaxy cross–correlation function and an Abell clustering classification both indicate that powerful FR II sources at intermediate redshifts ($z \sim 0.5$) belong to environments which are about three times richer than those nearby (Yates *et al.* 1989, Hill and Lilly 1991); (ii) the absolute K–magnitudes of the 3CR galaxies at redshifts $z \geq 0.7$ are comparable to brightest cluster galaxies (BCGs) at these redshifts (see Figure 3) ; (iii) powerful extended X–ray emission from the vicinity of a number of high redshift 3CR radio sources has been associated with cooling flows in relatively dense intracluster media (Crawford and Fabian 1996); (iv) [OII] 3727 emission line imaging has revealed companion galaxies near many of the radio sources (McCarthy 1988); (v) the large Faraday depolarisation and rotation measures of these radio sources require a dense, ionised surrounding medium (Carilli *et al.* 1997); (vi) spectroscopic studies of the fields of some distant 3CR galaxies show them to lie in at least moderately rich clusters (Dickenson 1997); (vii) finally, we have determined the infrared radial intensity profiles of our sample of 28 3CR galaxies with redshifts $0.6 < z < 1.8$ using the UK Infrared Telescope (UKIRT) (Best *et al.* 1997b). Their mean de Vaucouleurs radius is larger than that of a sample of nearby BCGs, suggesting that they must be dynamically highly evolved, even by a redshift of one. In addition, at radii $r \geq 35$ kpc the distant 3CR galaxies show excess emission as compared to a de Vaucouleurs profile, similar to the halos seen around cD galaxies at the centres of rich clusters (Oemler 1976).

All the evidence suggests that, whilst powerful FR II radio galaxies at low redshifts tend to lie in isolated environments or in small groups, the distant 3CR radio galaxies ($z \geq 1$) are associated with the brightest galaxies in rich cluster, or proto-cluster, environments. This dramatic change in the galactic environments of the 3CR radio galaxies casts doubt on the 'uniform population' interpretation of the K$-z$ relationship.

To produce a powerful radio source, three essential ingredients are required: (i) a supermassive central black hole; (ii) a plentiful supply of gas to fuel the black hole; (iii) a dense surrounding medium to minimise adiabatic expansion of the radio lobes. The 3CR radio galaxies at $z \sim 1$ release huge fluxes of kinetic energy in the form of relativistic jets, corresponding to the Eddington limiting luminosity of black holes with $M \sim 10^8 - 10^9\, M_\odot$ (Rawlings and Saunders 1991). Theoreticians have long argued that the masses of the central black holes should be roughly proportional to the masses of the host galaxies (Soltan 1982, Efstathiou and Rees 1988, Haehnelt and Rees 1993), and some evidence for such a correlation has recently been found for massive black holes in the nuclei of nearby galaxies (Kormendy and Richstone 1995). Massive central cluster galaxies would therefore be expected to host the most powerful central engines. In high redshift proto–cluster or young cluster environments, there will also be a plentiful supply of disturbed intracluster gas to fuel the central engine and confine the radio lobes. Furthermore, Ellingson *et al.* (1991) have found that the velocity dispersions of galaxies in the clusters around distant 3CR radio sources are significantly lower (400 to 500 $\mathrm{km\,s^{-1}}$) than those in comparably rich clusters at low redshifts (500 to 1000 $\mathrm{km\,s^{-1}}$). These smaller relative velocities increase the efficiency of galaxy mergers, which may trigger the onset of the radio source (Heckman *et al.* 1986). Thus, massive galaxies in young cluster environments at high redshifts possess all the necessary ingredients for producing the most powerful radio sources.

Why then are the most powerful FR II radio sources at the present epoch not also associated with central cluster galaxies? The radio sources hosted by nearby BCGs are almost invariably 'edge–darkened' FR I type radio sources; their radio jets are of much lower kinetic power than the FR IIs and, according to current ideas, are strongly decelerated in the inner kiloparsec by entrainment of surrounding material (Laing 1993 and references therein), producing lower luminosity radio sources. Since neither the black hole mass nor the density of the cluster environment can decrease with cosmic epoch, the availability of fuelling gas must be the factor which limits the jet–powers of these sources.

Rich regular clusters at low redshifts have had time to virialise the spatial distribution of the galaxies, and for the intracluster gas to take up an equilibrium configuration within the cluster gravitational potential. Studies of the Butcher–Oemler effect indicate that, compared to high redshift clusters, there is a dearth of gas–rich galaxies close to the centre of low redshift clusters which might merge with, and fuel, the central galaxy. Furthermore, the high velocity dispersion of the galaxies greatly reduces the merger efficiency. The evolution and virialisation of low redshift clusters is therefore likely to result in a more restricted fuelling supply for the central engines of BCGs. Instead, the most powerful FR II radio sources at small redshifts are generally found in small groups of galaxies: in these groups, velocity dispersions are small, mergers relatively frequent, and gas–rich galaxies plentiful. Occasionally, dramatic mergers will occur in low redshift BCGs, and we ascribe exceptional sources such as Cygnus A to this process.

These considerations suggest an explanation for the constancy of the stellar masses of the 3CR galaxies throughout the redshift range $0.03 \leq z \leq 1.8$. This mass may be interpreted as the typical mass which a central cluster galaxy attains before its galactic and gaseous environment is virialised. Galaxies of mass greater than a few times $10^{11}M_\odot$ will have undergone greater dynamical evolution, and generally live in virialised environments where the reduced supply of fuelling gas to the central regions results in FR I sources being formed. This leads naturally to a correlation between the redshift of a 3CR FR II galaxy and the richness of its environment: galaxies in highly overdense environments accumulate matter the fastest, and therefore reach this 'FR II upper mass limit' at early cosmic epochs, whilst those which lie in less dense environments evolve more slowly and can form powerful FR II radio galaxies at the current epoch.

In the light of these ideas, the K$-z$ relation of the 3CR galaxies can be compared with those recently derived for other samples of galaxies, such as the 6C radio galaxies, which are about a factor of five less powerful radio sources than the 3CR galaxies at a given redshift and a sample of BCGs. In these cases, the mean K$-z$ relations are broadly consistent with non–evolving stellar populations (see Figure 3). This is remarkable, since undoubtedly their stellar populations must have evolved significantly between redshift $z \sim 1$ and the present epoch. Indeed, the V$-$K colours of these galaxies are bluer at $z \geq 0.5$ than they are nearby, consistent with passively evolving stellar populations (Aragón-Salamanca *et al.* 1993).

At high redshifts ($z \sim 1$), the 6C radio galaxies have absolute K–magnitudes which are about 0.6 magnitudes fainter than those of the 3CR sample, whilst at low redshifts the two samples are equally luminous in the K–waveband. Eales *et al.* (1997) have suggested that this effect might be associated with emission of the active galactic nucleus contributing directly or indirectly to the K–band flux density of the distant 3CR sources. Analyses of our HST and UKIRT observations of a sample of distant 3CR galaxies have been used to exclude this possibility (Best *et al.* 1997b). Specifically, (i) the radial intensity profiles of all except two of the galaxies (3C22 and 3C41), show no evidence for a nuclear point source at $\geq 10\%$ of the total K–band flux; (ii) the contribution to the K–band light of the aligned blue component seen in these galaxies is generally small ($\leq 10\%$); (iii) on average, line emission contributes less than 5–10% of the total K–band flux density of these galaxies.

The greater K–band luminosities of the 3CR galaxies as compared with the 6C galaxies at high redshifts must indicate that they contain a greater mass of stars. It follows that, if there is a correlation between stellar mass and central black hole mass, the 6C galaxies possess less powerful central engines than the 3CR galaxies at high redshifts, thus accounting for their lower radio luminosities. At redshifts $z \leq 0.6$, the beam powers of all radio galaxies are well below the Eddington limit indicating that their radio luminosities are determined predominantly by the availability of fuelling gas in the host galaxy and its environment, rather than by the black hole mass. A weaker correlation would therefore be expected between galaxy mass and radio luminosity at these redshifts, accounting for the similarity of the K–magnitudes for the low redshift 3CR and 6C galaxies.

At redshifts $z \sim 1$, the 3CR galaxies and BCGs are equally bright, consistent with the proposition that the distant 3CR galaxies are amongst the most massive galaxies in clusters. At low redshifts, however, BCGs are up to a magnitude brighter in absolute K–magnitude than the 3CR galaxies. Since the stellar populations of the BCGs must have evolved over this redshift interval in a similar way to those of the 3CR galaxies, the shape of their K$-z$ relation must reflect the fact that BCGs continue to accumulate matter through mergers with massive cluster galaxies, and gas infall. Hierarchical clustering models for structure formation have suggested that the mass of BCGs increases by as much as a factor of 5 between a redshift of one and the present epoch (Kauffmann 1995).

For the BCGs, and also for the 6C galaxies, this growth of the mass of the galaxies between $z \sim 1$ and $z = 0$ increases their absolute magnitudes, thereby compensating for the dimming of their stellar populations — the two effects conspire to give rise to apparently simple 'no evolution' tracks. The 3CR galaxies at high redshifts also lie at the centre of galaxy clusters and so their masses would be expected to increase with cosmic epoch just like the BCGs; the 'passively evolving' K$-z$ relation of these galaxies suggests, however, that this does not occur. This, too, is a conspiracy. The galaxies sampled at high and low redshifts do not form a uniform population, as is indicated by the dramatic change in their galactic environments with redshift. The apparent passive evolution disguises an important result: the most powerful FR II radio sources at all redshifts $z \leq 1.5$ contain approximately the same mass of stars, a few times $10^{11} M_{\odot}$.

There are two important consequences of this discussion. First, we have presented evidence that powerful distant radio sources lie in (proto-)cluster environments, and so provide a unique tracer of these systems in the difficult redshift interval $1 \leq z \leq 3$. Studies of the evolutionary state of such massive collapsed structures at early epochs will provide important constraints upon theories of structure formation and galaxy evolution. Second, we have shown that the commonly–used 'closed–box' galaxy evolution models are not appropriate for interpretations of the K$-z$ relations. Instead, the shapes of these relations can be used to provide information about the merger histories of massive galaxies in clusters.

5 Concluding Remarks

I have told the story of the 3CR radio galaxies in some detail. For me, one of the important results of our analyses of the HST and infrared data is that that the key astrophysical issues are now much more closely linked to major problems in main-stream extragalactic astrophysics and cosmology. Whatever the ultimate explanation for the alignment effect, it seems essential that there be large quantities of cool gas distributed in clouds throughout the body of the galaxies to account for the aligned emission. This is an important result since it is consistent with recent analyses of the variation of the star-formation rates and abundances of neutral hydrogen clouds with cosmic epoch (Madau *et al.* 1996, Storrie-Lombardi *et al.* 1995). These provide a strong hint that the powerful evolutionary effects are associated with the abundance of cold gas in galaxies as they coalesce to become the galaxies we observe today. We have argued that, in the radio galaxies at redshifts $z > 1$, we are observing this process occurring for the most massive galaxies present at these epochs.

6 Acknowledgements

Parts of this text are based upon papers already submitted for publication. Specifically, Sections 2 and 3 are an abridged version of my introductory survey to the Tenerife meeting on *Observational Cosmology with the New Radio Surveys* held in January 1997. Section 4 is largely taken from a paper written in collaboration with Philip Best and Huub Röttgering which has been submitted for publication in *Nature*; I am most grateful to Philip and Huub for their splendid contributions to this work and for permission to use this material in this article.

1. Allington-Smith, J.R. (1982) *MNRAS*, **199**, 611
2. Allington-Smith, J.R., Perryman, M.A.C., Longair, M.S., Gunn, J.E. and Westphal, J.A. (1982) *MNRAS*, **201**, 331
3. Aragón-Salamanca, A., Ellis, R.S., Couch, W.J. and Carter, D. (1993). *MNRAS*, **262**, 764.
4. Best, P.N., Bailer, D., Longair, M.S. and Riley, J.M. (1995). *MNRAS*, **275**, 1171
5. Best, P.N., Longair, M.S. and Röttgering, H.J.A. (1996) *MNRAS*, **280**, L9
6. Best, P.N., Longair, M.S. and Röttgering, H.J.A. (1997a) *MNRAS*, **286**, 785
7. Best, P.N., Longair, M.S. and Röttgering, H.J.A. (1997b) *MNRAS*, (in press)

8. Carilli, C.L., Röttgering, H.J.A., Van Ojik, R., Miley, G.K. and van Breugel, W.J.M. (1997) *ApJSuppl.*, **109**, 1.
9. Chambers, K.C., Miley, G.K. and van Breugel, W.J.M. (1987) *Nat*, **329**, 604
10. Crawford, C.S. and Fabian, A.C. (1996) *MNRAS*, **282**, 1483
11. Dickenson, M. (1997) In *HST and the Large Redshift Universe*, (ed. N. Tanvir). Cambridge: Cambridge University Press (in press).
12. Downes, A.J.B., Peacock, J.A., Savage, A. and Carrie, D.R. (1986). *MNRAS*, **218**, 31
13. Dunlop, J.S. (1997). In *Observational Cosmology with the New Radio Surveys*, (ed. M. Bremmer). Dordrecht: Kluwer Academic Publishers (in press).
14. Dunlop, J.S. and Peacock, J.A. (1990) *MNRAS*, **247**, 19
15. Eales, S.A., Rawlings, S., Law-Green, D., Cotter, G. and Lacy, M. (1997). *MNRAS*, (in press).
16. Efstathiou, G. and Rees, M.J. (1988) *MNRAS*, **230**, P5.
17. Ellingson, E., Green, R.F. and Yee, H.K.C. (1991) *ApJ*, **378**, 476
18. Fabbiano, G., Miller, L., Trinchieri, G., Longair, M.S. and Elvis, M.1984. *ApJ*, **277**, 115.
19. Gunn, J.E., Hoessel, J.G., Westphal, J.A., Perryman, M.A.C. and Longair, M.S. (1980) *MNRAS*, **194**, 111
20. Heckman, T.M., Smith, E.P., Baum, S.A., van Breugel, W.J.M., Miley, G.K., Illingworth, G.D., Bothun, G.D. and Balick, B. (1986). *ApJ*, **311**, 526
21. Hill, G.J. and Lilly, S.J. (1991) *ApJ*, **367**, 1
22. Kaufmann, G. (1995) *MNRAS*, **274**, 153
23. Kormendy, J. and Richstone, D.O. (1995) *ARAA*, **33**, 581
24. aing, R.A. (1993) In *Astrophysical Jets* (eds. D. Burgarella, M. Livio and C.P. O'Dea), 95. Cambridge: Cambridge University Press
25. Laing, R.A., Riley, J.M. and Longair, M.S. (1983) *MNRAS*, **204**, 151
26. Lilly, S.J. and Longair, M.S. (1984) *MNRAS*, **211**, 833
27. Lilly, S.J. and Prestage, R.M. (1987) MNRAS, **225**, 531
28. Longair, M.S. (1965) *MNRAS*, **129**, 419
29. Longair, M.S. (1981). In *X-ray Astronomy with the Einstein Satellite*, (ed. R. Giacconi), 241. Dordrecht: D. Redel and Co.
30. Longair, M.S. (1989). *Alice and the Space Telescope.* Baltimore: Johns Hopkins University Press.
31. Longair, M.S. (1995) In *Twentieth Century Physics* (eds. L.M. Brown, A. Pais and A.B. Pippard), **3**, 1171
32. Longair, M.S. (1997) *Astronomy and Geophysics*, **38**, Issue 1, 10

33. Longair, M.S. and Willmore, A.P. (1974). *MNRAS*, **168**, 479
34. Madau, P., Ferguson, H.C., Dickenson, M.E., Giavalisco, M. Steidel, C.C and Fruchter, A. (1996) *MNRAS*, **283**, 1388
35. McCarthy, P. (1988) PhD Dissertation, University of California.
36. McCarthy, P.J., van Breugel, W.J.M., Spinrad, H. and Djorgovski, S. (1987). *ApJ*, **321**, L29
37. Miller, L., Longair, M.S., Fabbiano, G., Trinchieri, G. and Elvis, M. (1995). *MNRAS*, **214**, 109
38. Oemler, A. (1974) *ApJ*, *209*, 693
39. Peacock, J.A. (1985) *MNRAS*, **217**, 601
40. Peacock, J.A. and Gull, S.F. (1981) *MNRAS*, **196**, 611
41. Peacock, J.A. and Wall, J.V. (1985) *MNRAS*, **216**, 173
42. Prestage, R.M. and Peacock, J.A. (1988) *MNRAS*, **230**, 131
43. Ryle, M. and Sandage, A.R. (1964) *ApJ*, **139**, 419
44. Soltan, A. (1982) *MNRAS*, **200**, 115
45. Storrie-Lombardi, L.J., McMahon, R.G., Irwin, M.J. and Hazard, C. (1994) *ApJ*, **427**, L13
46. Yates, M.G., Miller, L. and Peacock, J.A. (1989) *MNRAS*, **240**, 129

CLUSTERS OF GALAXIES IN X-RAYS

A. CAVALIERE
Astrophysics, Dept. of Physics, 1 via Ricerca Scientifica,
Roma 00133, Italy
E-mail: cavaliere@roma2.infn.it

A personal cross section through 25 yrs of progress in the field.

1 Discovery

By the spring of 1971 *Uhuru* – the first step in the ladder of X-ray observatories planned and designed by Riccardo – had been in the sky for a few months, and the data from its survey were beginning to flow into the analysis rooms at American Science & Engineering.

I was lucky to have fled the "leaden years" then crushing Italy – egotistic riot leaders and fanatic mobs in a mix murderous at times even within the universities – and to have landed at MIT and then at nearby AS&E. There the center of the action was located, and one could truly feel the aura of intellectual drive and hectic exploration energized by Phil Morrison and by Riccardo.

At AS&E with Herb Gursky and Wally Tucker we noticed an unprecedented pattern in the quick-look data; that was a number of bright, high Galactic latitude sources correlated with clusters of galaxies. At the time, they came to 3 bona fide sources at nominal photon energies $2 \div 8$ keV with fluxes of some 10^{-10} erg/cm^{2}s, safely superposed to the clusters in Virgo, Perseus and Coma, and showing marginal evidence of an excess width relative to the point-source response of the 0.5^o passive collimators. So we started to ponder and argue over any meaning of all that.

Finally, we resolved to send in a Letter to Nature: Cavaliere, Gursky & Tucker (1971) *X-ray Sources and Galaxy Associations*, which made the cover of the June 18 issue, and also won a couple of sour columns in the News and Views section. In fact, in that Letter we were submitting two heretical propositions: i) a new class existed of extragalactic X-ray sources with powers $10^{42} \div 10^{44}$ erg/s, different from the active galaxies, and correlated instead with galaxy *associations*; ii) their X-ray emission was likely to be dominated by *hot intracluster gas*, and so truly extended and with thermal (though not necessarily isothermal) spectra.

All that proved true in due course (systematic detections of groups emitting 10^{42} erg/s came with the instrumentation of the '90, see Ponman et al. 1997;

complex thermal structure is just coming out), but at the time we were brushing quite against the grain of the prevailing wisdom. With the quasars known for some some 8 years, and the pulsars not yet 4 years old and still trendy, the astrophysical community was fascinated with the non-thermal compact emissions. If anything came next, it was another version of the inverse Compton emission, namely, photons of the cosmic microwave background scattered off an extended population of relativistic electrons.

In fact, what few extragalactic X-ray sources were then known, were apparently related with active galactic nuclei: M87, 3C273, and NGC4151. Still during the same 1971, papers were published by the NRL and by the *Uhuru* teams discussing the source in Perseus only in terms of NGC1275, and that in Virgo in terms of M87 plus M84. Moreover, the radioastronomers had just detected in Perseus and Coma head-tail radio sources and diffuse radio halos, the latter witnessing to a diffuse population of relativistic electrons possibly replenished by the former. To most, clusters – not to mention groups – were still nebulous entities if not suspicious constellations. They had been or were familiar only to a selected few: Zwicky, of course (I had just bought the last volumes of his *Catalogues of Galaxies and Clusters of Galaxies* with clusters mapped, to outline on the appropriate pages the *Uhuru* sources as they turned out), and G. Abell and a handful of believers, including G. Chincarini and H. Rood. Even a good introductory book like Motz & Duveen's *Essentials of Astronomy* had a section discussing the large velocity dispersions observed and the virial equilibrium in terms of "the problem of stability of clusters of galaxies"; this concluded with a punning quote from Ambartsumian: "many large clusters are disintegrating because they have large velocities of dispersion".

Actually, I was doubly lucky, since in Italy I had had a considerable research experience in plasma physics – which I had undertaken as the closest Ersatz for astrophysics then hardly existent in my country. So I happened to have some feeling of the macroscopic outputs produced at frequencies $h\nu \sim kT$ by the plasma bremsstrahlung emissivity $\ell \sim 10^{-27}\, n^2\, T^{1/2}$ erg/s cm^3. Now, an application to the hot intracluster gas of the controversial virial theorem was even then feasible on the back of a stamp, to find $kT \sim G\,M\,m_H/10R \sim 5$ keV for a total mass $M \sim 10^{15}\, M_\odot$. This is a particle kinetic energy truly huge compared with the electrostatic energy $e^2 n^{1/3} \sim 10^{-8}$ eV, for any reasonable density around $n \sim 10^{-3}$ particles/cm^3. So by the same token the plasma condition was satisfied exceedingly well (by far the best plasma in the Universe!), and the brems photons would fall smack in the *Uhuru* energy band. Ever since it looked appropriate to me naming this the *intracluster plasma* (ICP).

A few discussions had appeared before, concerning observations interpreted in terms of optically thin bremsstrahlung emission from intergalactic medium in thermal equilibrium. These were focused on the worrisome missing mass, and had been prompted by a claimed detection of the Coma cluster very strong at 25 keV, with an angular diameter of 4°. Felten et al. (1966) had noted the luminosity claimed would require a virial mass of gas at $T \simeq 2\,10^8$ K; but with a concentration similar to the galaxies, this should have been only $\simeq 2^o$ in size. On the other hand, Meekins et al. (1971) from new, if limited, rocket data had set a much lower upper bound to the emission from Coma and to any gas mass, obtaining for the latter $< 2\%$ of the virial value. So these early discussions overshot or missed the truth by a factor of 10 up or down.

A larger, uniform data base from *Uhuru* was eventually produced by Gursky et al. (1972) to comprise 5 clusters in the Abell (1958) catalog, plus Virgo, plus 3 more, all on power levels $L \sim 10^{43} \div 10^{44}$ erg/s. This paper confirmed that powerful X-rays were a systematic feature of clusters, while it still entertained the option of non-thermal emissions.

2 Corroborations

The debate over thermal bremsstrahlung vs. non-thermal emissions (either concentrated in active galaxies, or due to extended inverse Compton scattering) raged for several years on end. But in fact a new chapter in astronomy and in astrophysics had been opened up, with clusters first, and groups after, winning increasingly keen interest from all quarters.

It was soon clear that in the thermal option the mere detection of powerful X-rays implied that huge, hot ICP clouds had to be gathered and bound within real, long-lasting gravitational potential wells. But then the scaring virial velocities like 1000 km/s in the Coma cluster were vindicated by the very X-ray detection; they had to be straightforwardly related to the photon energies and to the temperature by $h\nu \sim kT \simeq m_H v^2/2$. In 1972 Gunn & Gott discussed how such large thermal energies could be obtained by spherical infall and compression of the gas into the potential wells; by the way, this paper treated in full the collapse of an isolated homogeneous sphere, an influential model that still shapes much of the current wisdom concerning the formation of clusters. All that contributed to steer the understanding of the virial velocities toward excess mass, unseen but real, then to become the "dark matter" (DM).

On the other hand, for bremsstrahlung luminosities Solinger & Tucker (1972) derived the X-ray-optical correlation $L \propto v_r^4$, and this was finding some support from the data, albeit with a wide scatter.

More direct ways to detect the ICP in other bands were pointed out. Sunyaev & Zel'dovich (1972) indicated that the hot ICP should cast a cold shadow (colder by about $-1/2$ mK) on the CMB in μw, but a warm glow longward of the CMB peak; this was the longed for inverse Compton scattering, only off the thermal ICP electrons. Miley (1973) proposed that the head-tail radio sources were swept back by ram pressure while ploughing through the ICP, and Jaffe & Perola (1973) gave and discussed the first quantitative models.

It took the spectral resolution achieved by Ariel 5 and OSO 8 to nail down the case for the thermal ICP. In 1976 Mitchell et al. detected in the Perseus cluster the first emission lines of very high excitation energies, like those of nearly stripped Fe at nearly 7 keV. Shortly after, Serlemitsos et al. (1977) detected the same lines also in the Virgo and Coma clusters, and Mushotzky et al. (1978) showed that the continuum was indeed exponential in shape, at temperatures around 5 keV.

Meanwhile, the list of identifications had overtaken the dozen in the census by Bahcall & Bahcall (1975). Time was ripe, then, to go *quantitative* concerning structure and profiles of the ICP clouds.

3 The ICP in the potential wells

The hot ICP within a gravitational potential well is expected to adjust to hydrostatic equilibrium, in fact over time scales somewhat shorter than the dynamical times; then its pressure gradient is to balance gravitation. The galaxies, and the DM, feel the same gravity and balance it by the gradient of the stress associated with their velocity dispersion v.

On this straightforward ground, Cavaliere & Fusco Femiano (1976) built up the handy description of the ICP disposition that became known as the standard β-model (we had used τ instead of β to denote the ratio of the scale heights $m_H v^2/2kT \lesssim 1$). In a nearly isotropic situation both the balances $d(2nkT)/nm_H\, dr = -d\phi/dr = d(\rho v^2)/\rho\, dr$ are to hold. To the extent that the state of the gas is approximately isothermal, and the galaxian v stays nearly constant along r, the link between their dispositions is simply given by

$$n(r) \propto \rho^\beta(r)\,. \tag{1}$$

The exponent has the value $\beta = 0.7\, v_8^2/T_8$ (in units of 10^8 cm/s and of 10^8 K, with the accurate mean molecular weight 0.6); this is expected to range between 0.5 and 1 on simple grounds of energy equipartition, with electrons partaking (as is permitted by the time scales) or not. The upshot is that isothermal ICP clouds tend to be more extended than the galaxy distributions.

The X-ray surface brightness is derived by column projection of the emissivity. For example, if the galaxies follow the (approximate) King distribution with core radius $r_c \simeq 0.13\, h^{-1}$ Mpc, then the surface brightness profile should be $S \propto [1+(r/r_c)^2]^{1/2-3\beta}$. This turned out, as found by Jones & Forman 1984, to be an efficient representation of the X-ray profiles even when the shape is not truly spherical but averaging over annuli is adopted – except for the central peak with minor output often caused by cooling flows, see Sarazin 1988.

Many of us (see Cavaliere & Fusco Femiano 1978, and Sarazin 1988 with its bibliography) scouted the area around the β-model, motivated by a couple of worries. What about other galaxy and DM distributions like the original King model with its $v(r)$ decreasing slowly from the center, or else DM $\rho(r)$ with a moderate central cusp? But these did not change materially the core-like structure of $n(r)$. What about the isothermal assumption (with the corresponding deep central minimum in the entropy profile), given that thermal diffusion is highly suppressed by even a minute magnetic field, conceivably tangled on galactic scales? So polytropic ICP equilibria were explored:

$$[n(r)/n_c]^{\gamma-1} = T(r)/T_c = 1 + \beta\, \frac{\gamma - 1}{\gamma}\, ln\, \rho(r)/\rho_c \tag{2}$$

with $1 < \gamma \leq 5/3$, corresponding to temperatures declining from the central T_c, and to flatter entropy profiles $s \propto ln\,(T/n^{2/3})$ up to the truly level one. But as moderatly resolved spectroscopy became feasible, the data retained consistency with rather flat temperature profiles, with γ in the range $1 \div 1.3$. The associated, rising entropy profiles should result from macroscopic events in the history of the clusters, as I will discuss later on.

So the β-model opened the way to all kinds of *quantitative* evaluations, and its extensions permitted to set bounds to them. It also allowed quantifying the low-brightness wings of the ICP clouds, which are important in several contexts.

The first context concerns the reconstruction of the full emitted flux when in the data only the central part stands out of the noise. This issue is important for the construction of source counts and of luminosity functions, and is discussed in this volume by P. Rosati.

Another point involves the Sunyaev & Zel'dovich effect. This (tiny) distortion of the CMB spectrum is proportional to the electron energy density integrated along the line of sight throughout the cluster, namely, $(\Delta T/T)_C \propto \int dl\, n\, T$; the profile here is $[1 + (r/r_c)^2]^{1/2-3\beta/2}$ on assuming again an approximated King distribution for the gravitational mass. The attractive possibility was soon raised by Cavaliere, Danese & De Zotti (1977), (1979) of using clusters as self-calibrating sources to derive the absolute distance scale and hence the Hubble constant; Silk & White (1978) extended this to deriving Ω_o. Schematically, from the expression for the flux $F \propto n^2 T^{1/2} R^3/D^2$ one may eliminate n using $(\Delta T/T)_C \propto n\, T\, R$ to obtain $D^2 \propto (\Delta T/T)^2_C\, R/F\, T^{3/2}$; since in turn the angular diameter reads $\theta \propto (1 + z)^2\, R/D$, the final expression contains only direct observables, and is linear in the luminosity distance D or in H_o^{-1}. The value for H_o is also proportional to the ratio of the shape factor for S over that for $(\Delta T/T)^2_C$. The SZ decrement in μw is delicate and painful to measure, as discussed by Birkinshaw (1990), but its integrated value has been reliably obtained for several clusters. Currently it is being resolved and mapped in a number of them, see Grainge et al. (1996) and the bibliography therein.

Finally, a sensitive issue concerns the total mass of the ICP, and the baryonic fraction in clusters. Cavaliere & Fusco Femiano (1978) had stressed that the cluster outskirts have low surface brightness (this declines with n^2), yet they contribute importantly to the plasma mass (which increases outwards with the volume integral of n). For a nearly isothermal ICP the mass sums up to some 30% of the virial value out to some 3 Mpc, and so they included it in their Poisson equation for a self-consistent potential. But the instrumentation of the time gave no chance to discern such wings unambiguously.

The data came in the early '90, and permitted White et al. (1993) to quantify the amount of baryons observed in the Coma cluster out to 1.5 h^{-1} Mpc, to the result that the baryonic fraction exceeds $M_B/M \simeq 0.15$, overwhelmingly more in ICP than in stars. White & Fabian (1995) probed some 20 clusters, deriving similar or larger baryonic fractions by comparison with the observed virial masses (see Biviano et al. 1993). The former authors stressed the lack of obvious ways to preferentially gather into clusters the baryons rather than the DM particles; if anything, the interacting ICP is expected to be inflated out relative to the collisionless DM by energy injections.

So these largest chunks of baryonic matter should sample fairly the universal ratio, to the effect that $M_B/M = \Omega_B/\Omega_o$ should hold. But then in high-density universes a baryonic fraction $\Omega_B \simeq 0.15$ is not easily made consistent with the abundances of light elements produced by the canonical cosmological nucleosynthesis; whence the slogan "baryonic crisis". The cheapest way out would be provided, of course, by a low-density cosmology with $\Omega_o \lesssim 0.3$, but a couple of stumbling blocks stand in this way, as discussed next.

4 X-raying cosmogony

Actually, data of the above quality and import had been made available thanks to the advent in 1978 with *Einstein* of the next-generation observatories, carrying out Riccardo's steady aim at implementing the grazing incidence optics. Such telescopes combine much higher signal/noise with resolution and imaging capabilities, so extending to the X-ray band the magic of *shapes*.

Shapes are particularly relevant for clusters that make extended, often highly complex sources, see Forman & Jones (1982) and fig. 1. Such shapes

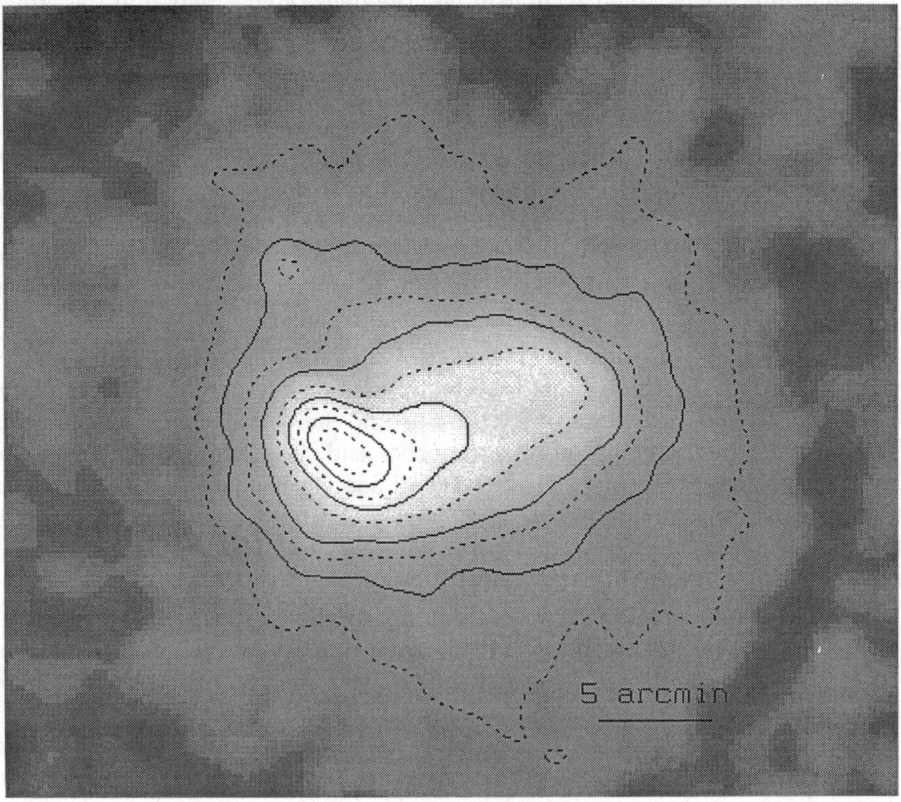

Figure 1: *ROSAT* map of A754, a cluster undergoing a major merging event with the subcluster on the left, see Henry & Briel (1995) and Zabludoff & Zaritsky (1995). Courtesy of S. De Grandi and S. Molendi.

are robust against superposition effects, once again because bremsstrahlung emits $L \propto n^2$. Most important, they offer telling signs of the formation process; actually, they fell into place, as it were, in the framework of hierarchical clustering scenarios.

These (see Peebles 1993) envisage structures forming from the bottom up, by repeated merging or aggregation of smaller into larger units. The canonical version is focused on direct collapse of density perturbations over the dynamical times $t_d \sim R/v$, due to gravitational instability. This sequentially involves perturbations with larger and larger scales but smaller and smaller amplitudes, which wrinkled the homogeneous background at the recombination.

Touching briefly upon technicalities, the density contrasts δ are assumed (consistent with the microphysics of the inflation) to have had initially a Gaussian distribution with variance $\sigma \propto M^{-a}$. The index a is related by $a = (n_e + 3)/6$ to the effective index $n_e(M) \lesssim -1.2$ (slowly decreasing with M) in the Fourier power spectrum of the perturbations $\langle|\delta_k|^2\rangle \propto k^{n_e}$. In the small-amplitude, linear stage the δ's increase slowly as self-gravity is opposed by expansion. In the critical universe, where it is fastest, the growth is given by $\delta(z) \propto (1+z)^{-1}$, and the transition across $\delta \simeq \sigma \simeq 1$ occurs at a redshift z for the mass $M_c(z) \propto (1+z)^{-1/a}$. Collapse and virialization of a condensed structure are assumed to follow. The rate of such a dynamical evolution is slowed down or even halted by a free Hubble expansion.

Expansion is restrained in the critical universe, and there rich clusters assemble today. So these constitute truly young structures in their making, which still feature remnants of the groups and subclusters they engulfed, or even merging events going on. At the other end, in very open universes the faster expansion freezes out for $z \lesssim 1/\Omega_o$ all nascent condensations. Whatever clusters can form would be old, with plenty of time for settling to relaxed, spherical shapes. The observed shapes (see, e.g., fig. 1 and Schindler 1996a) agree definitely better with the former picture.

Early N-body simulations sparsely sampling the dynamics of the gravitating mass as pioneered by White (1976) gave support to the general picture. An ensemble of simulated X-ray maps have been first produced by Cavaliere et al. (1986), on coupling to a still primitive N-body code a module computing density and emissivity of the ICP in hydrostatic equilibrium with the growing potential wells. In the critical cosmology these maps came out similar to the observed ones; at the present epoch, a sizeable fraction was caught in the stage of converging double clumps, or of a merging event in the act. Such simulations demonstrated the ability of the ICP to amplify and visualize the underlying dynamics of the DM.

Recent simulations including hydrodynamics (Mohr et al. 1995) took a step forward in quantifying how shapes depend on the cosmological framework. They indicated that morphologies as rich as observed require expansion either restrained by total density $\Omega_o > 0.3$, or tempered at the relevant $z \simeq 0.5$ by a cosmological constant $\Omega_\lambda = 1 - \Omega_o$. Incidentally, this means that the solution of the baryonic crisis is not cheaply found in an open universe. To proceed, two parallel lines are to be pursued: extensive (to now hardly adequate) simulations including from galaxies to ambient large scale structures as shown in fig. 2; a more objective assessment of type and amount of substructures in the observations. For this, a powerful tool is provided by the wavelets analysis as keenly developed by Lazzati (1996) under input of Riccardo.

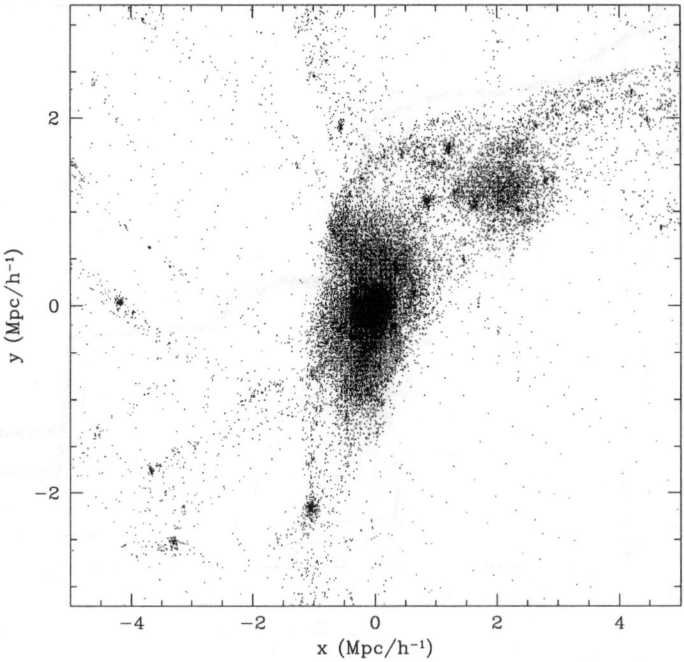

Figure 2: A complex merging event in the act at $z \simeq 0.5$ in the virtual universe of the N-body simulations. This involves condensations started long before as cold DM density perturbations in the critical cosmology. Note the prongs (just looming out in this image) of LSSs that guide such events. From N-body simulations by F. Governato at the HPCC, Seattle.

5 The punctuated equilibria of the ICP

Cluster *dynamics* and *baryon* content in the form of ICP are conceivably linked. The handy dynamical observable is the temperature $T \propto GM/R$, which directly probes the height of potential well dominated by the DM, with the baryons in the *passive* role of mere tracers. The luminosity, with its strong dependence on density, probes instead the baryonic content in its *active* role.

In fact, a definite L-T correlation is observed, but with considerable scatter. These features are understood on starting from the luminosiy factorized into dynamics $\times$ ICP in the form $L \propto T^2 \rho^{1/2} \int dV n^2(r)$, and tracing how $n(r)$ follows suit of the dynamically growing potential wells. We recently tackled a

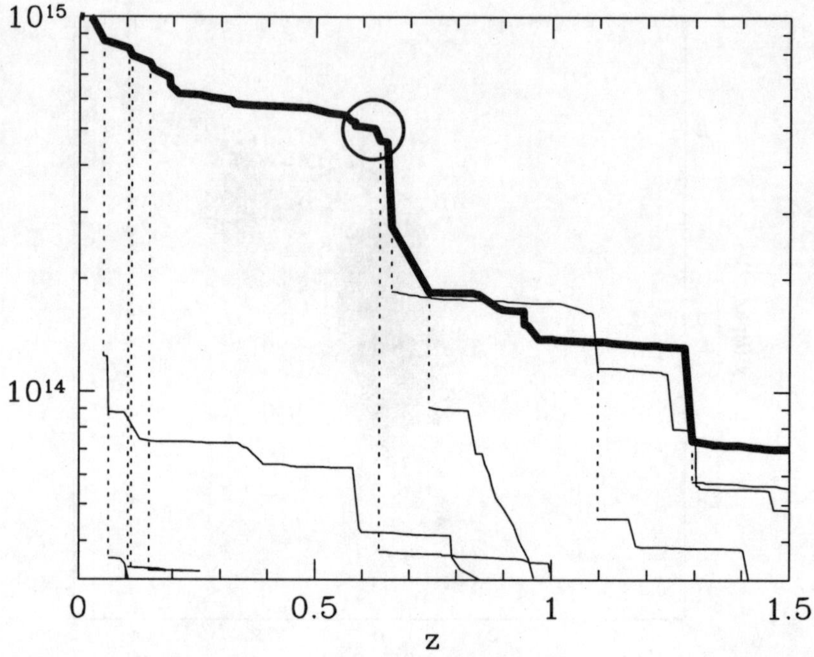

Figure 3: A realization of the merging history of the mass $M = 10^{15}\, M_{\odot}$ in the critical universe. The circle marks the accumulation of half the final mass. Monte Carlo simulation by P. Tozzi.

semi-analytical model of *punctuated equilibria* (Cavaliere, Menci & Tozzi 1997), which advances the β-model along two key directions: the boundary conditions to the ICP, and the hierarchical growth of groups or clusters.

The growth occurs by accretion of, or repeated merging with, further mass in various condensation stages, from comparable subclusters to nearly diffuse. The history of such episodes for a given final M is followed throughout the branches of the "merging tree" (see fig. 3) with Monte Carlo simulations, which are fast and easily produce a large statistics. The ICP is reset to a new equilibrium after each such episode that "punctuates" the cluster history.

The boundary conditions are set by the same accretion or merging process, which produces compressional shocks converting the inflow into thermal energy. The shock strength goes up with the ratio of the potential depth (wich sets the infall energy) to the temperature of the incoming gas. Strong shocks, producing density jumps up to 4 and temperatures up to the (conservative) value $T \simeq 3GMm_H/16R$, form close to the virial R. The strength is particularly large when rich clusters with deep potentials accrete small, cold groups or colder diffuse gas. Weaker shocks with smaller jumps or even closely adiabatic compressions always prevail in groups with virial $T \lesssim 1$ keV, since the gas to be accreted is already preheated to $T_1 \simeq 1/2$ keV by stellar feedbacks. These are due to energy inputs associated with star formation and evolution all the way to Supernovae. In groups such *nuclear* heating matches the *gravitational* virial heating, and tends to keep the baryons at bay from the shallower potential wells. The boundary conditions on the density at the radius R fix the constant in the proportionality of the β-models of eq. (1) or of eq. (2), given that the universal baryonic fraction is recovered outwards.

Using the set of merging histories we compute both the average values $\langle L\rangle \propto \langle n^2\rangle$ and the *intrinsic* dispersion, which is due to the different dynamical histories of the DM component but is *amplified* by the n^2 dependence of the emission. The result (fig. 4) is a behavior $L \propto T^3$ for clusters of richness class 1. But this steepens to $L \propto T^5$ for small groups with virial $T \simeq T_1$ (where the nuclear competes with the gravitational energy), and flattens to T^2 (the self-similar scaling of pure gravity) for very rich clusters with potentials so deep as to hold in the full universal baryonic fraction $\Omega_B/\Omega_o \simeq 0.15$. The dispersion is lower than the scatter observed when also the clusters with cooling flows are included, a point I shall comment upon later on.

Two points to note straightaway. Spherical shapes are not crucial in this context; what is relevant is the condition of small residual velocities inside the shocked region, as produced by aimed numerical simulations in all but the rare collisions of equals (see Schindler 1996b). At higher z the normalization of our $L-T$ relation increases but slowly, proportionally to $\rho_{eff}^{1/2}(z) = 1+z$; such z-dependence arises because clusters or groups grow within large scale structures (see fig. 2), where the effective density ρ_{eff} is currently higher than in the background but increases less with z.

We further argue that the modes of *entropy* production and distribution correlate with the global expansion. In the critical universe collapses and merging, with the shocks they induce, are currently ongoing so that strong shocks form close to the virial radius. Entropy is continuously generated in these outer regions, so that its inner distribution is forced to have a central minimum. Then the effective γ will be closer to 1 than to 5/3, leading to temperature profiles moderately decreasing out to the shock.

Conversely, in a very *open* universe all action is confined to early times: merging and mixing occurred early on, and then subsided; shocks had time

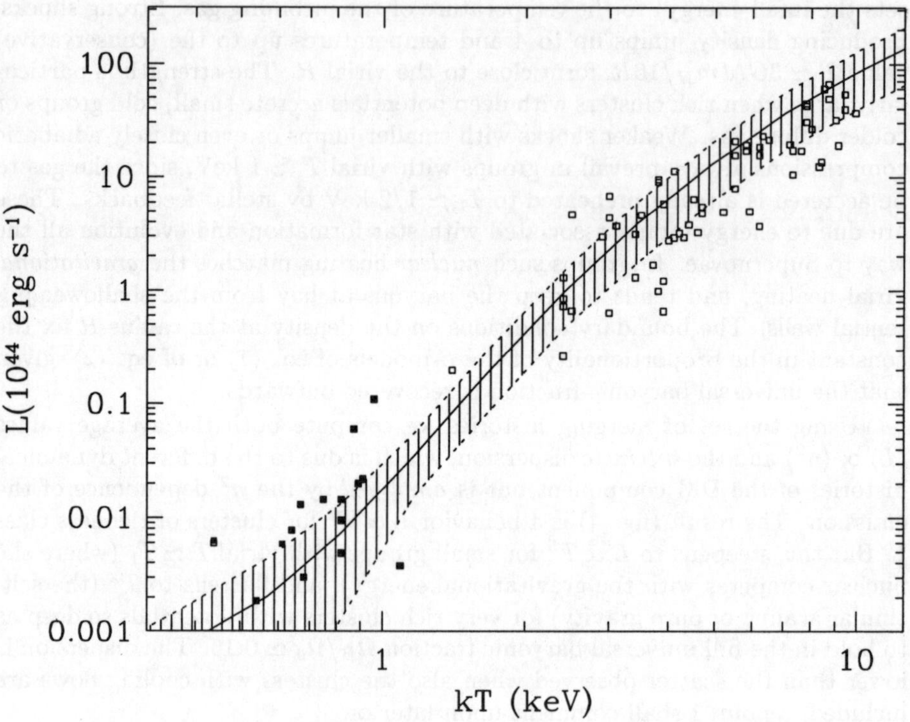

Figure 4: The $L-T$ relation derived by Cavaliere, Menci & Tozzi (1997) from the merging histories of the DM component of groups and clusters, with the ICP in punctuated equilibria. Bolometric luminosities. The strip outlines the 2-σ dispersion. The range $L_o \simeq 1.6\ h^{-2}\ 10^{44}$ erg/s $\pm 50\%$ corresponds to $T_o = 4.5$ keV. Data from Ponman et al.1997, David et al. 1993.

to expand beyond R and weaken. Correspondingly, the accretion petered off under nearly adiabatic conditions even for rich clusters; then the effective γ would be closer to 5/3.

6 Population statistics

The $L-T$ relation links, as pressed particularly by Oukbir & Blanchard (1996), the luminosity functions with the statistics of the dynamical variable $T \propto M^{2/3}\,\rho^{1/3}$, via the one-to-one connection $N(L)\,dL = N(M)\,dM$. The founding block of all such statistics is currently sought in the theoretical mass distribution proposed by Press & Schechter 1974. Normalizing $m \equiv M/M_o$ to the mass $M_o = 0.6\ 10^{15}\ \Omega_o\ h^{-1}\ M_\odot$ virializing now, this reads

$$N(m,z) = \sqrt{\frac{2}{\pi}}\ \frac{\delta_c\, a\, \rho(z)}{m_c^2(z)}\ \frac{m^{-2}}{\sigma}\ e^{-\delta_c^2/2\sigma^2}\ . \tag{3}$$

I used the shorthand $\sigma = \sigma_8\, m^{-a}\, \delta(z)$ for the evolved r.m.s. amplitude of the perturbations, with σ_8 expressing the normalization of their spectrum at the convenient scale $8\, h^{-1}$ Mpc; the collapse threshold δ_c is commonly taken at 1.68 and is discussed below. I expand on the mistery of this "P&S formula", in view of its central role in most current cosmogonical studies.

On the one hand, the simple and deft derivation originally devised by P&S has been examined and eventually pronounced dead (Bond et al. 1991). Their result, instead, is found to require an elaborate set of unrealistic simplifications and of ad hoc, even inconsistent, assumptions (hence its demotion to the status of Press & Schechter *formula*). These include: all perturbations are treated as independent, sharply defined spheres over a homogeneous, isotropic background; non-linear condensation and virialization is assumed, following the spherical collapse model of Gunn & Gott 1972, to occur invariably at a time corresponding to the threshold $\delta_c = 1.68$ in in an extrapolated linear growth; such condensations are assigned the mass of the largest sphere where the threshold is crossed at one point, with all substructures brushed off by fiat. Correspondingly, the formula contains, in addition to δ_c, only the linear dynamics in the form of the evolved σ.

On the other hand, the formula agrees (but over limited ranges of M and z) with the results from many N-body simulations, to the amazement of everybody. One amazing feature concerns the slope of the power law at the small-M end, which depends on brushing away all the correlations implied by realistic shapes of the perturbations, such as shown in fig. 2.

The other amazing issue concerns the large-M end, where one expects "cross-talks" with scales larger yet. In fact, groups and clusters are observed to concentrate in the ribs of the large scale structure network (Ramella et al. 1992). This deeply indents the mythical homogeneity and the isotropy of the background with large voids of several tens of Mpc, or swells it into filaments and sheets with contrasts of several units. Modern N-body simulations (e.g., Brainerd & Villumsen 1992, Bertschinger & Jain 1994, see also fig. 2) fulfill the views of Zel'dovich (see Shandarin & Zel'dovich 1989) and produce the network of non-linear LSS under generic conditions. They also display how clusters and groups form in the ribs and especially at their nodes; the LSSs apparently play an active role by channelling the velocities, and promoting aggregation and merging of condensed clumps which would never meet in a homogeneous isotropic background. All that complexity is apparently ignored by the formula.

To be fair with it, one has to realize that a simple univariate statistics will anyway require quite some pruning of the morphological richness of the sources, and some schematic rendition of the collapse process. Moreover, the formula is built upon sensible cornerstones: the mass scale $M_c(z)$ rooted in dimensional grouds; the faint end flatter than M^{-2}, implying a finite mass in all condensations (conserved by the formula at the value corresponding to all the available matter); the upper cutoff reflecting the Gaussian initial conditions. Finally, to the joy of the phenomenologists the formula is rather elastic, its cutoff being sensitively tuned in place and shape by the threshold δ_c and by the perturbation amplitude σ_8.

However, there are bounds to such tuning. The threshold δ_c may be adjusted, but at the cost of introducing perturbation shapes different from sharp spheres and so correlations which alter the faint slope of $N(M)$. The amplitude σ_8 is bracketed on starting from the CMB anisotropies measured by *COBE/DMR* on very large scales, and extrapolating down to $8\,h^{-1}$ Mpc in terms of the spectra of DM perturbations. For the time being, the current assessments (see Bunn & White 1997) contemplate for the large-scale end of the spectrum either the invariant form $n_e \to 1$ consistent with $\Omega_o < 1$, to yield $\sigma_8 = 0.8 \pm 10\%$ and $\sigma_8 = 1.1 \pm 10\%$ with the cosmological term $\Omega_\lambda = 1 - \Omega_o$; or the tilted form $n_e \to 0.8$ as required for $\Omega = 1$ (with h constrained to 0.5) to yield $\sigma_8 = 0.7 \pm 10\%$.

On its own, the P&S formula yields $n_{cl} \simeq 10^{-5}\ h^3\ \mathrm{Mpc}^{-3}$, the observed values of the local density of rich clusters, only if the rule $\sigma_8\ \Omega^{0.5} \simeq 0.5$ is closely obeyed. Another constraint is imposed by the density of distant clusters (see Bahcall, Fan & Cen 1997). As discussed below, in a few years the cosmogical parameters and the spectral amplitude will be pinned down more closely from the CMB anisotropies. Then the above degrees of flexibility will be fixed, and it will be more fruitful to discuss the numerology in terms of consistency or failure of the P&S formula.

Meanwhile, questions arise. Is anything amiss in the canonical treatment of structure formation? The correlations are certainly wanting. Cavaliere, Menci & Tozzi (1996) argue a two-fold way to hierarchical growth of structures: not only direct condensations from initial perturbations, but also non-linear *aggregations* of formed galaxies correlated by larger structures. They show how effectively aggregations work within groups to gradually erase galaxies over the time scale $1/n\,v\,\Sigma_{gal} \gtrsim t_d$, and to form by such coalescence the large, cD-like galaxies known to stand out from the Gaussian statistics. They expect groups and clusters to behave similarly within LSSs. After all, not much was perceived of the latter in 1974 when P&S devised their formula!

How the formula holds up when confronted with data over extended dinamic ranges? In the *virtual* sky of the simulations, there are signs (detected by Bertschinger & Jain 1994, and confirmed by Governato & Tozzi, in preparation) that P&S – however flexible – over-represents the small masses, and either under-represents the large ones or overdoes the evolution.

In the *real* sky, clusters and groups are felt to constitute a privileged, simpler testing ground for cosmogonic studies, being less affected than galaxies by the ill-understood processes of star formation and gas recycling. Here the X-ray *surveys* enter in full, because they are fast, virtually free from contamination, and with completeness easier to assess. They provide a local luminosity function with a dynamic range approaching 10^3 to check the shape, and – combined with optical redshifts – a range out to $z \simeq 1$ to test the evolution.

7 Evolution

The P&S formula as it stands – along with the derived luminosity functions – comprises two kinds of population evolution toward higher z: the normalization increases, like (1+z) in the critical universe, which may be termed a "number evolution" on borrowing the quasar parlance; meanwhile, the upper cutoff shifts toward smaller values, like $\exp -\delta_c^2(1+z)^2/2\sigma_8^2$ in the critical case, leading to "luminosity evolution".

Now, the statistically strong data by Rosati et al. (1997) from *ROSAT* indicate no evolution of the luminosity functions out to $z \simeq 0.8$ and up to a few times 10^{44} erg/s. This is only marginally consistent with previous evidence of negative evolution, which however was statistically weak as often stressed by Riccardo. So the subtle point arises how the P&S dynamical evolutions are modulated by cosmology (which enters through σ_8 and n_e, and again in the distances and volumes), and by the ICP evolution (which enters via the $L-T$ correlation), to end up in closely *constant* luminosity functions.

Can the very study of such luminosity functions at various z yield at one stroke also the parameters h, Ω_o, Ω_λ..., as often hoped for? To see through this, our approach (Cavaliere, Menci & Tozzi, in preparation) is to concentrate on *physical* models for the DM and for the ICP, discarding unphysical combinations or parametric mixes. The ensuing predictions are compared with two recent surveys from *ROSAT*, the local BCS by Ebeling et al. (1997), and the DCS out to $z \simeq 0.8$ by Rosati et al. (1997). So we select cosmologies that comply with all solid constraints, adopt physical perturbation spectra consistent with the former (and with the *COBE* amplitudes), and look for an entropy distribution in the ICP consistent with both. For example, strong shocks are

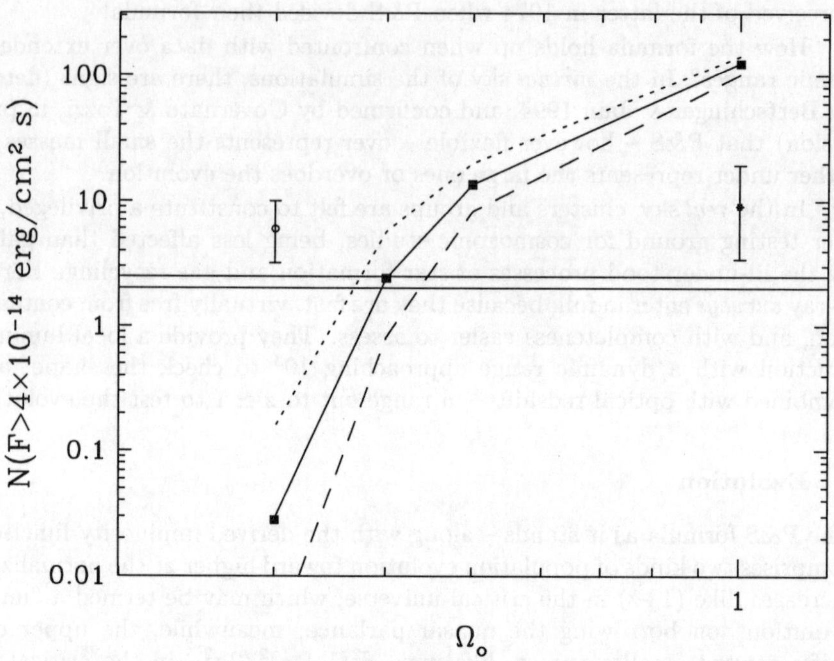

Figure 5: Grading cosmologies/cosmogonies in terms of faint counts: predicted by us, vs. observed by Rosati et al. 1997 (with uncertainty represented by the orizontal strip). The slanted line shows the results derived as reported in §7, using standard cold DM perturbations and shocked ICP in punctuated equilibria. The strip outlines the total variance comprised of the intrinsic dispersion in the $L-T$ relation, and of the uncertainty in the spectral amplitude σ_8. The isolated points represent: left, the case $\Omega_o = 0.3$, $\Omega_\lambda = 0.7$; right, tilted cold DM perturbations in the critical universe. Provided by N. Menci.

rare in very open cosmologies, and so are adiabatic compressions within rich clusters in the critical case. So we use the appropriate $L - T$ relation and, after checking the normalization on the local luminosity function, we predict the source counts at bright and faint fluxes and the full luminosity funtions.

Our results (an example is given in fig. 5) are as follows. Many scenarios are ruled out: so are the standard cold DM perturbations in the critical cosmology which overproduce the local clusters; but also the very open cosmologies with $\Omega_o < 0.3$, which underproduce them – by the way, this constitutes the other stumbling block for a cheap solution of the baryonic crisis. A small set of effectively disjoint cosmologies/cosmogonies survive, although none fares really well: the critical $\Omega = 1, h = 0.5$ universe, wrinkled by tilted cold DM perturbations and containing high baryonic abundance, combined with the shocked ICP in punctuated equilibria; the low density $\Omega_o = 0.3$ universe, but in flat geometry with $\Omega_\lambda = 1 - \Omega_o$, with weaker shocks; and (marginally) the open case $\Omega_o \simeq 0.4$. The upshot is that such set still spans quite some range, so the cosmological parameters can be constrained but only up *to a point*.

To what extent enlarging the data base on X-ray clusters will help in further constraining cosmology? Here I argue that the variance intrinsic to the hierarchical clustering, and amplified by the ICP emissivity, sets an effective *limitation* to clusters as cosmological signposts.

This should come to no surprise; in reducing the rich complexity of these sources to dry, univariate statistics of luminosities the "hidden parameters" are bound to re-emerge in the form of a wide *variance*. This is actually observed in terms of the scatter of the $L - T$ relation, see fig. 4; for example, the temperature $T_o = 4.5$ keV in fig. 4 is correlated with luminosities in the range $L_o \simeq 1.6\, h^{-2}\, 10^{44}\, (1 \pm 50\%)$ erg/s, where the dispersion still does not include the cooling flow clusters. This (minimum) intrinsic variance is used in fig. 5 to show that the present Poissonian error bars in the observed faint counts are already small. The current uncertainty in σ_8 contributes further variance; this is nominally smaller, but it is shown on the basis of eq. (3) to enter with comparable weight the overall uncertainty in the luminosity functions.

I conclude that large-area, bright surveys may be helpful in clarifying the issue of any negative evolution taking place at the bright end. But large-area, faint surveys will hardly provide a sharper insight into cosmology, unless one could previuosly *reduce* both the uncertainty in σ_8 and that in L_o.

8 What next

Down to what point are these reductions feasible? Theoretically, the variance $\Delta L_o/L_o$ of dynamical origin may be sharpened by Monte Carlo simulations so extensive as to provide the full scatter distribution. But then one has to include the extra emissivity associated with cooling flows: not only the direct central emission, but also the general density enhancement that triggers the short cooling times; related work is in progress with N. Menci and P. Tozzi.

Observationally, one needs a large statistics for the distributions of L and T, to build up narrower correlations for subsamples homogeneous in term of cooling flows strength, as started by Fabian et al. (1994) and by White, Jones & Forman (1997). A large effort is called for, aimed to truly spectroscopic measurements of T; these require *SAX* and *XMM*, or even *AXAF* to pin point the hot spots and structured thermal states of the ICP left behind by major merging events. However, I hasten to stress that such efforts will find soon a better aim than constraining Ω_o.

This is because of the other component of the uncertainty, namely, that related to the *linear* perturbation spectrum, and particularly to its amplitude. This is more directly fixed at scales quite larger than cluster's and still in their linear stage. This is the goal for the next missions to observe the CMB anisotropies: first *MAP*, and susequently *PLANCK*. But then these observations are also planned to pin down the global parameters h, Ω_o.. to better that 10%, thus superseding indirect constraints set at cluster scales gone *non-linear*.

With the cosmology so narrowed down, in a few years the study of groups and clusters in X-rays will resume its proper tack; this should be aimed, I maintain, at the physics of systems of intermediate *complexity* (now in its technical meaning), midway between elementary particles and biological systems. Again this should come to no surprise, reconsidering how many and diverse bricks contribute to build up the astrophysics of clusters. These include particle physics as for the nature of the DM, but comprise also cosmology, gravity, hydro- and thermodynamics, all in the highly structured combinations reviewed here. One manifestation of such complexity is constituted by the very morphological richness. To wit, I believe to be no coincidence that the parlance of the field candidly adopted from the beginning terms like "population" and "evolution".

I submit that quantifying such complexity has been the thread constantly underlying and linking all breakthroughs in the field. I maintain that the history of the cluster astronomy-astrophysics constitutes a major, quantitative achievement by the physics of complex systems over the last 25 yrs, made possible – once Bruno triggered it all – by Riccardo's vision of X-ray astronomy, by the just right observatories he designed and provided to the community, and by specific observations of his. I stress that the study of clusters as signposts for *cosmogony* constitutes the all-important, unfinished task. After all, shedding for a minute the complacent mood, one quails at realizing that most current cosmogony is based on DM whose nature is unknown, and is shaped by the P&S statistics which shouldn't apply!

I have argued the cluster complexity to be comprised of the cosmogonical sector related to the DM-dominated dynamics, and of the sector related to the ICP hydro- and the thermodynamics, with the latter sector amplifying and highlighting the former. The telling link is constituted by $L - T$ correlation, and this is one issue to probe locally and out to redshifts $z \simeq 0.5$ now sparsely sampled (see Mushotzky & Scharf 1997, Tsuru et al. 1997). Theoretically, one main issue is to fully uderstand its intrinsic scatter as discussed above. Near-future observations ought to clear the scatter of the large statistical uncertainties now affecting most measurements of T, and to assess any systematics from thermal inhomogeneities left by major merging events. In several cases mass measurements by gravitational lensing can already check the emission-weighted temperatures and/or the usual assumption of virial equilibrium for the ICP. In parallel, measurements of densities may be checked in many cases with the SZ effects, which are sensitive to the wings of the ICP disposition, but less so to the direct emissivity from cooling flows.

With the baryon history fully understood, the cosmology narrowed down, and the perturbation spectrum closely bracketed, cluster X-raying at last will expose the underlying process of *non-linear* condensation of the DM on scales $1 - 20$ Mpc in the active environment provided by larger-scale structures, so coming to grips with the pattern and the ways of structure formation.

Acknowledgments

I am indebted for many fresh data and informative discussions to S. De Grandi, D. Lazzati, and especially to S. Molendi, M. Ramella and P. Rosati. I owe the better part of my understanding to R. Fusco Femiano, C. Gheller, F. Governato, P. Monaco, O. Pantano, A. Trevese, and especially to N. Menci and P. Tozzi.

References

1. G.O. Abell, ApJ Suppl. **3**, 211 (1958).
2. J.N. Bahcall and N.A. Bahcall, ApJ **199**, L89 (1975).
3. N.A. Bahcall, X. Fan and R. Cen, preprint (1997).
4. E. Bertschinger and B. Jain, ApJ **431**, 495 (1994).
5. M. Birkinshaw, in *The CMB: 25 Years Later*, eds. N. Mandolesi and N. Vittorio (Kluwer, Dordrecht 1990), p. 77.
6. A. Biviano, M. Girardi, G. Giuricin, F. Mardirossian and M. Mezzetti, ApJ **411**, L13 (1993).
7. J.R. Bond, S. Cole, G. Efstathiou and N. Kaiser, ApJ **379**, 440 (1991).
8. E.F. Bunn and M. White, ApJ **480**, 6 (1997).
9. T.G. Brainerd and J.V. Villumsen, ApJ **394**, 409 (1992).
10. A. Cavaliere, H. Gurski and W.H. Tucker, Nature **231**, 437 (1971).
11. A. Cavaliere and R. Fusco Femiano, A&A **49**, 137 (1976).
12. ———————— A&A **70**, 677 (1978).
13. A. Cavaliere, N. Menci and P. Tozzi, ApJ **464**, 44 (1996).
14. ———————— ApJ 484, L1 (1997).
15. A. Cavaliere, L. Danese and G. De Zotti, ApJ **217**, 6 (1977).
16. ———————— A&A **75**, 322 (1979)
17. A. Cavaliere, P. Santangelo, G. Tarquini and N. Vittorio, ApJ **305**, 651 (1986).

18. L.P. David, A. Slyz, C. Jones, W. Forman, S.D. Vrtilek and K.A. Arnaud, ApJ **412**, 479 (1993).
19. H. Ebeling, A.C. Edge, A.C. Fabian, S.W. Allen, C.S. Crawford and H. Böhringer, preprint (1997).
20. A.C. Fabian, C.S. Crawford, A.C. Edge and R.F. Mushotzky, MNRAS **267**, 779 (1994).
21. J.E. Felten, R.J. Gould, W.A. Stein and N.J. Woolf, ApJ **146**, 955 (1966).
22. W. Forman and C. Jones, ARAA **20**, 547 (1982).
23. K. Grainge. M. Jones, G. Pooley, R. Saunders, J. Baker, T. Haynes, & A. Edge, MNRAS **278**, L17 (1966).
24. J.E. Gunn and J.R. Gott, ApJ **176**, 1 (1972).
25. H. Gursky, A. Solinger, E. Kellogg, S. Murray, H. Tananbaum, R. Giacconi and A. Cavaliere, ApJ **173**, L99 (1972).
26. J.P. Henry and U.G. Briel, ApJ **443**, L9 (1995).
27. W.J. Jaffe and G.C. Perola, A&A **26**, 423 (1973).
28. C. Jones and W. Forman, ApJ **276**, 38 (1984).
29. D. Lazzati, Thesis at the Univ. of Milano (1996).
30. J.F. Meekins, G. Fritz, T.A. Chubb and H. Friedman, Nature **231**, 107 (1971).
31. G. Miley A&A **26**, 413 (1973).
32. R.J. Mitchell, J.L. Culhane, P.J. Davison and J.C. Ives, MNRAS **178**, 29p (1976).
33. J.J. Mohr, A.E. Evrard, D.G. Fabricant and M.J. Geller, ApJ **447**, 8 (1995).
34. R.F. Mushotzky and C.A. Scharf, ApJ in the press (1997).
35. R.F. Mushotzky, P.J. Serlemitsos, B.W. Smith, E.A. Boldt and S.S. Holt, ApJ **225**, 21 (1978).
36. J. Oukbir and A. Blanchard, preprint (1996).
37. P.J.E. Peebles, *Principles of Physical Cosmology* (Princeton Univ. Press, Princeton 1993).
38. W.H. Press, and P. Schechter, ApJ **187**, 425 (1974).
39. T.J. Ponman, P.D.J. Bourner, H. Ebeling and H. Böhringer, preprint (1997).
40. M. Ramella, M.J. Geller and J.P. Huchra, ApJ **384**, 396 (1992).
41. P. Rosati, R. Della Ceca, C. Norman and R. Giacconi, preprint (1997).
42. C.L. Sarazin, *X-ray Emissions from Clusters of Galaxies* (Cambridge Univ. Press, Cambridge, 1988).
43. S. Schindler, MNRAS **280**, 309 (1996a).
44. S. Schindler, A&A **305**, 756 (1996b).
45. P.J. Serlemitsos, B.W. Smith, E.A. Boldt, S.S. Holt and J.H. Swank,

ApJ **211**, L63 (1977).
46. J. Silk and S.D.M. White, ApJ **220**, 390 (1978).
47. A.B. Solinger and W. H. Tucker, ApJ **175**, L107 (1972).
48. S.F. Shandarin and Ya.B. Zel'dovich, Rev. Mod. Phys. **61**, 185 (1989).
49. R.A. Sunyaev and Ya.B. Zel'dovich, Comm. Astrop. Sp. Sci. **4**, 173 (1972).
50. T. Tsuru et al., IAU Symposium 188 (1997).
51. D.A. White and A.C. Fabian, MNRAS 273, 72 (1995).
52. D.A. White, C. Jones and W. Forman, preprint (1997).
53. S.D.M. White, MNRAS **177**, 717 (1976).
54. S.D.M. White, J.F. Navarro, A.E. Evrard and C.S. Frenk, Nature **366**, 429 (1993).
55. A.I. Zabludoff and D. Zaritsky, ApJ **447**, L21 (1995).
56. F. Zwicky, E. Herzog, P. Wild, M. Karpowicz and C.T. Kowal *Catalogues of Galaxies and Clusters of Galaxies*, 6 Vols (Caltech, Pasadena 1961-68).

HOT GAS AND DARK MATTER IN GALAXIES AND CLUSTERS

W. Forman & C. Jones
Center for Astrophysics,
Cambridge, MA 02138, USA
E-mail: cjf@cfa.harvard.edu; wrf@cfa.harvard.edu

We summarize the pioneering work in the study of galaxies and clusters of galaxies made possible by the advent of imaging X-ray astronomy that was realized with the launch of the Einstein Observatory. We emphasize the uniqueness of the contributions from X-ray astronomy and focus on the discovery of hot coronae and dark halos around early-type galaxies, the determination of mass distributions and baryon fractions in clusters, and cluster structure. We discuss the cosmological implications of these investigations.

1 Hot Gas in Early-Type Galaxies and Clusters

Prior to the launch of the Einstein Observatory, virtually nothing was known about the X-ray emission from normal early-type (E and S0) galaxies. The first X-ray images of galaxies in the Virgo cluster showed extended X-ray emission from hot gas in several Virgo elliptical galaxies extending up to 100 kpc. The detection of significant amounts of hot gas in this class of galaxy led to the realization that early-type galaxies were not gas poor, as they were generally believed to be. Rather, the mass lost by evolving stars was deposited in a hot corona bound to the galaxy by a massive dark halo.

Einstein observations of clusters also radically changed our view of these systems. Before Einstein, clusters were modelled as symmetric, relaxed systems. However, the cluster X-ray images showed a variety of morphologies and demonstrated that many clusters were not relaxed systems, but actively undergoing mergers. For those clusters that are relaxed, the cluster imaging allows mapping the gravitating mass distribution which provided tests of cosmological models.

During the UHURU years, it was possible to review all X-ray research on galaxies and clusters in a single contribution such as this. Now, an entire proceeding is required to cover the wealth of observational and theoretical X-ray research related to galaxies and clusters. Therefore, in the sections which follow, we review major contributions to the study of hot gas and dark matter in galaxies and clusters from Einstein and bring these discussions up-to-date primarily in those areas in which Riccardo Giacconi was especially interested and in which he directly participated.

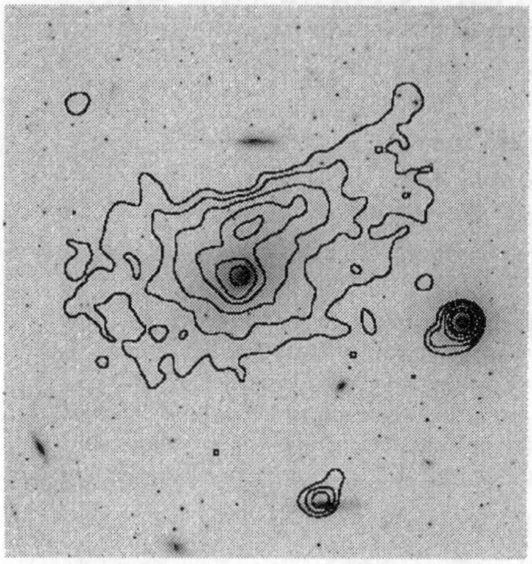

Fig. 1 shows the X-ray contours of the elliptical galaxy M86=NGC4406 superposed on an optical image. The galaxy to the west of M86 is M84=NGC4374.

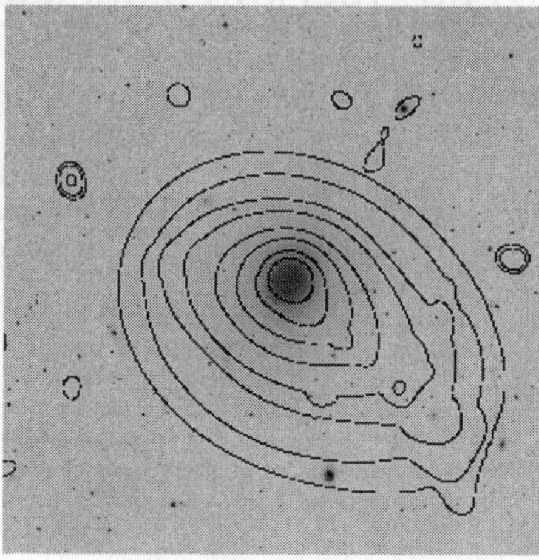

Fig. 2 shows the X-ray contour map superposed on an optical image of NGC4472, a nearly normal elliptical galaxy lying outside the Virgo cluster core.

2 Massive Dark Halos Around Early-Type Galaxies

The history of the study of gas in early type (elliptical and S0) galaxies began at least as early as 1957. In that year Sandage [1] estimated that during the course of normal stellar evolution, the constituent stars of a bright elliptical would shed $5 \times 10^9 M_\odot$ of gas. Sandage suggested that much of this gas could be in the form of neutral hydrogen, although a portion of it could be ionized and might explain the optical emission lines of forbidden oxygen, which were seen in 15% of the ellipticals observed by Humason, Mayall, and Sandage. [2]

In the 25 intervening years, the limits on various forms of gas in early type galaxies became increasingly more stringent and models were developed to explain the absence of gas in elliptical galaxies. In particular, Mathews and Baker [3] discussed the possibility of driving the gas from galaxies in galactic winds powered by supernova explosions.

With the launch of the Einstein Observatory and the ability to obtain spatially resolved X-ray observations on galaxy scales, it became apparent that E and S0 galaxies were not gas poor, but that the gas was stored in a reservoir of hot, ~1 keV (~ 10^7K) gas (Forman, Jones & Tucker [4]; see also Nulsen et al.[5] and Canizares et al.[6]). The gas mass was roughly that calculated by Sandage, 25 years earlier.

The first examples of hot gas surrounding elliptical galaxies were found in a survey of Virgo cluster galaxies (see Fig. 1). The measured redshift of M86 gives this galaxy a velocity of about $v \sim 1500$ km sec^{-1} with respect to the Virgo cluster. This high velocity implies that the galaxy is crossing the Virgo core, perhaps for the first time, and that accumulated gas is being ram pressure stripped from the galaxy by the ambient Virgo cluster gas seen around M87 (see Forman et al.[7], and Fabian et al.[8]). The unusual morphology of the emission around M86 provided strong evidence that the X-ray emission arose from thermal radiation from hot gas accumulated in the galaxy potential well when M86 was outside the cluster core region.

Observations of the X-ray emitting gas can provide fundamental information about galaxies — namely their gravitational mass and the shape of the underlying potential, which have been difficult to determine for early-type galaxies. In spiral galaxies, the presence of a cool interstellar medium has been used to systematically measure the dynamical masses at large radii with both optical and radio techniques. The absence of such tracers at large radii in early-type galaxies has made the determination of their mass distributions significantly more difficult. However, the observed X-ray coronae provide ideal test particles to trace the total mass distribution in individual galaxies, since the gas obeys well-understood laws and comprises a small fraction of the total galaxy mass.

The method used to derive the mass distribution in early type galaxies relies on two simple principles — the ideal gas law and hydrostatic equilibrium. The hydrostatic equation can be written as

$$\frac{dP}{dr} = -G\rho M(<r)/r^2 \tag{1}$$

where P is the pressure, ρ is the gas density, G is the gravitational constant, and $M(<r)$ is the total mass interior to the radius r. The mass in equation 1) is not just the mass in gas, but *all* the mass that contributes to the gravitational potential, whether it be stars, collapsed objects, or exotic particles. The ideal gas law allows us to eliminate the pressure, in favor of the gas temperature and density, and solve for the mass interior to r. We find:

$$M(<r) = -\frac{kT}{G\mu m_p}\left(\frac{d\ln\rho}{d\ln r} + \frac{d\ln T}{d\ln r}\right) r. \tag{2}$$

where T is the gas temperature at a radius r, μ is the mean molecular weight and m_p is the mass of a proton. Numerically this becomes:

$$M(<r) = 3\times 10^{12}(T/10^7K)\left(-\frac{d\ln\rho}{d\ln r} - \frac{d\ln T}{d\ln r}\right)(r/100kpc)M_\odot. \tag{3}$$

If one can measure the temperature and density gradients, then one can measure the total mass as a function of radius. This is a remarkably powerful technique and can be applied to clusters as well as galaxies.

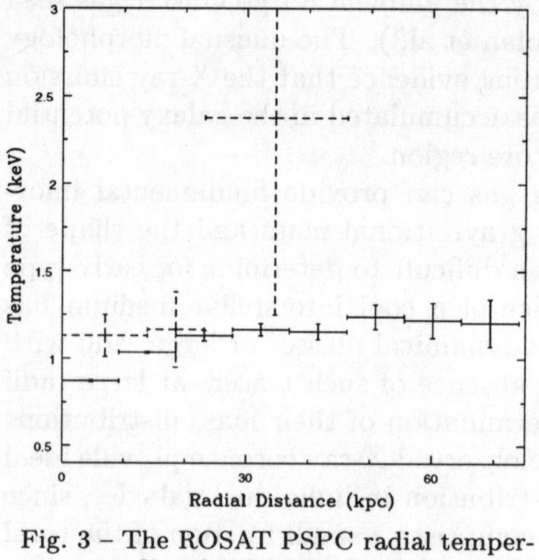

Fig. 3 – The ROSAT PSPC radial temperature profile of NGC4472 is shown (solid lines) along with the Einstein observations (dashed lines) which demonstrated the presence of a massive, dark halo.

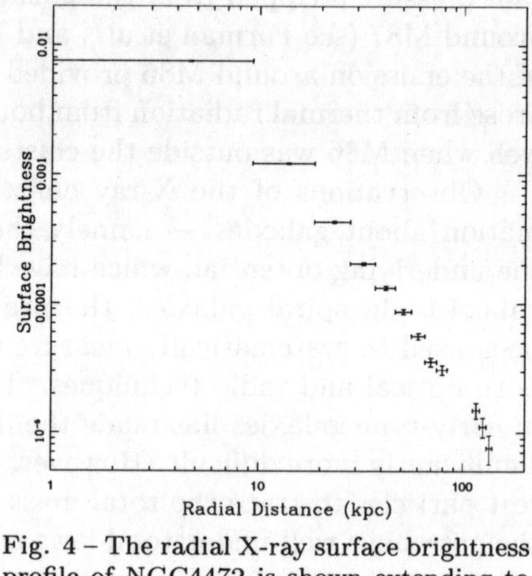

Fig. 4 – The radial X-ray surface brightness profile of NGC4472 is shown extending to large radii.

The first X-ray mass determination of the mass in an early-type galaxy was performed using Einstein observations of NGC4472. Fig. 2 shows an X-ray contour map superposed on an optical image. The X-ray emission extends well beyond the optical image of the galaxy. The gas temperature profile as measured from ROSAT PSPC images is shown in Fig. 3 along with the earlier Einstein temperature determinations. We see an isothermal corona with a temperature of about 1 keV (1.2×10^7K) measured to 16 arcminutes (74 kpc) from the galaxy center. Note that there is a small, but significant temperature decrease at small radii, where the gas density is high and the cooling time is short (see Thomas et al. [9] and Thomas [10] for a discussion of cooling cores in hot coronae).

The X-ray surface brightness, shown in Fig. 4, can be used to derive the gas density profile, which also is needed to calculate the gravitating mass. The surface brightness can be written as $S(r) = S_0 \int n^2 \Lambda(T) dz$ where $\Lambda(T)$ is the emissivity of the gas as a function of temperature and n is the gas density distribution, and z lies along the line-of-sight. Since the gas is essentially isothermal, $\Lambda(T)$ is a constant and the surface brightness distribution directly provides the gas density distribution required in equation 3.

With the measured gas density and temperature profiles, we can calculate the gravitating mass using equation 3 for NGC4472. The gravitating mass is $4.5 \times 10^{12} M_\odot$ at 70 kpc where the gas temperature is measured or $6.4 \times 10^{12} M_\odot$ at 100kpc if we assume that the gas temperature remains constant within the regions where the brightness profile is well-determined. The galaxy luminosity (in the blue band) is $L_B = 7.4 \times 10^{10} L_\odot$ for an assumed distance of 16 Mpc. Taking the uncertainties into account this yields a mass-to-light ratio of

$$M/L = \begin{cases} 78 - 93 M_\odot/L_\odot & \text{at 100 kpc} \\ 54 - 65 M_\odot/L_\odot & \text{at 70 kpc} \end{cases} \tag{4}$$

These mass-to-light ratios can be compared to the mass-to-light ratio where stellar matter dominates the mass. For example, Lauer [11] studied a sample of 14 elliptical galaxies. For NGC4472, Lauer found $M/L = 14 M_\odot/L_\odot$ within the galaxy core and $M/L = 11.9 M_\odot/L_\odot$ within 6.1 kpc. Thus, the outer regions of NGC4472 are dominated by dark matter. If we exclude the inner region and compute the mass-to-light ratio in the outer 16-70 kpc, we find a ratio of 96-118 $M_\odot/L_\odot$. Thus, the outer regions have a much higher mass-to-light ratio than the central regions.

Other bright galaxies have similar mass-to-light ratios – 97, 89, 110, and 50-90 (including diffuse light seen at large radii around the galaxy) for NGC507, NGC499, NGC5846, and NGC1399 respectively. For NGC4636, a massive halo with $M/L_B = 145$ at about 100 kpc was found (Mushotzky et al.[12]). A novel approach to determining mass distributions was developed by Buote and Canizares [13] using the geometric distributions of stellar light and the X-ray emission in a highly elliptical galaxy NGC720. They conclude that a dark halo must be present to generate the observed X-ray distribution. The most recent work of Buote & Canizares [14] accounts for the multi-component spectra in early-type galaxies observed with ASCA and confirms the existence of dark matter halos for NGC720 and NGC1332. As observations improve, it will be possible to measure the shape of the halo distributions for comparison with theoretical models of halo formation.

In conclusion, X-ray studies of early-type galaxies have demonstrated that these systems, like their sister spirals, are embedded in massive dark halos with $M/L_B \sim 100$. X-ray imaging observations provide fundamental insights into the mass distributions of early-type galaxies, not possible at any other wavelength.

2.1 Elemental Abundances

The hot coronae are the repositories of stellar mass loss and supernova ejecta and thus contain a record of past supernova activity in the galaxy. The heavy element abundances in the hot coronae are determined by the supernova yields of heavy elements, the initial mass functions which govern the frequency with which supernova progenitors are formed, and the mass fraction of primordial gas remaining after galaxy formation. The heavy elements presently seen in the hot coronae of early type galaxies arise from two sources – stellar mass loss from evolving stars and ejecta from type Ia supernovae which are the dominant type of supernova observed in elliptical galaxies. Because the radial flow time of the gas is billions of years, radial abundance measurements can provide estimates both of the supernova rate and its temporal evolution.

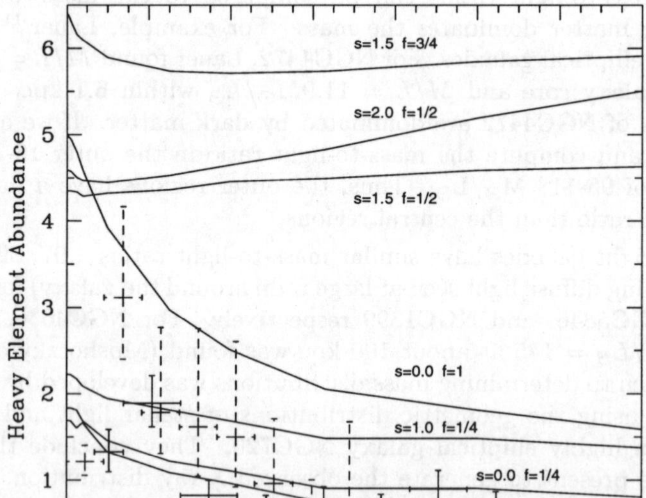

Fig. 5 shows the abundance distribution around NGC1399 (solid lines) and NGC4472 (dashed lines) compared to model calculations (Loewenstein and Mathews [15]). Note that for this comparison, the iron abundances are derived for Fe/H $= 3.16 \times 10^{-5}$ by (number).

A theoretical study of the dependence of the heavy element abundances on the supernova rate was made by Loewenstein and Mathews.[15] They characterized the time dependence of the type Ia supernova rate and its present epoch value as $R_{SN} = 2.2f(t/t_H)^{-s}$ where R_{SN} is the number of type Ia supernovae per 100 years per $10^{11}L_\odot$, t_H is the present time, f is the ratio of the present epoch supernova rate to that given by Tammann,[16] and s parameterizes the dependence of the supernova rate with time. Fig. 5 shows model calculations of the radial behavior of the heavy element abundance for galaxy models of a $10^{11}L_\odot$ galaxy with a heavy halo. Along with the models we show the abundance measurements derived from the ROSAT observations of NGC1399 and NGC4472 (Forman et al.[17]; Jones et al.[18]). As shown in Fig. 5 and also found from recent ASCA observations (Mushotzky et al. [12], Loewenstein et al. [19], Matsumoto et al. [20]), less than solar abundances of heavy elements are found over a wide range in radii.

As is apparent from Fig. 5, models with much higher supernova rates at earlier epochs are inconsistent with the observations. The models which adequately describe the observations have constant or slowly varying supernova rates ($|s| < 1$) and require relatively modest values of the present epoch supernova rate. In particular for the models with $s = 0$, the one with present epoch supernova values equal to Tammann's[16] rate ($f \approx 1$) produces iron abundances greater than observed, while the model with $f = 1/4$ gives iron abundances in good agreement with the observations. Thus, for models of elliptical galaxies with heavy halos, the observations suggest a constant or slowly varying supernova rate ($|s| < 1$) and a present epoch supernova rate significantly less than $f = 1$. Higher supernova rates would be possible, if the iron yields from type Ia supernovae are reduced or if the ejecta are not completely mixed into the hot ISM before cooling.

Another consequence of the low heavy element abundances is that the observed gas cannot be expanding in a galactic wind. If such a wind were occurring, then the gas would have to be rapidly resupplied from stellar mass loss. Since the stellar content has a high metallicity, this would have to be reflected in the gas. Hence, the low heavy element abundances argue that the gas cannot have high velocity and that the masses derived under the hydrostatic assumption must be reasonably accurate.

3 Clusters of Galaxies

Clusters are complex, multi-component systems with hundreds of galaxies, a hot (10^7 to 10^8K) intracluster medium (ICM), and dark matter evolving in tightly coupled ways. On scales of clusters of galaxies, most of the mass is dark. Only a few percent of the mass in clusters lies in the optical galaxies. In rich clusters, most of the luminous baryonic mass (about 25% of the total cluster mass) is in the hot ICM.

X-ray observations are of fundamental importance for cluster studies. First, the hot ICM is the dominant reservoir of *luminous* baryons in clusters, with gas mass exceeding stellar mass by factors of 3-5 (David et al.[21]; Arnaud et al.[22]). Second, X-rays are uniquely suited to defining unbiased cluster samples whose properties (e.g., mass function) are related to the underlying cosmology. Third, the hot ICM traces the total cluster mass as well as cluster morphology and the presence of substructure. By measuring the gas temperature and density distribution in relaxed clusters, we can measure the gas mass and, more importantly, assuming hydrostatic equilibrium, measure the mass and distribution of the dark matter (e.g., Bahcall & Sarazin[23], Mathews[24]). Simulations predict that different forms of dark matter should be distributed and evolve differently (e.g., Navarro et al.[25]; Kofman et al.[27]). Therefore, by comparing distributions of dark matter and luminous mass on cluster scales and measuring how these distributions change over time, one should find clues to the nature of the dark matter. In the next sections we describe some of the implications of X-ray imaging cluster studies.

Table 1: Frequencies of Cluster Morphological Classes

X-ray Morphological Class	Example	Percent	Mean L_x
S – single symmetric peak	A401	56	29.3×10^{43}
O – offset center	A2319	5	44.5
E – elliptical	A2256	14	30.8
C – complex, multiple structures	A514	13	9.6
D – double (roughly equal components)	A98	6	22.1
P – primary with a small secondary	A85	3	19.9
G – primarily galaxy emission	A2666	3	0.4

3.1 *Cluster Morphology and Substructure*

For the first time with Einstein, cluster morphologies for a large sample of clusters could be studied. The Einstein images showed that in the centers of many clusters, the gas density is sufficiently high that the radiative cooling time of the gas is short compared to the age of the Universe. A cooling flow is established as the gas density rises to support the weight of the overlying cluster gas. The sharply peaked X-ray surface brightness at the centers of many clusters as well as the cool temperature components seen at the centers of several clusters from Einstein FPCS and SSS measurements are indicative of cooling flows. Even at the spatial limits of the Einstein images, nearly half of all clusters have central cooling flows. Cooling flow rates of 50 to 100 solar masses per year are typical, with several extreme cases exceeding 500 solar masses per year. Inhomogeneities in the gas may lead to distributed mass deposition over the entire cluster core. For a detailed discussion of cooling flows and their properties see Fabian [26].

As part of the Einstein cluster survey, we generated X-ray iso-intensity contour maps to investigate cluster structure. We used these to define seven morphological cluster classes (see Jones & Forman [28]) which are summarized in Table 1 along with their relative frequency in the sample, typical examples, and mean luminosity. The classification sample included 208 clusters which were at sufficiently low redshift ($z \leq 0.15$) and with a sufficient number of counts to allow an accurate classification.

The morphological classification of clusters is closely related to the study of substructure. All except the "single" and "galaxy" classes contain clusters with appreciable structure. While the categories of "double", "complex", and "primary with small secondary" show obvious structure, the "elliptical" and "offset center" clusters also are structured systems as demonstrated by detailed X-ray and optical studies of clusters in these categories including Coma, A754, and A2256 (e.g., Fabricant, Kent, and Kurtz [29]; Fitchett and Webster [30]; Briel et al. [31]; White, Briel & Henry [32]; Mohr et al. [33]; Davis and Mushotzky [34]; Vikhlinin et al. [35], [36]).

Clusters exhibiting clear substructure (double "d", primary with small secondary "p", and complex "c") comprise 22% of the classifiable sample. Including elliptical ("e") clusters (e.g. A2256-like) brings the fraction of clusters with substructure to 36%. Finally, including clusters with offset centers ("o" e.g. A2319-like) gives a substructure fraction of 41%. The limited resolution of the IPC and the bias in classifying faint or poorly observed clusters as centrally peaked and symmetric suggest 41% is a lower limit to the number of clusters with substructure. There is little or no change in the percentage of clusters with substructure as a function of X-ray luminosity.

In addition to the Einstein imaging survey, a wide variety of approaches have been used to characterize and identify substructure in clusters of galaxies. Analysis of the spatial distribution of cluster member galaxies was used to show that 40% of a sample of 65 clusters exhibited significant substructure (Geller and Beers [37]). Statistical tests utilizing both galaxy positions and velocity data for 15 clusters detected substructure in 30-40% of the sample (Dressler and Shectman [38]). A correlation analysis of the galaxy positional data for the same set of clusters showed that more than 50% of the clusters contain significant substructure (Salavador-Sole, Sanroma, and Gonzalez-Casado [39]).

Wavelet techniques, applied to both optical and X-ray data, also show high fractions of substructure in cluster samples (Escalera et al.[40]; Slezak, Durret, and Gerbal[41]). Velocity offsets of bright cD galaxies from the cluster mean were used to infer that 21 of 25 clusters were not relaxed (Bird[42]). Finally, analysis of variations in the low order moments of X-ray cluster images showed that a large fraction of clusters were not dynamically relaxed systems (Mohr, Fabricant, and Geller[44]). In summary, both optical and X-ray studies suggest a high fraction of substructure in rich clusters with typical fractions exceeding 40%. As sensitivity improves and detection techniques become more sophisticated, the fraction of clusters with significant substructure can be expected to increase.

The morphology of X-ray cluster samples is a powerful discriminator between cosmological models (Richstone, Loeb & Turner[43], Mohr et al.[44], Buote & Xu[45]). It is well-known that the structure growth rate (and hence, cluster formation rate) strongly depends on the underlying cosmology: in a flat, $\Omega = 1$ universe, there is no preferred time scale and clusters should be actively forming at the present and evolve self-similarly; in a matter-dominated, $\Omega < 1$ universe, the cluster formation rate is effectively terminated at $z = \Omega^{-1} - 1$ (White & Rees[46]), while in a flat, cosmological constant dominated universe ($\Omega_M + \Lambda = 1$), this occurs at $z = \Omega_M^{-1/3} - 1$ (Carroll, Press & Turner[47]). Qualitatively, a high substructure fraction should argue for a high value of Ω, assuming that substructure from mergers is quickly erased.

While initial substructure studies seemed to indicate a high Ω universe, Jing et al.[48] found a larger substructure fraction in simulations for low Ω models than did Evrard et al.[49] (see also Nakamura et al.[50]). In addition, Kaufman and White[51] and Lacey and Cole[52] argued that the timescales over which substructure is erased are uncertain and that the magnitude of the substructure is not precisely quantified for comparison with numerical predictions. In particular, Lacey and Cole found that if the time to erase substructure following a merger is 20% of the present age of the Universe, then the lower limit on Ω is 0.4 to 0.6, depending on the initial fluctuation spectrum. If the erasure time for substructure is as long as half the Hubble time, then Ω is only constrained to be greater than 0.1. Taken together, these arguments would allow a smaller value of Ω to be consistent with the observed substructure fractions.

Another approach to constraining cosmological parameters is to directly compare an ensemble of cluster images with those derived from cosmological simulations (Cen and Ostriker[53]). This approach was applied to a sample of 65 Einstein cluster observations by Mohr et al.[44]. This study, which analyzed the shift of the emission centroid, the rotation of the isophotes, and properties of the surface brightness profiles argued that a universe with a high

Ω best matched the X-ray cluster properties. However, simulations used in this analysis, to maintain sufficient spatial resolution, do not include large scale structures such as filaments which could serve to enhance the structure predicted for low Ω model universes. The tendency of cluster ellipticities to lie along the direction toward neighboring clusters (West et al. [54]), suggests that clusters grow through infall along filaments and that some of the cluster structure may arise from alignments along filamentary structures.

In conclusion, while the fundamental idea of using cluster morphology to constrain Ω appears sound, the exact correspondence remains in doubt. With only the present epoch "snapshot" of substructure in clusters, our confidence in the comparison of observation and theory remains uncertain. Future observations of clusters at higher redshifts will allow a comparison of the models to "snapshots" at different epochs and will provide a much more confident estimate of the cosmological density parameter from substructure frequency.

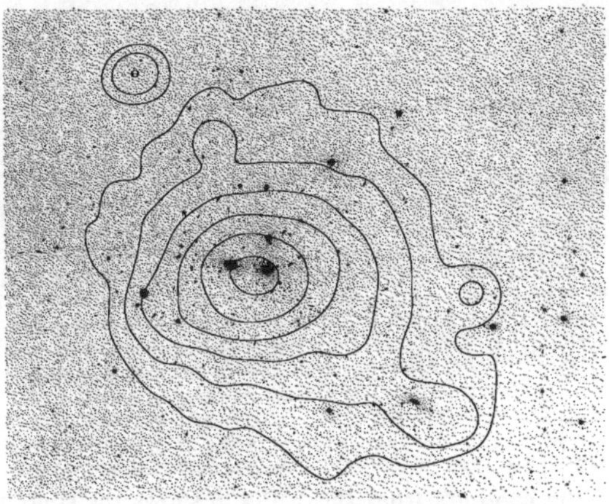

Fig. 6 shows the background-subtracted, vignetting and exposure corrected image of the Coma cluster in the energy band 0.5-2.0 keV. The merging subcluster is seen to the south west.

Substructure in the Coma Cluster

Increases in sensitivity and new analysis techniques have continued to demonstrate the complexity of the cluster environment. The Coma cluster, long thought to be the prototype for a relaxed cluster, is in fact a remarkably complex system with structure on many different scales. X-ray studies have shown its remarkable complexity (Briel et al. [31]; White, Briel & Henry [32]; Vikhlinin, Forman & Jones [35,36]) including a merging subcluster and a variety of smaller scale features in the cluster including two concentrations around the central cluster galaxies (see Fig. 6 for an overall view of the cluster). Optical indications of substructure come from the work of Fitchett & Webster [30] who demonstrated the presence of substructure and Mellier et al. [55] who were able to derive the mass to light ratios for the core substructures surrounding the two dominant core galaxies. Most recently, Colless and Dunn [56] used a sample of 552 galaxy redshifts to further clarify the merger history of the Coma cluster. Biviano et al. [57] combined ROSAT and optical redshift and photometric samples to identify the main body of the cluster and suggest that the cluster was undergoing rotation. They also concluded, based on the difference in velocity between the NGC4889 and NGC4874 groups and the cluster mean, that these groups only recently arrived within the cluster core.

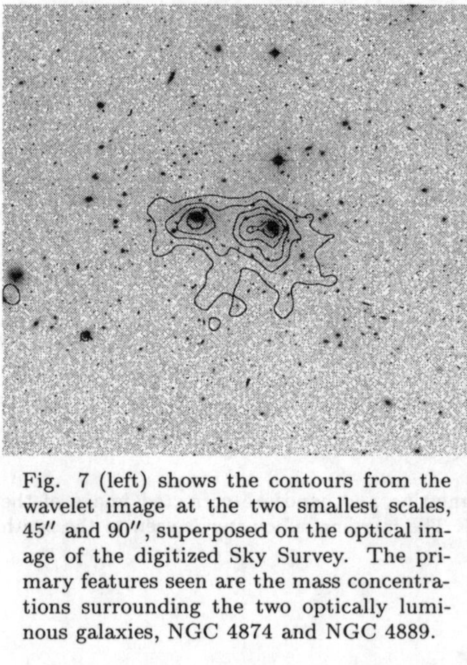

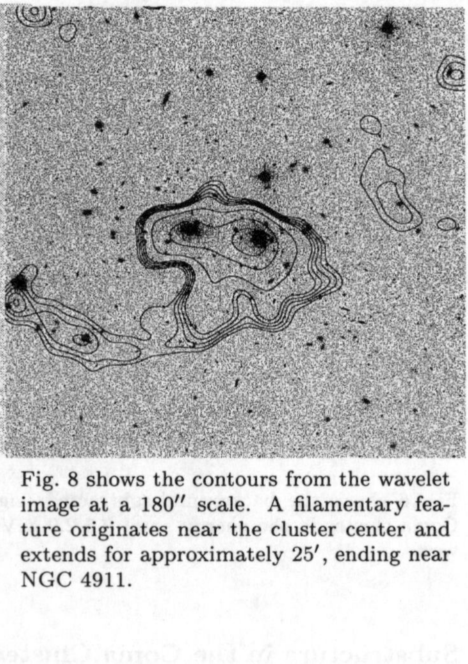

Fig. 7 (left) shows the contours from the wavelet image at the two smallest scales, 45″ and 90″, superposed on the optical image of the digitized Sky Survey. The primary features seen are the mass concentrations surrounding the two optically luminous galaxies, NGC 4874 and NGC 4889.

Fig. 8 shows the contours from the wavelet image at a 180″ scale. A filamentary feature originates near the cluster center and extends for approximately 25′, ending near NGC 4911.

To study the small scale structure in more detail, a wavelet decomposition analysis of the ROSAT Coma cluster images was performed (Vikhlinin et al. [36]). Contour maps of wavelet images generated for small scales are shown in Fig 7 and 8. On the smallest scales, 60″, the contours show the extended sources surrounding the two brightest cluster galaxies NGC 4874 and NGC 4889 first detected in X-rays by White et al. [32] (see also Vikhlinin et al. [35]). On larger scales, 120″, the wavelet analysis reveals an X-ray filament, approximately 1 Mpc (25′) in length with a luminosity of 3×10^{42} ergs $s^{-1}$and a gas mass of 5×10^{11} $M_{\odot}$. The filament originates near the cluster center, curves to the south and east for $\sim$ 25′ towards the galaxy NGC 4911, and ends near the galaxy NGC 4921.

We consider two possible scenarios for the origin of this filamentary feature. First, this feature could arise from gas that has been ram pressure stripped from a merging group since the observed gas mass is well within the reasonable range for the gas mass in the core of a group. An alternative is that the observed filamentary feature arises from cluster gas compressed in the potential well of a dark matter perturbation formed by a tidally disrupted infalling galaxy group which provides a gravitational potential perturbation and hence enhances the surface brightness. For the observed properties of the filament, and assuming a gas temperature that of the cluster, we find a mass of $M = 6 \times 10^{12}$ $M_{\odot}$, consistent with the mass in the core of a group.

We mention that, with the removal of the small fraction of counts associated with the tail, the remaining Coma X-ray emission is remarkably regular. Thus, even though the Coma cluster contains significant structure, the cluster core appears relaxed – although this appearance may change as the merging group penetrates into the main body.

The Complex Substructure of A1367

A1367 is a nearby, X-ray bright cluster of galaxies which was extensively studied with both the Einstein IPC and HRI (Jones et al. [58]; Bechtold et al. [59]). The cluster is characterized by a cool gas temperature and high fraction of spiral galaxies, typical of a dynamically young system. In a detailed analysis, Bechtold et al. found fifteen 1′ scale emission features in the central region of the cluster, nine of which were associated with cluster galaxies.

To explore the nature of the X-ray emission and to better understand the small scale features in A1367, we analyzed a ROSAT PSPC observation using a wavelet transform to detect and determine sizes of features in the X-ray images (Grebenev et al. [60]). The angular extents (FWHM) of all the sources detected in the A1367 field are shown as a function of off-axis angle in Fig. 10. The figure demonstrates that there is a class of sources located chiefly near the cluster center which appear extended in the PSPC image. Many of these sources coincide with those detected in the original Einstein analysis.

If the extended features are associated with individual early-type galaxies in A1367, these galaxies are unusually luminous in X-rays compared to galaxies in less dense environments (Canizares et al. [6]). Alternatively, these features could be remnants of groups of galaxies that have not been fully absorbed by the cluster. In fact, the luminosities are very like the emission reported for such groups by Ponman & Bertram [61]. Although the number of systems detected in the central region of A1367 seems large, even such relaxed systems as the Coma cluster exhibit multiple extended features. A1367 has a gas temperature roughly half that of the Coma cluster and hence may have a longer dynamical timescale and be less evolved. Thus, it may not be surprising to find a large number of merger remnants.

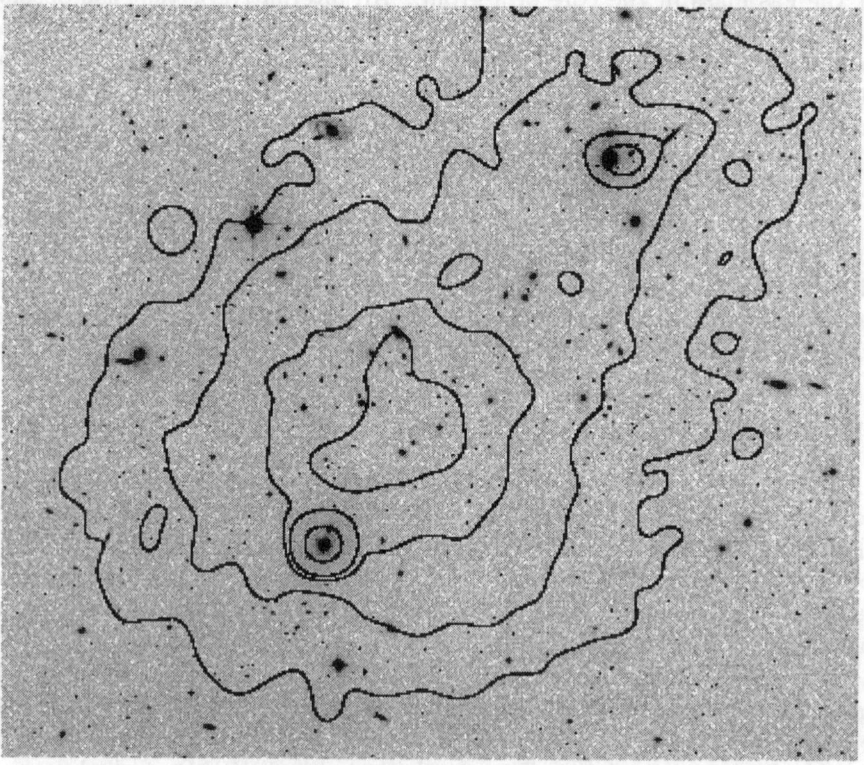

Fig. 9 shows the ROSAT X-ray contours of A1367 superposed on an optical image. The overall elongation of the cluster and the impression of a merging subcluster approaching the main concentration from the north west is suggested.

The extensive small scale structure in A1367 remains somewhat unique – no other cluster contains a comparable number of small scale features, many of which are associated with individual galaxies. AXAF images with good spatial and spectral resolution should tell us much about these interesting features and indicate whether they are present in other clusters.

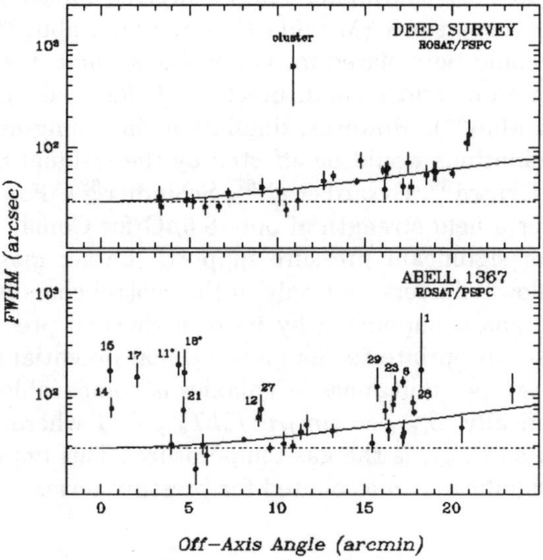

Fig. 10 shows the distribution of angular extents (FWHM) of X-ray sources detected with the ROSAT PSPC in a Deep Survey (top) and A1367 (bottom) as a function of off-axis angle. The solid lines (fit to the Deep Survey point source extents) show the degradation of the angular resolution at large off-axis angles.

3.2 Cluster Masses and Baryon Fractions

Clusters are unique because of their size. A massive, rich cluster with $M_{total} \sim 10^{15} M_\odot$ forms from a density perturbation of radius $\sim$ 15Mpc. Clusters are sufficiently large that an inventory of their components should be unbiased, as shown by CDM simulations (e.g., Cen & Ostriker [62]; Frenk et al. [63]). Hence, determining baryon fractions in clusters yields the cosmological baryon fraction, provided the processes responsible for the hot gas distribution are modeled adequately. Combining measurements of the cluster baryon fraction with nucleosynthesis calculations provides an upper bound on the total mass density in CDM cosmologies.

Measurement of the gas density and temperature distribution allows determination of the total cluster mass through the hydrostatic equation (Bahcall & Sarazin [23], Mathews [24]) as described above for early-type galaxies. Since the sound crossing time is short compared to the dynamical timescale for most

X-ray luminous clusters, hydrostatic equilibrium appears applicable over the central few Mpc. However, the assumption of hydrostatic equilibrium has been questioned, motivated by the higher masses derived for some clusters from lensing or velocity dispersions (Miralda-Escude and Babul [64]). The hydrostatic assumption could be violated for two reasons - first, the gas could have substantial bulk motions and second, magnetic fields could provide pressure support (Loeb and Mao [65]). However, simulations including mergers show the X-ray mass determinations would be affected by the residual bulk flows by no more than ≈30% (Evrard [66], Evrard et al. [67], Schindler [68]). For magnetic fields, Feretti et al. [69] infer a field strength of only 8.5μG for Coma, a small fraction of that required for significant pressure support. Larger magnetic fields are found in "cooling flow" clusters, but only in the central cores (Ge [70]). Another argument that the gas is supported by its own thermal pressure is that the gas temperature is appropriate to the gravitational potential in which it lies. Generally, the energy per unit mass in galaxies is comparable to or less than that in gas. Specifically, $\beta_{spec} = \mu m_H v^2 / 3kT_{gas} \sim 1$ where v is the cluster velocity dispersion and T_{gas} is the gas temperature. This implies that for the gas, there are no significant unaccounted for pressure terms.

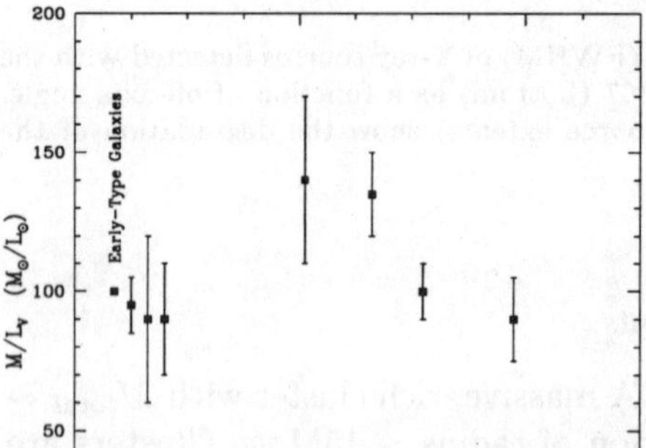

Fig. 11 – The mass-to-light ratios for systems ranging from early-type galaxies (within about 100kpc) to rich clusters (measured within 1 Mpc).

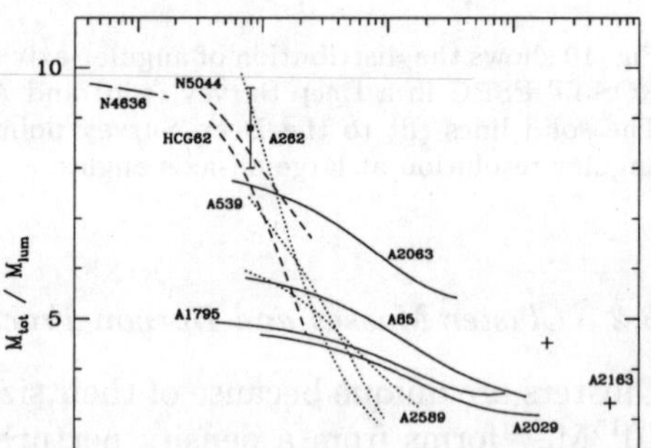

Fig. 12 – The ratio of total gravitating mass to luminous mass as a function of total gravitating mass for an elliptical (long dash), groups (short dash), cool clusters (dotted), and hot clusters (solid). A2163 is shown with "+" signs.

While there were significant systematic disagreements between early lensing studies and X-ray mass measurements for some clusters, recent work has shown better agreement (Squires et al.[71],[72]; Allen et al.[73]). As Daines et al.[74] showed for A1689, clusters with discrepant X-ray and lensing mass estimates can arise because of projection effects due to cluster substructures or larger scale filaments. In conclusion, while it now appears that cluster mass measurements from X-ray observations are likely to be accurate, any differences between the X-ray and the lensing determinations will provide new insights into the physical processes in clusters.

Since the discovery of a hot ICM, it has been recognized that the gas is a significant contributor to the mass of a cluster. Particularly in rich clusters, the ICM is the primary luminous baryonic component (Blumenthal et al.[75], Arnaud et al.[22], David et al.[76], White & Fabian[77]). In galaxies, the gas represents only a few percent of the total mass and the stellar light dominates the observed baryonic matter. In rich clusters, the hot intracluster gas is the dominant baryonic component. With the ability to determine the gas, stellar, and gravitating masses, we can compare the observable (luminous) baryonic matter to the total gravitating matter.

The results of applying hydrostatic equilibrium to a set of groups and clusters and assuming that these systems are essentially isothermal is shown in Fig. 11 and 12 (see David et al.[76]). Fig. 11 shows a remarkable constancy of the ratio of dark to luminous matter on scales from galaxies to rich clusters. The linear scales range from 100kpc for the galaxies to 1 Mpc for the clusters. This suggests that the dark matter in clusters is the same as that in galaxies. The dark matter in clusters may be only the dark matter stripped from galaxy halos. No new dark matter component is required in clusters.

Fig. 12 shows the variation of the ratio of total mass to luminous mass (the inverse luminous baryonic fraction). From a high value of 10 for galaxies, the value decreases by more than a factor of two. Thus, as one goes to larger scales, the universe is *brighter* (as long as one takes a panchromatic view). Thus, while there is much dark matter at large scales, it is a smaller fraction of the total gravitating mass than on smaller scales.

We can extend the analysis of David et al. to determine a mean baryon fraction for the large ensemble of Einstein clusters, again assuming clusters are isothermal. Fig. 13 shows the measured gas mass fractions (essentially the baryon fraction) as a function of bolometric luminosity. In general, the gas masses are directly measured within one Mpc, while at three Mpc, the gas masses are extrapolations of the observed X-ray surface brightness profile, which probably accounts for the relatively larger scatter. We find that the average gas mass fraction at 1 Mpc is $\sim$15%. This fraction increases with the X-ray luminosity from an average of $\sim$10% for clusters with X-ray bolometric

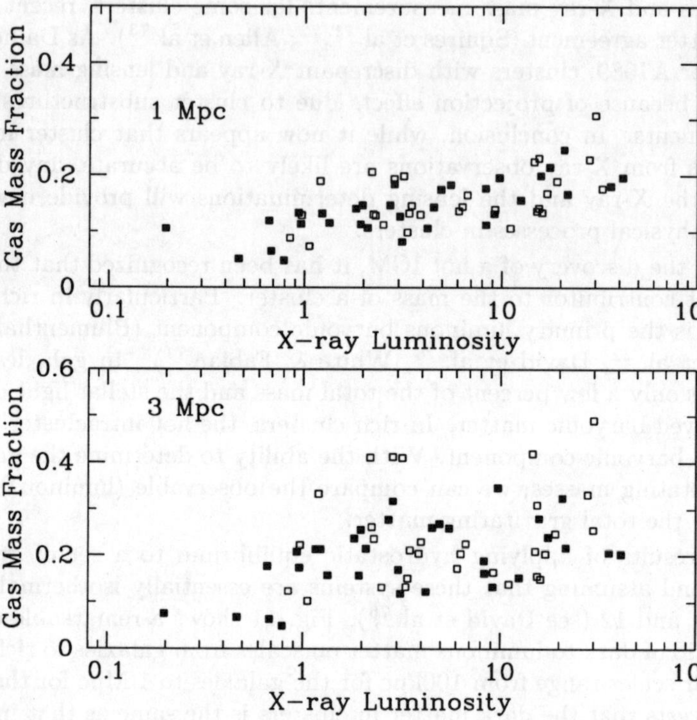

Fig. 13 shows the gas mass fraction vs. X-ray luminosity ($\times 10^{44}$ ergs s^{-1}) within 1 and 3 Mpc for clusters with little substructure (open symbols) and without substructure (solid symbols).

luminosities less than 10^{44} ergs s^{-1} to ~15% for clusters with luminosities between 10^{44} and 10^{45} ergs s^{-1} to ~20% for clusters with X-ray luminosities exceeding 10^{45} ergs s^{-1}. Thus, while the gas mass fraction increases with cluster radii, these results show that this fraction also increases with cluster luminosity at fixed metric radii. It seems likely that the increasing gas mass fraction can be explained by the greater ability of more massive clusters to gravitationally bind hot gas, while the less massive systems lose some gas to the intracluster medium at early epochs when supernova driven winds are enriching the intracluster medium with heavy elements. The observed gas mass fractions yield a lower limit to the baryon fraction, since some gas could be lost from a cluster and some baryons reside in galaxies or may be "dark".

In CDM cosmologies, the observed large baryonic fraction, combined with

"Big Bang" nucleosynthesis calculations, suggests that Ω may be significantly less than one (White et al. [78]). Since Big Bang nucleosynthesis calculations constrain the baryon contribution to the mass density to be $\Omega_b h_{50}^2 = 0.05 \pm 0.01$ (see Walker et al. [79]), then the lower limit on the baryon fraction translates into an upper limit of $\Omega < 0.3$ (see for example, White et al. [78]). This assumes that cluster mass components are representative of the Universe as a whole.

Numerical simulations for CDM cosmologies show that cluster baryon fractions are representative (e.g., Cen & Ostriker [62]; Frenk et al. [63]). Furthermore, the isothermal assumption has been shown to provide an unbiased estimator of the total mass with a dispersion of 50-60% around the "true" value (see Evrard et al. [67]). Hence, the mean baryonic fraction from the Einstein sample provides a strong lower limit to the baryonic fraction of the Universe for CDM models and a strong upper limit on Ω (from Big Bang nucleosynthesis).

Although most of the results above assume that clusters are isothermal, recent results suggest that the actual temperature profiles only further lower the limit on Ω. A detailed study of A2256 shows that the observed ASCA temperature profile implies a gravitating mass, at the virial radius, which is 1.6 times smaller than that derived assuming isothermality (Markevitch & Vikhlinin [80]. Hence, the observed baryon fraction is larger than that derived with the isothermal assumption. It seems difficult to escape the "baryon catastrophe" for CDM cosmologies. However, the numerical simulations, upon which argue that clusters contain representative mass fractions, do not account for all the observed cluster properties. For example, the increasing baryon fraction with radius or luminosity is not predicted Hence, there may yet be physical effects that will alter the prevailing view.

4 Future Prospects

While we have reviewed some key properties of clusters at low redshifts, there are other research areas in which X-ray emission from clusters is important. X-ray observations play a key role in the study of distant clusters, primarily because X-ray surveys provide a powerful technique for selecting well-defined distant cluster samples. These samples are ideal for comparing cluster properties as a function of redshift with predictions from numerical simulations.

Also important, distant cluster observations can be combined with microwave and radio observations to study the Sunyaev-Zel'dovich effect. Although, subject to projection effects and possible substructure, the S-Z effect allows the determination of the Hubble constant, and possibly q_0, free of the distance ladder. As we better understand cluster evolution, the ensemble effects of clusters on the CMBR will provide important tests of cosmological models.

With the launch of large, new X-ray observatories in the coming few years, the study of distant clusters will provide definitive results. Larger samples will be observed and clusters at high redshift will be studied in detail. We will be able to compare cluster properties at different epochs and therefore more strongly test predictions of numerical simulations over cosmological times. We will compare detailed X-ray cluster mass determinations to those from gravitational lensing to better understand any discrepancies and probe the physics of the intracluster medium to determine the magnitude of non-thermal pressure support.

With the launch of the next generation of X-ray observatories, we will achieve a Golden Age in X-ray astronomy. With AXAF, XMM, SRG, and ASTRO-E, we will have telescopes of large area, high angular resolution, and high spectral resolution. We will answer many of the questions posed by the first imaging observations pioneered with Einstein, ROSAT, and ASCA. For galaxies and clusters, we can look forward to a rich symbiosis with optical and radio observations as we probe in X-rays into new regimes of angular and spectral resolution.

References

1. Sandage, A. *ApJ* **125**, 422 (1957).
2. Humason, M.L., Mayall, N.U., and Sandage, A.R. *AJ* **61**, 97 (1956)
3. Mathews, W. and Baker, J. *ApJ* **170**, 241 (1971).
4. Forman, W. Jones, C. and Tucker, W. *ApJ* **293**, 102 (1985).
5. Nulsen, P.E.J., Stewart, G. & Fabian, A.C.*MNRAS* **208**, 185 (1984).
6. Canizares, C. R., Fabbiano, G. and Trinchieri, G.*ApJ* ,, 312 (503) 1987.
7. Forman, W., Schwarz, J., Jones, C., Liller, W., and Fabian, A. *ApJ* **234**, L27 (1979).
8. Fabian A.C., Schwarz J. and Forman W. *MNRAS* **192**, 135 (1980)
9. Thomas, P., Fabian, A., Arnaud, K., Forman, W., and Jones, C. *MNRAS* **222**, 655 (1986)
10. Thomas, P. *MNRAS* **292**, 104 (1988)
11. Lauer, T. *ApJ* **292**, 104 (1985)
12. Mushotzky, R., Loewenstein, M., Awaki, H., Makishima, K., Matsushita, K., Matsumoto, H. *ApJ* **436**, 79 (1994)
13. Buote, D. & Canizares, C. *ApJ* **427**, 86 (1994)
14. Buote, D. & Canizares, C. *ApJ* **474**, 650 (1997)
15. Loewenstein, M. and Mathews, W. *ApJ* **373**, 445 (1991)
16. Tammann, G. in Supernovae and their Remnants ed. Cosmovici, (Dor-

drecht: Reidel) (1974)
17. Forman, W., Jones, C., David, L., Franx, M., Makishima. K., and Ohashi, T. *ApJL* **418**, L55 (1993)
18. C. Jones, C. Stern, W. Forman, J. Breen, L. David, W. Tucker Franx, M. *ApJ* **in press**, (1997)
19. Loewenstein, M. et al. *ApJL* **436**, 75 (1994)
20. Matsumoto, H., Koyama, K., Awaki, H., Tsuru, T., Loewenstein, M. & Matsushita, K. *ApJ* **482**, 133 (1997)
21. David, L., Arnaud, K., Forman, W., Jones, C. *ApJ* **356**, 32 (1990)
22. Arnaud, M., Hughes, J. P., Forman, W., Jones, C., Lachieze-Rey, M., Yamashita, K., Hatsukade, I. *ApJ* **390**, 345 (1992)
23. Bahcall, J. and Sarazin, C. *ApJ* **213**, 699 (1977)
24. Mathews, W. *ApJ* **219**, 413 (1978)
25. Navarro, J., Frenk, C., and White, S. *ApJ* **462**, 563 (1996)
26. Fabian, A. *A&AR* **32**, 277 (1994)
27. Kofman, L. Klypin, A., Pogosyan, D. Henry, J. P. *ApJ* **470**, 102 (1996)
28. Jones, C. and Forman, W. , Proceedings of NATO ASI, Cambridge, England, ed. A. Fabian, 1992
29. Fabricant, D., Kent, S., and Kurtz, M. *Ap. J.* **336**, 77 (1989)
30. Fitchett, M. and Webster, R. , *ApJ* **317**, 653 (1987)
31. Briel, U., Henry, J.P., and Bohringer, H. *A&A* **259**, L31 (1992)
32. White, S., Briel, U., and Henry, J.P. *MNRAS* **261**, 8 (1993)
33. Mohr, J., Fabricant, D. & Geller, M. *ApJ* **413**, 492 (1993)
34. Davis, D., and Mushotzky, R. , *AJ* **105**, 409 (1993)
35. Vikhlinin, A., Forman, W., and Jones. C. *ApJ* **435**, 162 (1994)
36. Vikhlinin, A., Forman, W., and Jones. C. *ApJ* **474**, 7 (1997)
37. Geller, M.J., and Beers, T.C. *PASP* **94**, 421 (1982)
38. Dressler, A. and Shectman, S. *AJ* **95**, 985 (1988)
39. Salvador-Sole, E. Sanroma, M., Gonzalez-Casado, G. *ApJ* **402**, 398 (1993)
40. Escalera, E. et al. *ApJ* **423**, 539 (1994)
41. Slezak, E., Durret, F., Gerbal, D. *AJ* **108**, 1996 (1994)
42. Bird, C. *AJ* **107**, 1637 (1994)
43. Richstone, D., Loeb, A., and Turner, E. L. *ApJ* **393**, 477 (1992)
44. Mohr, J., Evrard, A., Fabricant, D., and Geller, M. *ApJ* **447**, 8 (1995)
45. Buote, D. & Xu, G. *MNRAS* **284**, 439 (1997)
46. White, S. D. M. & Rees, M. *MNRAS* **183**, 341 (1978)
47. Carroll, S. M., Press, W. H., & Turner, E. L. *ARA&A* **30**, 499 (1992)
48. Jing, Y. P., Mo, H. J., Borner, G. & Fang, L. Z. *MNRAS* **276**, 417 (1995)
49. Evrard, A., Mohr, J., Fabricant, D., Geller, M. *ApJ* **419**, L9 (1993)

50. Nakamura, F. E., Hattori, M., Mineshige, S. *A&A* **302**, 649 (1995)
51. Kaufman, G. and White, S. *MNRAS* **261**, 921 (1993)
52. Lacey, C. and Cole, S. *MNRAS* **262**, 627 (1993)
53. Cen, R., and Ostriker, J. *ApJ* **429**, 4 (1994)
54. West, M., Jones, C., and Forman, W. *ApJ* **451**, 5 (1995)
55. Mellier, Y., Mathez, G., Mazure, A., Chauvineau, B. & Proust, D. *A&A* **199**, 67 (1988)
56. Colless, M. & Dunn, A. *ApJ* **458**, 435 (1996)
57. Biviano, A., Durret, F., Gerbal, D., Le Fèvre, O., Lobo, C., Mazure, A., & Slezak, E. *A&A* **311**, 95 (1996)
58. Jones et al. *ApJ* **234**, L21 (1979)
59. Bechtold, J., Forman, W., Giacconi, R., Jones, C., Schwarz, J., Tucker, W., and Van Speybroeck, L. *ApJ* **265**, 26 (1983)
60. Grebenev, S., Forman, W., Jones, C., and Murray, S. *ApJ* **445**, 607 (1995)
61. Ponman, T. and Bertram, D. *Nature* **363**, 51 (1993)
62. Cen, U.and Ostriker, J. *ApJ* **417**, 404 (1993)
63. Frenk, C. S., Evrard, A. E., White, S. D. M., & Summers, F. J. *ApJ* **472**, 460 (1996)
64. Miralda-Escude, J. and Babul, A. *ApJ* **449**, 18 (1995)
65. Loeb and Mao, S. *ApJ* **435**, L109 (1994)
66. Evrard, A. *ApJ* **363**, 349 (1990)
67. Evrard, A., Metzler, C., and Navarro, J. *ApJ* **469**, 494 (1996)
68. Schindler, S. *A&A* **305**, 756 (1996)
69. Feretti, L., Dallacasa, D., Giovannini, G., and Tagliani, A. *A&A* **302**, 680 (1995)
70. Ge, J.-P. , Ph.D. thesis, New Mexico Inst. Mining Tech. (1991)
71. Squires, G. et al. *ApJ* **461**, 572 (1996)
72. Squires, G., et al. ApJ 482 648 1997
73. Allen, S., Fabian, A. C., and Kneib, J. P. *MNRAS* **279**, 615 (1996)
74. Daines, S., Jones, C., Forman, W., and Tyson, A. *ApJ* **in press**, (1997)
75. Blumenthal, G., Faber, S., Primack, J., and Rees, M. *Nature* **311**, 517 (1984)
76. David, L., Jones, C., and Forman, W. *ApJ* **445**, 578 (1995)
77. White, D., and Fabian, A. *MNRAS* **273**, 72 (1995)
78. White, S. D. M., Navarro, J. F., Evrard, A. E., & Frenk, C. S. *Nature* **366**, 429 (1993)
79. Walker, T., Steigman, G., Schramm, G., Olive, K., and Kang, H. *ApJ* **378**, 186 (1991)
80. Markevitch, M. & Vikhlinin, A. *ApJ* **submitted**, (1997)

Clusters of Galaxies: A Commentary

Guido Chincarini
Universita` degli Studi di Milano & Osservatorio Astronomico di Brera

The Master said:
At 15 years of age my goal was knowledge
At 30 years of age my thinking was firm
At 40 years of age I had no doubts
At 50 years of age I knew the law of God
At 60 years of age I got to know to listen and when I became 70
I could follow the guidance of my heart without offending the law of Nature.
[A personal translation from " I dialoghi di Confucio (LUN YÜ)" Libro II versetto II.]

I. Introduction

When Herbert Rood and I started to collaborate on the studies of Clusters of Galaxies in 1970 the goals were to some extent very simple. At that time the catalogue by George Abell and the papers in Volume 66 (1961) page 533 - 636 of the Astronomical Journal were carefully studied together with the milestone work by Humason Mayall and Sandage which contained also the list of redshifts of galaxies available at that time. In these early studies we could fully appreciate the need to study the dynamics and the importance of studying the clustering of clusters. This was a matter of debate especially between Abell and Zwicky who had a strong difference of opinion on the argument. While Abell defended the existence of Clusters of Clusters of galaxies, Zwicky would insist that it did not make sense. He could not detect, as he was stressing, a cluster of clusters with the characteristic relaxation which was observed, for instance, in the Coma cluster of galaxies. That was, from the empirical point of view, only matter of semantic and of definition of clustering and implied, however, a different theoretical approach to cluster formation and evolution. We know, nowadays, that clusters are clustered without forming relaxed systems of clusters.

We did not really know whether or not it made really sense to apply the Virial Theorem to Clusters and yet it was more or less the only sensible investigation to be done considering the need to understand the dynamics and to investigate the problems raised by the early investigation by Zwicky and his considerations about the unseen mass as discussed also in his book Morphological Astronomy (1957). Indeed it is in part thanks to that and to long discussions I had with Zwicky that we started to ask about the size of the Coma Cluster and about the discrepant redshifts. It is thanks to the follow up of redshift surveys aimed to estimate the size of the Coma Cluster of galaxies and to explain the discrepant redshift in the Seyfert Sextett that we came to the discovery of the distribution of galaxies on Large Scale and to the definition of the Voids. Later on the existence of the hot intracluster medium addressed part of my work to the interaction of the intracluster medium with the galactic interstellar medium and that is how the work at Arecibo started in collaboration with Giovanelli and Haynes. Meanwhile Neta Bahcall investigated and properly illustrated the reality of clustering of clusters indicating its importance in cosmology.

An extensive discussion on Clusters of Galaxies was held also in Erice in 1979 following the success of HEAO-2 (Einstein Observatory) launched at the end of 1978.

In parallel the theory made a tremendous progress both in the understanding some of the processes of cluster formation and relaxation, in determining in a simple and efficient way the equation of hydrostatic equilibrium (Cavaliere and collaborators) and in detailing the mechanism of emission and galaxy - galaxy and galaxy - intracluster medium interaction. A fundamental review on clusters of galaxies came out in 1986 in Reviews of Modern Physics by Sarazin and that became a reference point for planning research and educating students in the field. Since this early work the study of Clusters of galaxies became, thanks also to the observations in the X-ray band and at radio wavelengths, a very mature field of endeavor where clusters of galaxies were studied, both observationally and theoretically, in all of their aspects and in relation to various branches of cosmology, astronomy and astrophysics.

In the last few years the role of clusters became more and more fundamental in the detection of high z galaxies via their lensing properties (see, for instance, the recent detection by Marijn Franx of the galaxy at z = 4.92 lensed by the cluster Cl 1358+62), in the understanding of formation and evolution and in the determination of the characteristics of the Large Scale Structure at various z. Indeed Clusters have an unique role in Cosmology and Astrophysics and constitute a fantastic tool to understand our Universe.

II. Why do we need catalogues of clusters of galaxies

The Universe is a fantastic large laboratory where we can not control experiments, but where past and present can be observed. It is therefore a matter of being able to reconstruct evolution with models and theories based on the observed facts rather than build and experiment to follow the evolution of a phenomenon in its details. The observation, therefore, either as simple as the darkness of the sky or as sophisticated as those requiring the most advanced facilities to observe extremely high redshift galaxies, is the very fundamental basis of our quest. The interplay between theory, prediction and facts being most rewarding toward the building of our knowledge about the architecture of the Universe, its formation and evolution. To be specific with the argument I am dealing with, and however this applies to almost any category of cosmic objects, we need to discover where clusters of galaxies are if we want to study their distribution in space and time, their characteristics and infer form these data information on their formation, evolution and fate.

As recently as two years ago, however this happened in the past as well, I was told that it was'nt necessary to build a satellite to look for distant, high z, clusters since there were indication, we now all agree those were false indications, that X-ray clusters do not exist (or more precisely are extremely rare) even at moderately high redshift, $0.6 < z < 1.2$. Not much later I asked a post-doc, expert in cluster simulations, whether or not it was worth to spend about 100 Million dollars to build a satellite to search, among other things, for distant clusters of galaxies. I admit he had no much time to think about, in any case the answer was that simulations already show that at fairly large redshifts, let' s say $z > 1$, clusters are very rare so that it probably would have been a wasting of money to look for them at such an high cost. I obviously made the story very simple and extreme; we should not, however, forget what we learned since Galileo: to understand Nature we must ask Nature about its content (that is we must observe and interpret it) and we will, soon or later, get the answer.

Undoubtedly clusters of galaxies form and evolve, we must find out when and how, and to do that we need to have deep surveys. It has been now shown that the tentative negative evolution of clusters as detected in the EMSS was overestimated and indeed we have no sign of evolution in the recent determination of the X-ray Luminosity Function estimated in the redshift range $0 < z < 1$, Rosati (1997). Clearly massive clusters might evolve faster and large area surveys are needed to evidence, or negate, that. At the moment, however, we have no evidence of very fast evolution and deep and large areas surveys are needed.

Surveys in the visible light to detect clusters are seriously limited and difficult. The detection of a distant cluster over a background of galaxies is progressively confused at larger z by the increase of the background and the simultaneous decrease of the number of cluster galaxies which is determined by the very tip of the Luminosity Function, Cappi et al. (1989). At these wavelengths we are limited, based on the experience of the survey conducted by Postman et al. (1996), to compile more or less complete sample up to $z \sim 1.2$. Quite likely this will be the most distant clusters we will be able to detect with the EIS survey at

ESO and other similar surveys which are now in progress. Such samples, furthermore, will be selecting mainly massive clusters and present a considerable degree of contamination due to projection effects.

Radio observations are also very useful and optical follow up will certainly enable cluster detection. I see that, however, only as a good complement to cluster searches since it is not clear, yet, how biased such a sample could be and how far we could search. The detection of a cluster via radio observations requires a radio galaxy moving, with its bent ejects, in a fairly dense ICM, a classical example being NGC 1265 in the Perseus cluster. The detection occurs through the observation of a secondary phenomenon which is fascinating under many aspects and which is however not a primary cluster indicator as could be the over density of galaxies or the signature of the hot ICM. Such signature is clearly evident in the X-ray band where the gas emits, quite clearly, brehmstrahlung radiation. Indeed having enough collecting area, spectral resolution and spatial resolution we could not only detect clusters but also estimate their redshift, temperature, luminosity and mass. In the future and when the Planck satellite will fly, microwave observations will clearly supply fundamental catalogues of clusters because of their detection via the Sunayev-Zeldovich effect. These observations and catalogues, however, will still require observations in the X-ray band and likely at visible wavelengths so that it does not pay to delay or cancel searches in the X-ray, on the contrary these should be planned as soon as possible.

As I said to be very efficient and to be complete we must have high spatial resolution, good spectral resolution and large field of view in order to make the survey cheap in time and cost. Ultimately we might have to push the observations beyond the X-ray background flux limit in order to detect forming clusters and see if the expected merging of substructures, assuming they are already lighted, is at work. If gas is present we should be able to detect at least the shock waves originated by collisions and interactions. Such endeavor, but that is a requirement also for the study of high z galaxies and distant star forming galaxies, will ultimately ask for very large efficient area wide field of view X-ray telescopes which we should be studying and planning from now. The technology is partly at hand and could be acquired in the coming years, it is only matter to work in view of the future and the future is now.

A first step in this direction consists in a wide angle X-ray survey that we propose to carry out at soft energies using a wide field X-ray telescope. I mentioned this project not only to advertise it, but also because the idea was triggered many years ago by Riccardo Giacconi who was pursuing it in relation to his research and deep scientifical interest in clusters of galaxies. While in Baltimore he was working with Burrows and Burg on the design of a wide field X-ray optics and now and then I would see he was upset (I doubt I ever saw Riccardo depressed) because they could not get the solution. But then, one day, he called me in Merate because they succeeded in the design (and later found out, as often happens in science, that earlier Werner had found a similar solution) and since then we started to improve our technology in order to make a prototype and possibly get a mission funded.

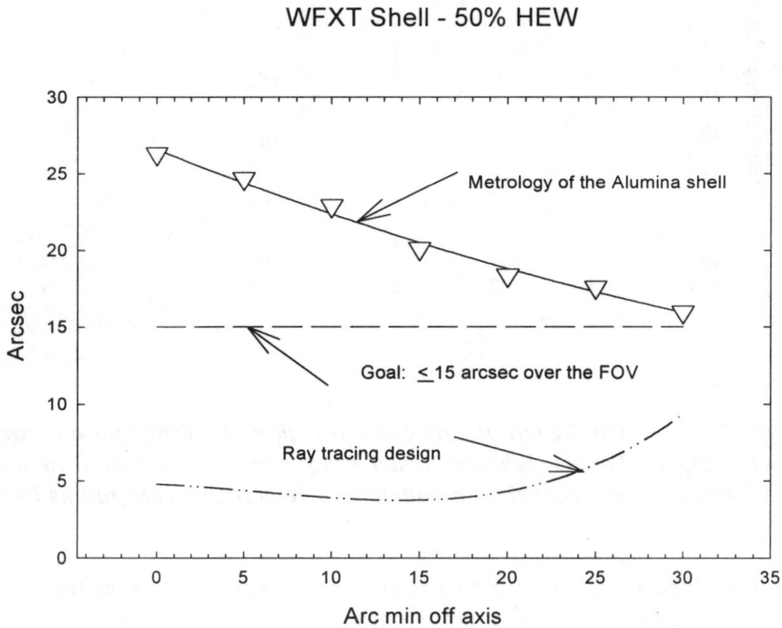

Figure 1. Result of the first 60 cm prototype Alumina shell

The new ceramic material, Alumina and Silicon Carbide, not only are very rigid compared to Nickel, but also much lighter. In our case this allows to double the number of shells, from about 10 as it was planned for the Telescope made by Nickel electroformed mirror shells to about 20. That is we can now reach, without increasing the weight, an efficient area of the Telescope, that is without accounting for the detector sensitivity, of about 350 cm^2. That is very important to reach the faint flux from high z objects, Figure 2.

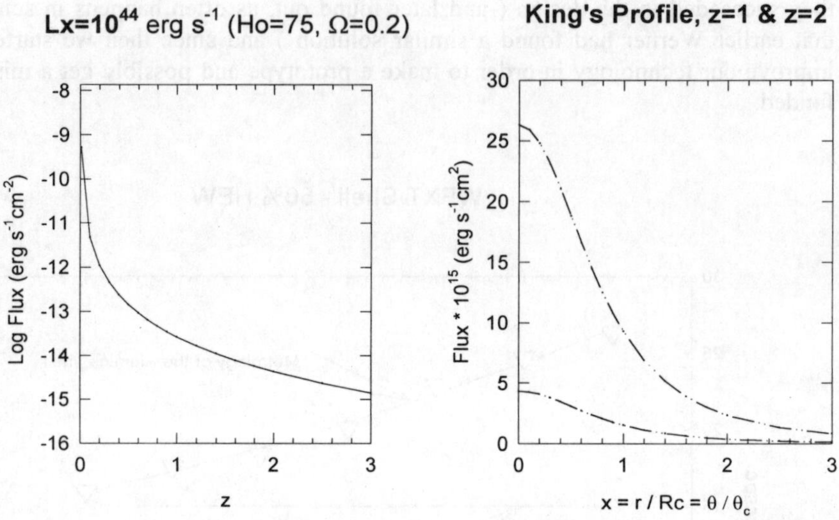

Figure 2. ***On the left the flux as a function of redshift for a cluster with Lx = 10^{44} erg s^{-1}. To the right the cluster King's profile observed from a cluster at z =1 and z = 2 respectively. I assumed as a reference a core radius Rc = 0.25 Mpc.***

Here I used the King' s profile as a reference. It really does not matter, for what I am discussing here, whether or not the distant clusters depart from it. A cluster at z =1 will have a core radius, θ_c, of 36.6 arcsec (33.1 at z=2) in the geometry we used so that the central element of resolution (assuming for the instrument a resolution of only about 15") will have a radius about x = $15/\theta_c \sim 0.41$ with a flux of ~ 1. 10^{-14} erg $s^{-1}cm^{-2}$ (1.74 10^{-15} erg $s^{-1}cm^{-2}$ for z=2 within x ~ 0.45). We can do an excellent and sensitive survey.

We now have a first prototype which is rather satisfactory, we demonstrated that the design and the technology work and we are confident that we can improve the quality to the specified goal as requested by the science drive. This first step would open the way to a much more ambitious project with the goals outlined above.

III. Where we are going.

It is only matter to look at the recent literature to know what is going on and to realize how active is the field and how many new and old ideas are getting planned and implemented. The ROSAT survey is poring a wealth of data which are being used both to ensemble complete sample of nearby and distant clusters aimed to estimate the Luminosity Function, its evolution and the Large Scale distribution on clusters. Such complete samples will allow a detailed study of physical properties and a better understanding of cluster substructure evolution and allows firm constraints on the value of Ω. But one of the most fascinating endeavor toward which a large effort is already initiated is the study of the mass distribution on large scales, and especially about clusters and galaxies, using the weak lensing effect. The very first ideas on lensing, it is worth remembering it, came mainly from the work by Einstein (1936), Zwicky (1937) and later by Zwicky in Morphological Astronomy (1957) and in more details from a paper by Bourassa, Kantowsky and Norton in 1973. The field developed tremendously following the first detection and it is now becoming a beautiful, albeit complicate, machinery thanks to a) the fast improvement in the instrumentation where high resolution large collecting area and field of view are required, b) the highly sophisticated analysis and the advanced theoretical interpretation. Indeed we are moving toward the making of "gravitational" instrumentation, that is instrumentation which is capable of mapping the distribution of mass independently of its Nature, whether barionic or else. Soon enough we should be preparing for programming observations with the NGST.

AXAF will undoubtedly supply us with unprecedented data also on clusters of galaxies and quite likely will open unforeseen windows of knowledge. Following that, one of the most ambitious project which is already under study is the plan to put in orbit the High Throughput X-ray Telescope (HTXS) which with its high spectral and reasonably high spatial resolution, very large efficient area will allow to study in very high details the physics of the hot cluster gas. It is for instance tremendously important to have the capability, as pointed out by Bregman, to estimate with the accuracy of the order of 100 km/sec the motion of the gas in the outskirts of the cluster in order to properly estimate the contribution to the kinetic energy, have better estimate of the cluster mass (the hydrostatic approximation must be modified) and track down the motion of the components at different temperatures. Abundance will be estimated with high precision.

One notice, however, in this field and other field of astronomy that quite often, even if for good reasons, the role of magnetic fields is disregarded. These can not be measured directly, are generally expected to be very weak, both in galaxies and in the intracluster medium, so that any effect might be insignificant at least to the first order. On the other hand the more we refine our observing tools and improve the accuracy and resolution of the data, the more we might get close to the threshold where the presence of magnetic fields might change, or improve, the interpretation of the phenomenon we are observing. Valinia et al. (1995), for instance, show that the magnetic field anchored to the gravitational potential of galaxies can sensibly influence the temperature distribution in the ICM.

IV. More

The recent literature does not deal much with the use of cluster of galaxies to measure the geometry of the Universe.
We all remember the Abell's Luminosity Function and his proposal to use the Luminosity of the break, the point where the slope changes, as a distance indicator. That did not work out. In preparation of my Erice lecture in 1979, I spend a considerable amount of time, partly in collaboration with Materne, to reduce all the published counts on clusters of galaxies to see, among other things, if I could say something about the use of the core radius as a standard length for cosmological tests. The paper by Chincarini and Materne referenced (in preparation) in that article was never published not only because the two of us got busy with something else, but above all because I felt I had not full control of that parameter and the data were not yet as clean as they should be. Something better could be done, perhaps, using the X-ray images. That might be worth trying as soon as more complete and accurate data are becoming available.
It is very well known that each time we did try to use standard candles, or metric lengths, to estimate the geometry of the Universe (Sandage dedicated a large fraction of his life to this problem) we discovered that we had first to understand how objects evolve with time since evolution dominates at high redshifts. Indeed the understanding of the evolution of cosmic objects at high z became one of the most important and fascinating problems of today cosmology. It seems evident, however, that evolution of Luminosity, obviously related to the time scale of stellar evolution and eventual environment effects, act on shorter time scale than the metric (as referred to the structure of the cosmic objects) since after the objects reach equilibrium the time needed to change the structure, dynamics time, is generally larger. We also know that the cooling time of the cluster emission is rather large so that once in equilibrium the luminosity would change only on a very large time scale. Other phenomena and interactions, as shown by the high percentage of substructures and merging, could also play the role of the evil and , of course, it all depends on the time of cluster formation and on how the gas formed and was heated. In a few words clusters, and their medium, could also be strongly affected by evolution in the observable range of redshifts (but it must be proved first).

But we can look at it from a different point of view. Let's do our best to use clusters to measure the geometry of the Universe and study related cosmological problems, then either we succeed or we fail because of evolution or other phenomena and we will have new knowledge on these. Even if we fail the main goal we will learn a lot on important issues. The goal then could be simply to investigate the structural properties at various z and find out what happens. The suggestion is then to observe a sample of galaxies in a reasonably large range of redshifts having as a goal, as it has been done for galaxies particularly by Sandage and Perelmuter (1990) and more recently by Pahre et al. (1996), the Tolman test on the expansion of the Universe.

The Petrosian metric radius is defined as that radius where the ratio of the average surface brightness ratio within that radius to the Surface Brightness (SB) at that radius has a particular value, $\eta(r)$. For a cluster assuming for simplicity $\beta=2/3$ in the King's profile and $x = r/Rc$, the radius normalized to the core radius, we have:

$$\eta = \frac{2\left[1-\left(1+x^2\right)^{-0.5}\right]}{x^2}\left(1+x^2\right)^{1.5}$$

This, being a ratio of SBs, is independent of redshift, absorption and luminosity evolution as long as the variations of SB due to evolution are independent of the radius. It is therefore a very powerful tool.
$\eta(r) = 4.8$ for $x=3$ and $\eta(r) = 3.09$ for $x = 2$. That is taking the range of core radii measured by Jones and Forman (1984) we will be able to measure the running of the <SB> as a function of redshift to medium-large distances with high accuracy, Figure 2.
Moreover, since in an expanding Universe $\langle SB\rangle \propto (1+z)^{-4}$ compared to $\langle SB\rangle \propto (1+z)^{-1}$ for a non expanding Universe, the test is very robust limiting the analysis to a relatively limited range of moderately low redshif clusters and avoiding therefore complication due to evolution.

V. Final remarks.

Astronomy is undoubtedly changing in the sense that the need of managing large projects and large amount of data is becoming essential. It is not something new, large and fundamental projects have been handled in the past, see the catalogue of

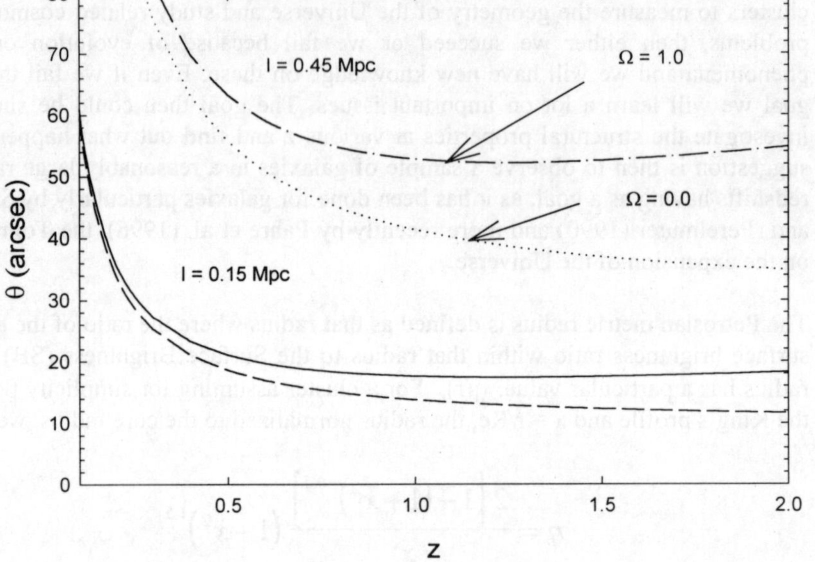

Figure 3. Angle subtended by a linear size of 0.15 and 0.45 Mpc as a function of redshift.

galaxies by Shane and Wirtanen to name one among a few. This effort lasted years and I still recall seeing, at the time I was a post-doc at Lick Observatory, Mr. Wirtanen very busy in measuring the position of galaxies on the large photographic plates obtained with the 20-inch Carnegie Astrograph of Lick. What is new is that now we are dealing with many large projects acquiring very accurate data, where data flow rate is very high thanks to the development of technology and with scientists that feel the need to have the scientific results in a very competitive way and in a short time. The future of any research is always more tantalizing than the past so that we are eager to pass to the next challenge.

To a rather large extent the research must, therefore, be based on teams which collaborate in a very effective way and which must be formed keeping the science goals and deadlines in mind. This must be kept in mind when we educate students, hire people and set up facilities. In a way the approach to science is more challenging with more degrees of freedom based partly on a different thinking and, perhaps, with more fun because of the faster rate of knowledge albeit the danger of big blunders. We must remember, however, that this is only part of the story. Even if a large project takes off having a well defined science goal in mind and must be carried out by creative minds, we always have the need to have also those original thinkers and very creative persons that uneasily adapt to a large team and rather work as singles, this being especially true for theoreticians, or as part of small groups. It could be a difficult equilibrium hard to evaluate but again very important and challenging. Having this at hand we can tackle the future with collaborative competition and solve many of the questions we have today, create many ideas and new projects at an higher level of knowledge and fully enjoy the process of getting to know first new wonders of Nature. Nature has the capability to satisfying every person of good will so that we do not have to worry if we have as a goal a genuine desire of knowledge rather than power and manipulation over others.

And since this is a contribution to honor Riccardo Giacconi I will simply refer to Figure 1. He triggered this idea of a wide field optics and I am glad to have shown that at the Observatory of Brera we did pick it up, had many interactions with him and did a lot of work to make it work. Now we hope to fly the optics because of the deep interest we have in the science we can do with it since I believe this could make a breakthrough in the understanding of clusters of galaxies, in the study of the Large Scale Structure and in many other fields of astronomy. This however much depends on the financing agencies and on those scientists who want to keep control of science. That not always goes in the direction of the best science. What I know is that if it does not happen now it will happen soon or later because the steps of knowledge are, in some sense, due in a well defined sequence. If it will not be us, it will be somebody else since it might be true, even if I do not like it, that ultimately all scientists are anonymous and irrelevant in the sense specified by Lewis Wolpert (1996).

REFERENCES

Bourassa, R.R., Kantowski, R. and Norton, T.D., 1973, Ap.J. 185, 747

Cappi, A., Chincarini, G., Conconi, P., Vettolani, G., 1989, A&A, 223,1

Einstein,A., 1936, Science, 84, 506

Jones, C. and Forman, W., 1984, Ap. J. 276,38

Pahre, M.A., Djorgovski, S.G. and Carvalho, R.R., 1996, Ap. J. 456, L79

Postman, M., Lubin, L.M., Gunn, J.E., Oke, J.B., Hoessel, J.G., Schneider, D.P. and Christensen, J.A., 1996, A.J., 111, 615

Rosati, P., 1997, ***X - RAY SURVEYS WORKSHOP***, Postdam June 18 - 20 1997

Sandage, A. and Perelmuter, J.M., 1990, Ap.J., 350, 481

Valini, A., Tajima,T., Matsumoto, R., Makishima, K. and Shibata, K., 1996, in ***Clusters, Lensing, and the future of the Universe***, Edited by V. Trimble and A. Reidenegger, ASP series Vol. 88.

Wolpert,L., 1996, in "***The future of science has begun, ….***", Prodeedings of the V International Conference, Fondazione Carlo Erba.

Zwicky, F., 1937, Phys. Rev. Letters, 51, 290

Zwicky, F., 1957, ***Morphological Astronomy*** (Berlin: Springer-Verlag), page 215.

Evolution of the Abundance of X-ray Galaxy Clusters out to $z \sim 1$

P. ROSATI

European Southern Observatory, Karl-Schwarzschild Str. 2
D-85748 Garching, Germany
E-mail: prosati@eso.org

I review recent observational progress made in the search for and study of distant galaxy clusters, with particular emphasis on the evolution of the abundance of X-ray clusters out to $z \sim 1$. The very existence of massive collapsed objects at high redshift challenges theories of structure formation. New deep X-ray surveys are now revealing that the bulk of the population of galaxy clusters is not evolving since at least $z \simeq 0.8$, with possible mild evolution limited to only the most luminous, presumably most massive systems. Thus far, a well defined sample of high redshift ($z \gtrsim 1$) clusters has been difficult to assemble and represents one of the most challenging observational tasks for the next decade. The next generation of X-ray instruments in conjuction with the new imaging and spectroscopic capabilities offered by 8-meter class telescopes will open up a new window on the physical understanding of the evolutionary processes involved.

1 Introduction

Clusters of galaxies are powerful tools for addressing a broad range of astrophysical and cosmological problems. The temporal evolution of the cluster abundance is a sensitive probe of the underlying cosmological model and can be used to determine the value of the cosmological density parameter, Ω_0, with a minimum of assumptions (e.g. Eke *et al.* 1996 [12], Bahcall *et al.* 1997 [1]). The spatial distribution of galaxy clusters is a direct tracer of the large scale structure and reflects the spectrum of initial density perturbations. Clusters themselves serve as laboratories to investigate galaxy evolution in a variety of rich environments over a substantial fraction of the Hubble time. The distribution and the evolution of the chemical abundance (particularly iron) of the intra-cluster medium, directly measured with X-ray observations, can be used to trace the baryon circulation and the star formation history in the Universe (e.g. Renzini, 1997 [24]). In addition, the method of using clusters as 'lensing telescopes' is now well established for studying both the mass distribution in rich clusters and for constraining the properties and distances of faint background galaxies magnified serendipitously in the cluster field (e.g. Luppino & Kaiser, 1997 [19]). Thus, the construction of large homogeneous samples of galaxy clusters, spanning a wide range of redshifts, is a fundamental goal of observational cosmology.

2 Searches for high-z clusters

The constraining power of the abundance of massive clusters and their properties on cosmological models becomes increasingly important at larger redshifts. Unfortunately, finding clusters at cosmologically interesting lookback times, i.e. $z > 0.5$, not to mention defining a complete sample, is a time consuming and difficult task. As a result, the much-needed observational constraints for theories of cluster formation have not been forthcoming. In accomplishing this task the selection methods which are briefly described below and the observational strategy have to be judiciously planned.

2.1 *Optical and near-IR Surveys*

Significant progress has recently been made in optical surveys of distant galaxy clusters with the advent of large format CCDs (Postman *et al.* 1996 [23], Lidman & Peterson, 1996 [18], Zaritsky *et al.* 1997 [32]). These surveys utilize objective finding algorithms on homogeneous CCD imaging data and have shown that rich clusters out to $z \sim 1$ can be identified, although at these redshifts it becomes extremely difficult to detect enhancemnets in the galaxy surface density against the overwhelming field galaxy population. In order to increase the contrast of the cluster against the field multi-color and narrowband imaging techniques are extremely useful.

At very large redshifts ($z>1$) imaging in near-infrared bands is essential to beat down the k-correction which dims significantly at optical wavelengths the dominant population of early-type cluster galaxies. Near-IR color techiniques can greatly facilitate the serendipitous detection of clusters beyond $z = 1$ (e.g. Stanford *et al.* 1997 [30]), a method which will come increasingly popular with the advent of large format near-IR imagers. Moreover, studies of spectrophotometric properties at optical *and* near-IR bands provide insights on the evolution of cluster ellipticals and their early formation [30]. A method which has had some success for very high-z cluster identification is the one of targeting high redshift radio-loud AGN (e.g. 3C 324 at z=1.2, Dickinson 1996 [9]). Half a dozen of these candidates have been discovered so far at redshifts $1<z<2$ and studied in near-IR bands, however, most of these systems still need spectroscopic confirmation with 8-meter class telescopes. Unfortunately, this method, although convenient from the observational point of view, does not seem to be suitable for assembling a *representative* samples of high-z clusters and for mapping the large scale structure.

At even larger redshifts ($z > 2$) the search for 'protoclusters' has been pursued with narrow-band techniques, particularly Lyman α in emission (e.g Pascarele *et al.* 1996 [22]). More recently, the first serendipitous concentration of galaxies at $z \simeq 3$ has been identified and spectroscopically confirmed among a large population of primeval galaxies selected with the Lyman break technique (Steidel *et al.* 1997 [31]). It has been argued that this structure may represent the progenitor of a rich, present-day Abell cluster.

2.2 X-ray Surveys

Although optically selected clusters, first and foremost the Abell catalogue, have been the basis of innumerable studies in the low redshift Universe, optical surveys are well known to have serious statistical shortcomings, such us contamination by chance coincidendes of unvirialized systems, which becomes increasingly important at high redshifts. The main limitation inherent in the optical selection, however, is the difficulty in defining the completeness of the sample due to the lack of a physical cluster observable parameter which can be used to define the survey limit. The cluster richness is not a well-defined observable and volume completeness is not attenable at high redshifts. These difficulties can be significantly alleviated by selecting clusters via their X-ray emission which provides a unique tracer of cluster potential wells. Pure X-ray selection leads to a selection function which can be modelled in a relatively straightforward way, being essentially that of a flux limited sample. As a result, distribution functions of observables, such as the number counts, the X-ray Luminosity Function (XLF) and the redshift distribution, can be directly compared with theories of structure formation.

Until very recently, the only available sample of X-ray selected clusters at high redshifts was the one compiled from the Einstein Extended Medium Sensitivity Survey (EMSS) (Gioia *et al.* 1990[13], Henry *et al.* 1992[15]). For years this survey has been the basis for several follow-up studies (e.g. lensing (Gioia & Luppino 1994 [14]), CNOC survey (Carlberg *et al.* 1997 [4])). The unprecedented spatial resolution provided by the ROSAT-PSPC detector, coupled with its high sensitivity and low background, make clusters high contrast, extended objects in the X-ray sky, thus providing an excellent opportunity to efficiently construct large cluster samples which are deep enough to probe cosmological distances.

3 The ROSAT Deep Cluster Survey

In this spirit, we embarked some time ago on the ROSAT Deep Cluster Survey (RDCS), a project aimed at constructing a large homogeneous sample of distant galaxy clusters selected solely on the basis of their X-ray properties (Rosati *et al.* 1995 [26] (R95)). A full discussion of the selection technique for the RDCS sample can be found in R95. Cluster candidates have been selected from a serendipitous search for extended X-ray sources in deep pointed observations drawn from the ROSAT-PSPC archive. A wavelet-based technique (Rosati *et al.* 1993 [25]) is used to detect and characterize X-ray sources. The multi-scale analysis performed by the the wavelet algorithm, besides reducing biases towards low surface brightness features, is particularly efficient at reducing the effects of confusion which is the main limiting factor in deep X-ray imaging. Cluster candidates are selected as those sources extended at 99% confidence level (via a statistical comparison with the local PSF), with flux $f_{-14} = f_X[0.5-2.0]\mathrm{keV}/(10^{-14}\ \mathrm{erg\ cm^{-2}\ s^{-1}}) > 1$ and off-axis angle $\theta < 18$ arcmin. This selection technique yielded 160 cluster candidates over an area of about $\sim 50\ \mathrm{deg}^2$, drawn from 180 X-ray fields scattered across the two galactic caps ($|b| > 20°$). The selection function of the survey and the completeness level at different fluxes are well understood.

A large optical identification program of RDCS cluster candidates has been carried out over the last two years. The imaging survey in V and I bands is now 90% complete and has revealed a very high success rate of identification, with about 115 new clusters (or groups) confirmed to date, i.e. displaying a significant overdensity of galaxies around the peak of the X-ray emission. The spectroscopic follow-up work has secured 95 cluster redshifts so far, spanning the range 0.045–0.83. A significant fraction of the newly discovered clusters lie at high redshift, about one-third at $z > 0.4$ and a quarter at $z > 0.5$ (two of which are at $z > 0.8$), making the RDCS the largest sample of spectroscopically confirmed distant clusters compiled to date.

3.1 The evolution of the X-ray Luminosity Function

The study of the evolution of the cluster XLF began with the compilation of the EMSS sample [13]. ¿From a complete subsample of 67 EMSS clusters at $0.14 < z < 0.6$ Henry *et al.* (1992) found evidence of a statistically significant negative evolution of the XLF at $z \gtrsim 0.3$. Earlier claims of strong evolution at lower redshifts (Edge *et al.* 1990 [11]) have been ruled out by a large compilation of X-ray clusters from the ROSAT All-Sky Survey (Ebeling *et al.* 1997 [10]). Covering a very large solid angle (735 deg^2), but being relatively shallow, the EMSS sample can probe only the bright end of the XLF and cannot provide strong constraints on the evolution of the XLF at higher redshifts since only 6 EMSS clusters lie at $z > 0.5$. On the other hand, the RDCS reaches considerably fainter fluxes, but covers a much smaller solid angle and therefore probes a complementary region in the $L_X - z$ plane to the EMSS. Surveys similar to the RDCS, utilizing the PSPC archival data but somewhat different selection techniques, are now well underway (RIXOS (Castander *et al.* 1995 [5]), SHARC

(Collins *et al.* 1997 [7]), WARPS (Scharf *et al.* 1997 [29])).

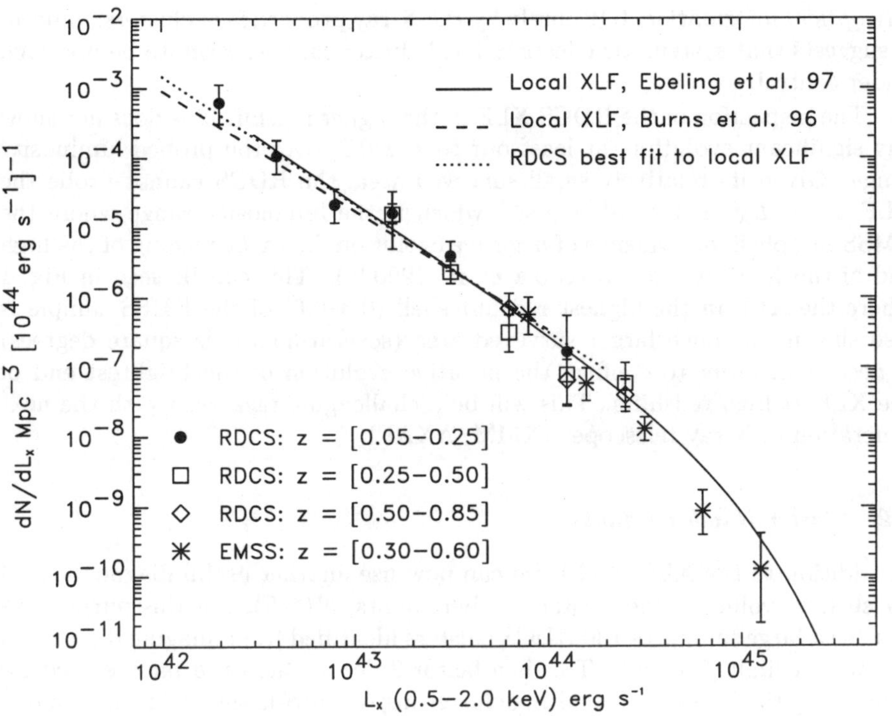

Figure 1: The X-ray Luminosity Function for the RDCS sample in three redshift shells. Also plotted two independent determinations of the local XLF from ROSAT All-Sky Survey data and the EMSS XLF at high redshfts (Rosati *et al.* 1997 [27]).

The imaging and spectroscopic identification of the RDCS cluster candidates is complete down to $f_{-14} = 4$, hence a complete flux limited subsample of 70 clusters with measured redshifts can be used to derive the XLF over a large redshift range (Rosati *et al.* 1997 [27]). In Fig. 1 we show the derived XLF in three redshift shells, [0.045–0.25], [0.25–0.5], [0.5-0.85]. The luminosities in the observed [0.5–2.0] keV band range from 3.5×10^{44} to 1×10^{42} erg s^{-1} and thus probe the region of moderately rich clusters to poor groups. We point out the excellent agreement between the local XLF of the RDCS, i.e. in the lowest redshift bin, and two independent determinations of the local XLF from Ebeling *et al.* 1997 [10] and Burns *et al.* 1996 [3]. Both these surveys use ROSAT

All-Sky Survey data but completely different selection techniques. The agreement found between these independent determinations is by no means trivial since the RDCS is the only sample of the three whose selection *is not driven by any optical information*, but purely by the X-ray properties of clusters. Indeed, it suggests that systematic effects in local cluster samples seem to be now well under control.

The inspection of the RDCS XLF in the higher redshift bins does not show any significant evolution, at least out to $z \simeq 0.8$, over the probed luminosity range. Given its relatively small surveyed area, the RDCS cannot probe the XLF *above* $L_X \simeq 3 \times 10^{44}$ erg s^{-1} which is the luminosity range where the EMSS sample finds evidence of negative evolution, i.e. a steepening of the high end of the XLF at $z \gtrsim 0.3$ (Gioia *et al.* 1990 [13]). This can be seen in Fig. 1 where the XLF in the highest redshift shell (0.3–0.6) of the EMSS sample is also shown. A much larger surveyed area (several hundreds square degrees) is needed in order to confirm the negative evolution of the brightest end of the XLF at high redshifts. This will be a challenging task even with the next generation of X-ray telescopes (XMM, AXAF).

3.2 *Cluster Number Counts*

In addition to the XLF, $N(L)$, we can now use another useful diagnostic tool for cluster evolution, the cluster number counts, $N(>S)$. For this purpose we can use a large sample of 130 RDCS clusters identified from imaging data down to the flux limit $f_{-14} = 2$. This is a factor 2 deeper than the sample used for computing the XLF. The resulting cluster $\log N$–$\log S$ is shown in Fig. 2 along with previous determinations. This observed curve can be integrated to provide the cluster contribution to the soft X-ray background, $I_{CXB}(0.5-2\ \mathrm{keV}) \simeq 1.5\,\mathrm{keV\,cm^{-2}s^{-1}sr^{-1}}$ down to $f_{-14} = 1$, and is found to account for about 10% of the observed soft extragalactic X-ray background.

Given the accurate determination of the local XLF (Fig.1), we can now compare our cluster $\log N$–$\log S$ with the predicted number counts for a non-evolving XLF. In Fig. 2 we show that these predictions, obtained by integrating the BCS XLF over the luminosity range $1 \times 10^{42} - 5 \times 10^{45}$ erg s^{-1} out to $z = 1.1$ for two values of q_0, are very close to the observed counts. A significant steepening (or flattening) ($|\Delta\alpha| > 0.2$) of the faint end of the XLF at $z > 0.3$, not directly probed by the RDCS, would lead to an overprediction (or underprediction) of the observed faint number counts outside the error bars, for any value of $0 \leq q_0 \leq 0.5$. This further strengthens the evidence that the XLF does not evolve, within the present uncertainties, over a wide range of luminosities ($2 \times 10^{42} \lesssim L_X(\ \mathrm{erg\ s^{-1}}) \lesssim 3 \times 10^{44}$). An independent study by

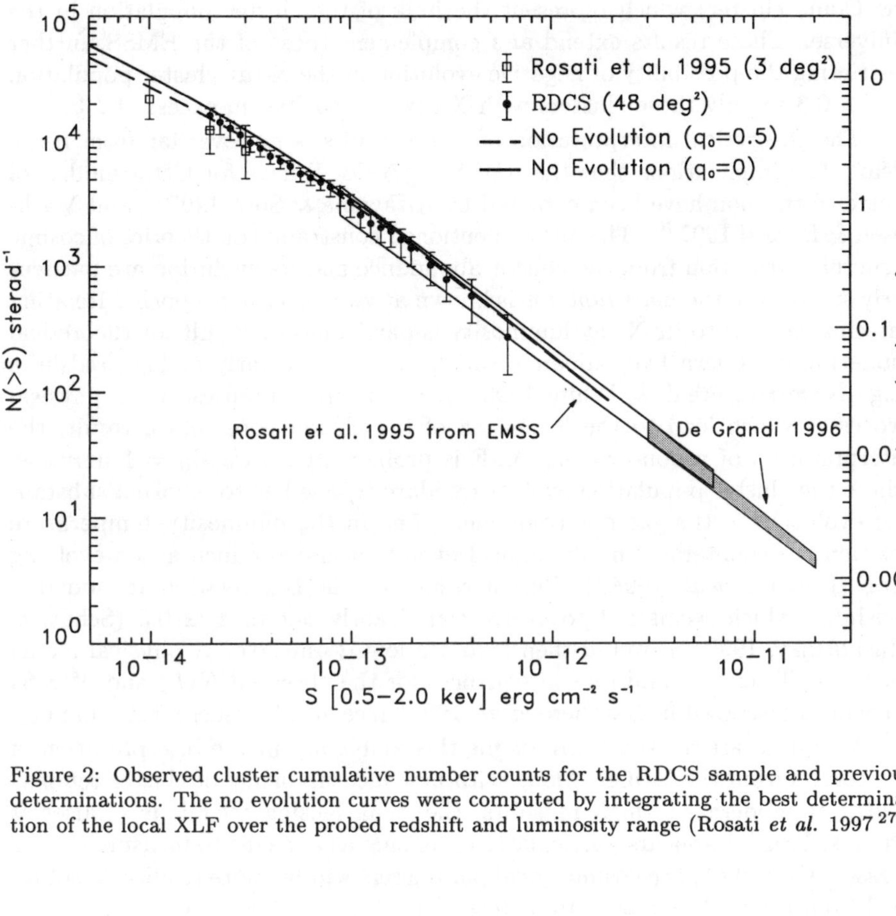

Figure 2: Observed cluster cumulative number counts for the RDCS sample and previous determinations. The no evolution curves were computed by integrating the best determination of the local XLF over the probed redshift and luminosity range (Rosati *et al.* 1997 [27]).

Collins *et al.* [7] also found no evidence of significant evolution of the cluster abundance at intermediate X-ray luminosities.

¿From these results a better observational picture of the evolution of X-ray clusters is now emerging. The large size of the RDCS sample and its wide redshift baseline provides an accurate assessment of the evolution of the space density of galaxy clusters over a considerable range in X-ray luminosities and out to the highest redshifts probed so far in X-ray surveys. By combining the RDCS XLF with an improved determination of faint cluster number counts, we find that the XLF does not evolve significantly at $z \lesssim 0.8$, from luminosities typical of poor groups up to the local $L^*_{[0.5-2]\mathrm{keV}} \simeq 4 \times 10^{44}$ erg s^{-1} (roughly the Coma cluster) which represent the bulk of the cluster population in the Universe. These results extend and complement those of the EMSS, further restricting the possibility of negative evolution in the X-ray cluster population at $z \gtrsim 0.3$ to only those clusters with X-ray luminosities in excess of L_X^*.

The theoretical interpretation of these results is however far from being clear. The implications that the RDCS $\log N$–$\log S$ have for CDM models of cluster formation have been explored by Kitayama & Suto 1997 [17] and Mathiesen & Evrard 1997 [20]. The above mentioned constraints on theories of cosmic structure formation from the cluster abundance and its evolution are particularly stringent *if* the *mass function* is known *at varying cosmic epochs.* Relating the cluster mass to its X-ray luminosity is particularly difficult for theoretical models and modern hydrodinamic simulations, particularly at high redshifts (e.g. Navarrro, Frenk & White 1995 [21]), due to uncertainties in the physical processes which lead to the formation of the cluster core. As a result, the interpretation of a non-evolving XLF is problematic. In a $\Omega_0 = 1$ universe, where the cluster population evolves rapidare ly, one has to invoke a substantial evolution in the gaseous component, i.e. in the luminosity–temperature relation, to counteract the dynamical evolution and produce a non-evolving XLF (Borgani *et al.* 1998 [2]). Recent results on the L-T relation at moderate redshifts, which seems not to evolve significantly out to $z \simeq 0.5$ (Scharf & Mushotszky, 1997 [28]), would then favour a low-Ω universe. A physical model for the L–T relation and its consequences for the observed $N(L)$ and $N(>S)$ have been discussed in Cavaliere *et al.* 1997 [6] (see also Cavaliere, this volume).

New data are needed to overcome this ambiguity in the interpretation of the results from the RDCS. Along with new measurements of cluster temperatures at moderate-to-high redshifts, the weak lensing technique (Kaiser & Squires, 1993 [16]) appears particularly promising as a means to measure cluster masses. Only then, the cosmological parameters will be more readily disentangled from the evolutionary parameters.

4 Future Prospects

The next decade promises to be a crucial one for the study of the formation and evolution of high redshift galaxy clusters from both theoretical and observational viewpoints.

Several upcoming X-ray missions will figure prominently. XMM, with its unprecedented collecting area, will push X-ray selected cluster surveys to higher redshift and will make it possible to measure cluster temperatures at intermediate redshifts. AXAF, with its unprecedented spatial resolution, will shed new light on the physics of the intra-cluster medium of nearby clusters.

The ABRIXAS All-Sky Survey will further increase the sample of low-z clusters allowing an accurate measurement of the local cluster temperature function and, in conjuction with the Sloan Digital Sky Survey, will map the nearby large scale structure in great detail. N-body and hydrodynamic simulations on the latest generation of parallel supercomputers will be *quantitatively* compared with these new multi-wavelength datasets.

The new generation of optical and near-IR mosaic imagers and highly efficient multiplexing spectrographs on 8-meter class telescopes (Keck, VLT, GEMINI) will provide a wealth of complementary information which may be difficult to handle. Large multi-color surveys in the optical and near-IR bands will cover a region of the parameter space of redshift, solid angle and limiting flux completely unexplored with 4m-class telescopes and conventional detectors. This will allow us to study the fundamental plane of cluster galaxies and cluster velocity dispersions in a large number of clusters at $z > 0.5$. The Advanced Camera aboard HST (1999) will add sub-kpc morphological information to this multi dimensional data set and will permit detailed studies of their lensing patterns. The impact that surveys based on Sunyaev-Zeldovich effect will have on cluster astrophysics is still difficult to predict.

Finally, we will be able to utilize $\sim$one massive ($L_X > 10^{44}$ erg/s) *gravitational telescope* per square degree out to $z \sim 1$ to infer *directly* the mass distribution of galaxy clusters and probe the faintest, otherwise unaccessible population of background galaxies in the early Universe.

Acknowledgments

I am grateful to my collaborators Roberto Della Ceca, Colin Norman and Riccardo Giacconi who contributed to the work presented here. Particularly, Riccardo Giacconi has inspired the RDCS project with enlightening discussions over the years on the technique and the methodology to pursue. Remo Ruffini and Luigi Stella are also acknowledged for the early encouragment on this work.

References

1. Bahcall, N.A., Fan, X., Cen, R., *ApJ* **485**, L53 (1997)
2. Borgani, S., Rosati, P., Tozzi, P. , in preparation (1998)
3. Burns, J.O. et al. *ApJ* **467**, L49 (1996)
4. Carlberg, R.G., Morris, S.L., Yee, H.K.C., Ellingson, E. *ApJ* **479**, L19 (1997)
5. Castander, F.J. *et al. Nat* **377**, 39 (1995)
6. Cavaliere, A., Menci, N., Tozzi, P. *ApJL* **484**, L21 (1997)

7. Collins, C.A., Burke, D.J., Romer, A.K., Sharples, R.M., Nichol, R.C. *ApJ* **479**, L117 (1997) *ApJ* **479**, L19 (1997)
8. De Grandi, S. Proceedings of Röntgenstrahlung from the Universe, MPE Report 263 , 577 (1996, ed. Zimmermann H.U., Trümper, J., Yorke, H.)
9. Dickinson, M. in Fresh Views of Elliptical Galaxies , 283 (1995, ed. A.Buzzoni, A.Renzini, A.Serrano, (ASP: San Francisco))
10. Ebeling H., Edge, A.C, Fabian, A.C., Allen, S.W., Crawford, C.S., Böhringer, H. *ApJ* **479**, L101 (1997)
11. Edge, A.C., Stewart, G.C., Fabian, A.C, Arnaud, K.A. *MNRAS* **245**, 559 (1990)
12. Eke, V.R., Cole, S., Frenk, C.S., *MNRAS* **282**, 263 (1996)
13. Gioia, I.M., Henry, J.P., Maccacaro, T., Morris, S.L., Stocke, J.T., Wolter, A. *ApJ* **356**, L35 (1990)
14. Gioia I.M. & Luppino, G.A. *ApJ Suppl.* **94**, 583 (1994)
15. Henry, J.P., Gioia, I.M., Maccacaro, T., Morris, S.L., Stocke, J.T., Wolter, A. *ApJ* **386**, 408 (1992)
16. Kaiser, N. & Squires, G. *ApJ* **404**, 441 (1993)
17. Kitayama, T. & Suto, S. *ApJ* **490**, 557 (1997)
18. Lidman C.E. & Perterson C.A. *AJ* **112**, 2454 (1996)
19. Luppino, G.A. & Kaiser, N. *ApJ* **475**, 20 (1997)
20. Mathiesen, B. & Evrard, A.E. *MNRAS* , in press (1997)
21. Navarro, J.F., Frenk, C.S., White, S.D.M *MNRAS* **275**, 720 (1995)
22. Pascarelle, S.M. *et al*, *ApJ* **456**, L21 (1996)
23. Postman, M., Lubin, L.M., Gunn J.G., Oke, J.B., Hoessel, J.G., Schneider, D.P., Christensen, J.A., *AJ* **111**, 615 (1996)
24. Renzini, A. *ApJ* **488**, 35 (1997)
25. Rosati, P., Burg, R., Giacconi, G., 1993, *AIP Conference Proc.* **313**, 260, eds. Schlegel & Petre (1993)
26. Rosati, P., Della Ceca, R., Burg, R., Norman, C., Giacconi, R. *ApJ* **445**, L11 (1995)
27. Rosati, P., Della Ceca, R., Norman, C., Giacconi, R. *ApJ* , in press (1997)
28. Scharf, C.A. & Mushotzky, R.F. *ApJL* **482**, L13 (1997)
29. Scharf, C.A., Jones, L.R., Ebeling H., Perlman, E., Malkan, M., Wegner, G. *ApJ* **477**, 79 (1997)
30. Stanford, A. *et al*, *ApJ* , in press (1998)
31. Steidel, C.C. *et al*, *ApJ* , in press (1997)
32. Zaritsky, D., Nelson, A.E., Dalcanton, J.J., Gonzales, A.H. *ApJ* **480**, L91 (1997)

LIFE AND TIMES OF THE X-RAY GAS IN ELLIPTICAL GALAXIES

ALVIO RENZINI
European Southern Observatory
Garching bei München, Germany

The global gas flows in elliptical galaxies are initiated by stellar mass loss and their diagnostics rely on X-ray observations. The flows are controlled by a number of factors, including supernova heating, the depth and shape of the potential well as determined by the amount and distribution of bright and dark matter, AGN fueling and its feedback effects, interaction with the intracluster medium, and star formation. As a result no steady-state solution can satisfactorily describe the complex, *evolutionary* behavior of the gas flows, which can experience supersonic wind, subsonic outflow, and inflow phases, and transitions between one such flow regime to another. Having identified heating by Type Ia SN's as one of the key factors controlling the flows, constraints on its evolution with cosmological time are derived by considering the total amount of iron contained in whole clusters of galaxies, while the iron abundance in individual galaxy flows can set constraints on the present rate of SNIa's in ellipticals. The central issue of the problem remains the fate of the gas. It is argued that in one way or another, via SN-driven winds, ram pressure stripping, or AGN violent ejection, most of the gas is ultimately expelled from galaxies thus joining the intracluster medium.

1 Introduction

In the late sixties Riccardo Giacconi was preparing for the launch of one his first satellites: UHURU. At that time my activity was very far from space-based astronomy: I was making stellar models trying to understand the horizontal branch (HB) of globular clusters, and concluded that stars have to lose some $\sim 25\%$ of their mass prior to the HB phase. Otherwise all globular clusters would have stubby red clumps in their color-magnitude diagram, rather than the well known extended feature we call the HB.

Evolving stars in elliptical galaxies have nearly the same mass and age as those in globular clusters, hence should lose a similar fraction of their mass. This implies a global present rate $\dot{M}_* \simeq 1.5\times10^{-11}L_{\rm B} \quad M_\odot{\rm yr}^{-1}$ (e.g., Faber & Gallagher 1976; Renzini & Buzzoni 1986), where $L_{\rm B}$ is the B-band luminosity of the galaxy. When allowing for the predicted increase back with cosmological time ($\dot{M}_* \propto\sim t^{-1.3}$ for a Salpeter IMF) it turns out that over one Hubble time the stellar population of an elliptical galaxy sheds 20-50% of its initial mass, or several $10^{10}M_\odot$ in a giant elliptical.

Early attemps at detecting diffused gas in ellipticals, especially through the 21 cm emission, failed to reveal any signal of the gas. Upper limits were some 100 times lower than the expected amount, and many called the HB argument into question: perhaps stars were not losing mass, after all. The case was finally vindicated by another Riccardo's creature, *Einstein*, which revealed that elliptical galaxies – no matter in which environment – can be powerful X-ray emitters, and the emission is dominated by a hot diffused gas at $kT \sim 1$ keV: the result of stellar mass loss was finally found (Forman, Jones & Tucker 1985; Canizares, Fabbiano & Trinchieri 1987; Fabbiano 1989).

A posteriori both the absence of HI gas and the X-ray emission appeared obvious, as ellipticals are dynamically *hot* stellar systems, hence the cool stellar winds emanating from individual red giants are quickly heated to nearly the same dynamical *temperature* of the stars as such winds collide with each other or with the ambient ISM. This implies $T \simeq m_{\rm p}\sigma_\star{}^2/k \simeq 10^7$K ($\sigma_\star$ being the stellar velocity dispersion), and the gas shines in X rays.

While there is now general agreement that stars do indeed lose mass, the two simple questions *"What happens to the gas being lost by the stars?"* and *"Where is it ultimately disposed?"* have not yet been answered with general satisfaction. Early suggestions proposed stationary *cooling flow* solutions (Nulsen, Stewart, & Fabian 1984; Sarazin and White 1987), as galactic wind solutions (Mathews & Becker 1971) underproduce the X-ray emission by a large margin. One of the problem with the albeit elegant cooling flow solution is that it does not really answer the second question. Since no star formation is observed, most of the cooling gas is postulated to form very low mass objects, such as brown dwarfs or Jupiters. Yet, this is not the only problem.

Fig. 1 shows the X-ray luminosity of elliptical galaxies vs their optical luminosity. Data refer to an homogeneous set of objects observed with *Einstein* (Fabbiano, Kim, & Trinchieri 1992). Galaxies at any given optical luminosity exhibit a very wide range of X-ray luminosities, and Fig. 2 demonstrates that this is not a result of distance errors (i.e. this is a distance-independent diagram). So, the X-ray luminosity of galaxies with the same optical luminosity and/or the same $\sigma_\star$ may differ by up to a factor ~ 100, hence their hot gas content by more than a factor of 10.

The lines in Fig. 1 represent the predicted X-ray luminosity for the two classical steady state solutions for the gas flow. In a supersonic wind very little gas accumulates, and $L_{\rm X}$ reduces to $\sim L_{\rm discr}$, the contribution of discrete sources. In a "cooling flow" a dense ISM is in place, and $L_{\rm X} \simeq L_{\rm in} + L_{\rm sn} + L_\sigma$ (where $L_{\rm in}$, $L_{\rm sn}$ and L_σ are the rates of energy deposition from the gravitational PdV work, from supernova explosions, and from the thermalization of stellar winds, respectively). Clearly, neither of the two steady state solutions accounts for the large scatter of the observed X-ray luminosity. Yet, some galaxies are faint enough in X-rays for their gas flow to be in a supersonic wind regime,

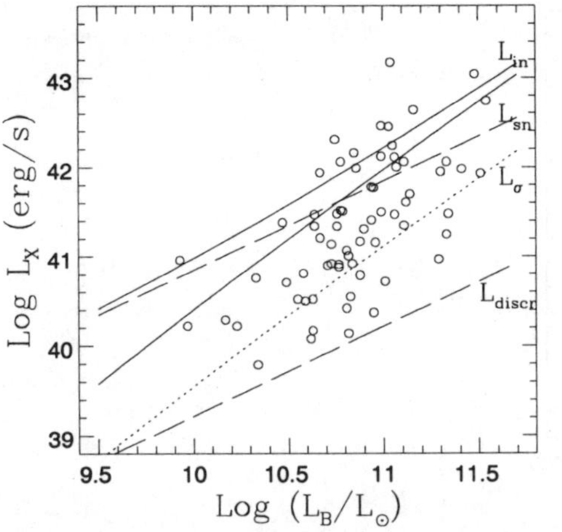

Figure 1: The X-ray luminosity of elliptical galaxies observed with the Einstein Observatory vs their optical luminosity (From Fabbiano *et al.* 1992). $L_{\rm discr}$, L_σ and $L_{\rm sn}$ are the expected contributions from discrete sources, from the thermalization of stellar winds, and from SN explosions (with $\vartheta_{\rm SN} = 1$), respectively. The expected X-ray luminosity of galaxies in the inflow regime ($L_{\rm in}$) is also shown, for $\vartheta_{\rm SN} = 0$ and 1 (adapted from Ciotti *et al.* 1991).

and indeed their X-ray spectrum requires very little hot ISM contribution in excess of the emission from low mass X-ray binaries and other stellar sources (Pellegrini & Fabbiano 1994). Meanwhile, some galaxies are bright enough in X-rays for being in a cooling flow regime. However, most galaxies exhibit X-ray luminosities that are intermediate between these two extremes. Hence, the majority of *Einstein* galaxies cannot be in the inflow neither in the wind regime, and other alternatives need to be explored.

In an attempt to save as much as possible of the "cooling flow" models a distibuted mass dropout from the hot ISM has been postulated (e.g., Sarazin & Ashe 1989; Kritsuk 1992; Bertin & Toniazzo 1995). In this way the X-ray luminosity of cooling flow models can be somewhat reduced, as only part of the flow reaches the center and the PdV work done by the gravitational field is correspondingly reduced. Yet, this is not enough to account for the galaxies with $L_{\rm X} < L_{\rm sn} + L_\sigma$. Indeed, these energies have to be radiated away in order to ensure the "cooling flow" itself. There are also good theoretical reasons for dismissing mass dropout driven by local thermal instabilities. Low amplitude thermal perturbations are generally stable (Balbus 1991), or oscillate radially without producing mass dropout (Loewenstein 1989). Even if instabilities were possible, the overdense regions would rapidly dissolve as a result of Rayleigh-Taylor instabilities without leading to mass deposition (Hattori & Habe 1990;

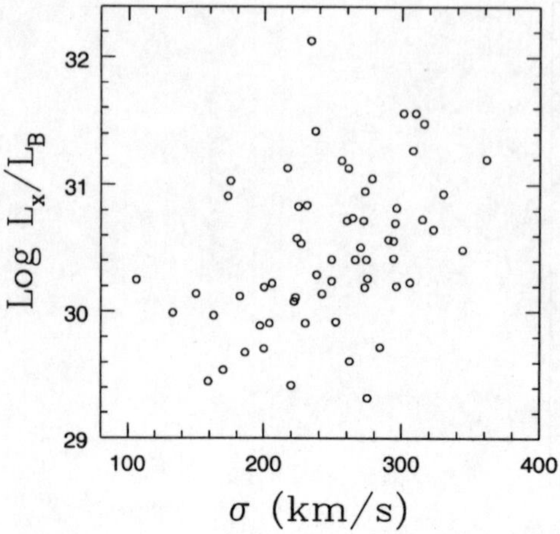

Figure 2: The X-ray to optical luminosity ratio for the elliptical galaxies in Fig. 1 as a function of the central velocity dispersion. Luminosity units are as in Fig. 1.

Malagoli, Rosner, & Fryxell 1990). In conclusion, there are more minuses than pluses in the mass dropout postulate.

A firm conclusion is that steady state models of any kind can hardly account for the observed distribution of X-ray luminosities. On the other hand, both the rate of stellar mass return and the supernova (SN) heating are expected to exhibit a strong secular evolution, and therefore *evolutionary* – as opposed to time-independent – solutions are to be sought.

To prevent cooling flows an energy source has to operate inside galaxies. Type Ia SNs are known to occur in ellipticals, hence they represent the most obvious source. Activation of an AGN by accretion on a central black hole (BH) is an additional option. These two possibilities are explored next.

2 Models with Type Ia Supernovae Leading the Play

SNs heat the gas and if their energy input is large enough they can drive the gas out of the galaxy. This is contrasted by the galaxy potential well, hence by the amount and distribution of dark matter that one assumes for the galaxy. Unfortunately, both the current SNIa rate in ellipticals and the dark matter parameters are poorly constrained by observations. Therefore, both need to be parameterized and the corresponding parameter space has to be explored.

Evolutionary models of gas flows that include dark matter and SN heating meet several of the observed X-ray properties of ellipticals. There are basically two families of such models, depending on the assumed past evolution of the

Type Ia SN heating, which can be conveniently parameterized assuming a power-law evolution with time (SN Rate $\propto t^{-s}$). Also the evolution of the rate of stellar mass return is well represented by a power-law ($\dot{M}_\star \propto\sim t^{-1.3}$), hence, for $s\gtrsim1.3$ the specific heating of the stellar ejecta secularly *decreases* with time (the SN heating was more efficient in the past). Galactic flows correspondingly evolve from early supersonic winds (W), to subsonic outflow (O), to an inflow (I) regime (Ciotti *et al.* 1991, hereafter WOI models). The O–I transition takes place when SN heating finally falls short of compensating for the radiative losses, and a central cooling catastrophe takes place.

Coversely, for $s\lesssim1.3$ the specific SN heating *increases* with time, and models evolve in the reverse sequence of flow regimes, from early inflows, to subsonic outflows, to late supersonic winds (Loewenstein & Mathews 1987; David, Forman, & Jones 1991, hereafter IOW models).

The X-ray luminosity is very low during wind (W) phases, as the hot gas rapidly leaves the galaxy and does not accumulate. During inflow (I) phases galaxies are very bright in X-rays, as the hot gas remains confined to the galaxy and cools down. During outflow (O) phases $L_{\rm X}$ is intermediate between the two previous extremes, increasing with time in the WOI models as gas accumulates, decreasing with time in the IOW models. Thus, the large dispersion in the $L_{\rm X}/L_{\rm B}$ ratio can be easily understood in the frame of either the WOI or the IOW models, and follows from relatively small galaxy-to-galaxy differences in e.g. the depth and shape of the potential well, which can cause relatively large differences in the epoch of the O–I or I–O transition. In the frame of WOI models galaxies evolve upward in Fig. 1, from near the bottom line $L_{\rm discr}$ when they are in the W phase, to the top line $L_{\rm in}$ when they finally reach the I phase. IOW models instead drip from the top line, evolve downward, and finally land on the bottom line. In both cases, the observed X-ray luminosities indicate that most of present-day ellipticals are in the O phase.

How can we distinguish between these two opposite options? The most direct way would be to look at high redshift ellipticals, with $z\gtrsim1$, hence $t\lesssim5$ Gyr. In the case of IOW models high-z E galaxies are predicted to be brighter in X-rays compared to low-z galaxies, while the reverse applies to WOI models. This test is curerently unfeasible, as e.g. ROSAT can hardly see individual galaxies beyond the distance of the Coma cluster, but should become feasible by a next to the next generation of X-ray observatories (Elvis & Fabbiano 1997).

The present and past rate of SNIa's in ellipticals is a crucial ingredient that direct observations should better constrain in the future. Current estimates of the present rate remain uncertain, as based on a fairly small number of events. Cappellaro *et al.* (1997) most recent estimate is based on 7 SNIa events in ellipticals, and gives a rate $\vartheta_{\rm SN} = 0.25 \quad (\pm \sim 0.1)$, $\vartheta_{\rm SN}$ being the present rate in units of 2.2×10^{-13} SNs $\rm yr^{-1} L_\odot{}^{-1}$ (i.e., in units of the rate as estimated by van den Bergh & Tammann (1991). Very little is currently known about the SNIa rate in high-z ellipticals, but current efforts at detecting and studying high-z SNIa's may soon start setting useful constraints (e.g. Leibundgut 1998; Perlemutter *et al.* 1997).

3 The Iron Discrepancy

Way of setting constraints on the present and past integral rates of SNIa's have been explored (Ciotti *et al.* 1991; Renzini *et al.* 1993), based on the iron limes in the X-ray spectrum of the ISM and in the intracluster medium (ICM), respectively. Within galaxies, one expects the iron abundance in the hot ISM to be $X_{\rm Fe}^{\rm ISM} \simeq < X_{\rm Fe}^* > + 5\vartheta_{\rm SN}$, thus reflecting the average stellar abundance ($< X_{\rm Fe}^* >$) further enriched by Type Ia SNs. Using Cappellaro *et al.* latest estimate one therefore expects $X_{\rm Fe}^{\rm ISM} \gtrsim 2X_{\rm Fe}^\odot$. From the iron-L complex at ~ 1 keV very low values are instead inferred from ASCA data, $X_{\rm Fe-L}^{\rm ISM} \lesssim 0.3 X_{\rm Fe}^\odot$ (e.g. Matsushita *et al.* 1994). Formally, the ASCA results requires a vanishing SNIa activity in present day ellipticals. This *Iron Discrepancy* is extensively discussed by Arimoto *et al.* (1997), where a series of possible solutions are explored. These include having SNIa delivering much less iron than currently believed, or hiding it in flakes or Jupiters, or diluting it with infalling intracluster medium (ICM). The former possibilities are excluded by Arimoto *et al.* on astrophysical grounds, while dilution by the ICM may work in some cases, as early explored by Renzini *et al.* (1993) for the case of NGC 1399 and NGC 4472. However, this is a viable solution for just a few, exceptional galaxies such as cD's which are surrounded by a dense ICM which would suffocate any attempt at establishing a galactic wind (D'Ercole *et al.* 1990). Brigenti & Mathews (1997) argue that accretion from a surronding ICM is necessary to account for the temperature profile in NGC 4472 (with T increasing outwards), and suggest accretion as a general explanation for both the low iron content *and* of the large scatter of $L_{\rm X}/L_{\rm B}$. Overluminosity compared to the $L_{\rm in}$ line (cf. Fig. 1) is a natural product of accretion from the ICM, as more gas flows in and radiates in addition to that made available by stars. It may indeed be required to account for the few galaxies above the $L_{\rm in}$ line (see Fig. 1), but it can hardly work as a general solution to the other problems. The issue is in fact to account for so many galaxies being fainter than $L_{\rm in}$, while galaxies exist with apparently very low iron that are not overluminous. I conclude that accretion from an ICM (or a remnant intragroup medium) can play a role in only a small minority of elliptical galaxies. The opposite process, ram pressure stripping as galaxies plough through a cluster ICM, can also be at play and help accounding for the L_X dispersion (Renzini 1994, and references therein), but no clear correlation of L_X/L_B with the environment has yet been reported.

Having excluded these and other solutions to the iron discrepancy, Arimoto *et al.* (1997) suggest to explore the possibility that some atomic physics problem may be hidden in the diagnostic tools currently used to analyze X-ray spectra. Indeed, in ellipticals and groups temperature is too low for detectable emission by electronically simple H- and He-like iron (Iron K), and the iron abundance can only be inferred from the Iron-L emission at ~ 1 keV which is due to much more complex iron ions. Called into question are the theoretical collisional excitation probabilities starting from the L-shell of iron ions with 4 or more electrons. While waiting for new quantum mechanical calculations, Arimoto *et al.* suggest a mere empirical check of the iron-L diagnostics. If a problem really exists, it should show up not just for ellipticals, but also for all sort of X-ray objects, such as starburst galaxies, supernova remnants, X-ray stars, etc., for which there may be independent evidence concerning their iron content. For several of these objects *ASCA* data appear to indicate unconfortably low iron abundances from iron-L (Fabbiano 1995; Arimoto *et al.* 1997). As more and more *ASCA* data become available we may expect the situation to fully clarify in the near future.

Fabbiano (1995) also pointed out that most of the very low iron abundances come from fitting a one-temperature isothermal model to the observed X-ray spectra, while allowing for two-temperature fits higher iron abundance should be found. Follow up of this suggestion reveals that indeed higher iron abundances are derived. For example, Boute & Fabian (1997) analyze *ASCA* data relative to 20 ellipticals, and find an average metal (i.e., not just iron!) abundance $<Z>= 0.19 \pm 0.12$ solar from single-temperature fits, while the derived abundance goes up almost a factor of five using two-temperature fits, i.e. to 0.9 ± 0.7 solar. Still, using two-temperature fits (but a different code) Arimoto *et al.* find iron 0.25 solar for a sample of seven *ASCA* ellipticals. With a two-temperature fit (the hot component fixed at 10 keV) Matsushita *et al.* (1997) find iron to be solar in NGC 4649, with evidence of ISM-ICM mixing in the outer regions. These examples illustrate current uncertainties in iron-L diagnostics, hence the factor of ~ 2 between the expectation (using the SNIa rate from Cappellaro *et al.* 1997) and the result of Boute & Fabian (1997) or Matsushita *et al.* (1997) may not be alarming. The residual difference may well come from imperfect atomic physics having been used for iron-L electronic configurations (Fabian *et al.* 1994, first questioned iron-L atomic physics), though Boute & Fabian (1997) don't find evidence for a major failure.

4 Constraints from the Intracluster Medium Archive

The total amount of iron and other elements in clusters of galaxies (ICM plus galaxies) represents a record of the overall past SN activity, setting a constraint on the past evolution of the SN rates. The total iron mass to cluster optical luminosity ratio appears to be constant among clusters, irrespective of their richness, with $M_{\rm Fe}/L_{\rm B} \simeq 0.03 \pm 0.01$ $(M_\odot/L_\odot)$. To make all this iron requires either a much stronger SNIa activity in the past ($s \gtrsim 1.3$), or a much stronger Type II SN activity at early times than expected from a Salpeter IMF (Ciotti *et al.* 1991; Arnaud *et al.* 1992; Renzini *et al.* 1993; Elbaz, Arnaud, & Vangioni-Flam 1995). The elemental abundances in the ICM – namely the α-elements over iron ratio – offer an attractive way to distinguish between these two alternatives (Renzini *et al.* 1993), hence providing indirect constraints on the history of gas flows in ellipticals. Again, early ASCA observations seemed to require very little SNIa contribution to the ICM (Mushotzky 1994), with $[\alpha/\mathrm{Fe}] \simeq +0.4$, which would favor SNII over SNIa production, hence IOW over WOI models. If true, this result would reduce SNIa's to a very marginal role in the chemical evolution of galaxies, thus rising the puzzling question as to how the solar elemental proportions have been established in our own Galaxy, while non-solar proportions may prevail at galaxy cluster scales. Subsequently Mushotzky *et al.* (1996) revised somewhat down the α-element abundances, to a modest $<[\alpha/\mathrm{Fe}]> \simeq +0.2 \pm 0.2$. Moreover, Ishimaru & Arimoto (1997) have recently pointed out that the small residual α-element enhancement in the ICM comes from Mushotzky *et al.* (1996) having assumed the solar iron abundance from "photospheric" model atmosphere analysis. The result is different if one uses "meteoritic" abundances instead, because the meteoritic iron abundance is ~ 0.16 dex lower than the photospheric value (Anders & Grevesse 1989). After noting that the meteoritic iron is now more generally adopted, Ishimaru & Arimoto conclude that the α-element enhancement in the ICM virtually disappears, going down from $<[\alpha/\mathrm{Fe}]> \simeq +0.2$ to only $\sim +0.04$ (consistent with no enhancement or even small depletion, given the reported errors).

In conclusion X-ray derived cluster abundances are consistent with no α-element enhancement, and therefore with a combined SNIa and SNII enrichment giving near solar elemental proportions on the whole cluster scale, as predicted by the standard model for the Galactic chemical evolution (Renzini *et al.* 1993). In particular, there appears to be no need to invoke a special IMF or a suppression of SNIa's in clusters, compared to our own Galaxy (Renzini 1997). This requires a fairly high value of the exponent s in the parametrized SNIa rate, but abundances are certainly still too uncertain to risk saying whether s is larger or smaller than 1.3.

5 Feeding a Central Black Hole

WOI models solve only in part the problem of the final disposal of of the gas. Most of the stellar mass loss goes indeed out of galaxies during W and O phases, but the transition to the inflow phase is soon or later unavoidable. The question of what happens to the cooling gas remains unanswered also in WOI models, where the sinking gas just disappears past the innermost mass zone (e.g., Ciotti *et al.* 1991). Ciotti *et al.* mention in fact two astrophysically attractive alternatives to Jupiters: 1) the central cooling catastrophe may result in a major starburst with several $10^8 M_\odot$ of gas turning into stars over a short time scale, and 2) the material lost by stars and flowing to the center of the galaxy can feed a central black hole (BH) that may be lurking there. Both options may actually co-exist in real galaxies, and both may provide additional heating to the ISM, whenever SNIa's fail to maintain an outflow.

Ciotti & Ostriker (1997) have recently explored the BH option by means of 1D hydrodynamical simulations, and a brief *digest* description of some of their main results is given here. During inflow phases up to several $10 M_\odot \mathrm{yr}^{-1}$ sink towards the center of the galaxy, and either IOW or WOI models simply postulate that this mass silently disappear. However, if instead this mass flow is accreted by a central BH the luminosity of the so powered AGN is expected to be in the range:

$$L_X^{AGN} \simeq \epsilon \dot{M}_{\mathrm{SINK}} c^2 \simeq 10^{44} - 10^{47} \quad \mathrm{erg\,s^{-1}}$$

for $10^{-3} < \epsilon < 0.1$, i.e., for 0.1 to 10% mass-energy conversion efficiency, with most of this luminosity being emitted in the hard X-ray range.

In the specific model show in Fig. 3, driven by SNIa'sthe evolution proceeds through the wind phse during the first ~ 5 Gys, then followed by the outflow phase up to the cooling catastrophe at $t = 9.4$ Gyr. At this point the central BH starts accreting at a very high rate ($\sim 60 M_\odot \mathrm{yr}^{-1}$), and the BH luminosity L_{BH} is activated (see Fig. 3a). Absorption of part of this radiation suddenly heats the gas in the central regions to the Compton temperature associated to L_{BH} ($\sim 10^9$K), driving an outgoing shock wave that leaves behind a hot bubble a few hundred parsecs in diameter. As the bubble expands the galaxy X-ray luminosity drops (Fig. 3b), the inflow is discontinued as so is accretion onto the BH: L_{BH} is shutoff (Fig. 3a). Then a new outflow regime is established, in this particular case lasting for ~ 1 Gyr, until a new central cooling catastrophe reactivates the BH, and so on. Even in this 1D approximation the physics of the process is rather complex. In a realistic case, with disks, jets, and all the AGN paraphernalia one may recover much of at least some classes of radiogalaxies hosted by giant ellipticals, in particular those that need a dense ISM to confine jets and radiolobes (see also Binney & Tabor 1995).

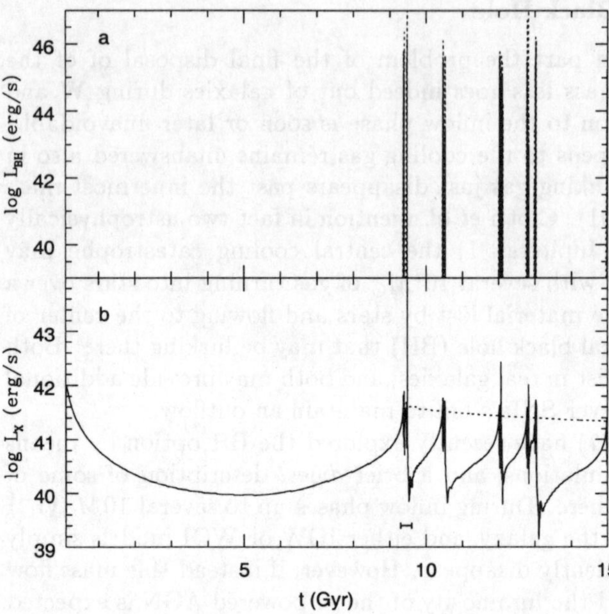

Figure 3: (a) The time evolution of the BH luminosity $L_{\rm BH}$ for a model with $\epsilon = 0.1$. (b) The time evolution of the galaxy X-ray luminosity for the same model, and for a model with $\epsilon = 0$ (unimpeded cooling flow). Model parameters are specified in Ciotti & Ostriker (1997).

The evolution of the gas flow then proceeds through a series of outflow phases separated by brief AGN phases during which the ISM of the galaxy is expelled. After each expulsion the AGN dies out of starvation, and stellar mass loss starts replenishing the ISM again until the next central cooling catastrophe takes place. The duration of the outflow phases depends on several model parameters, and can be as long as several Gyr provided SN heating delays the next central cooling catastrophe. The X-ray luminosity evolution of the galaxies (excluding AGN phases) is confined between the upper and the lower lines in Fig. 1, and therefore this model quite naturally accounts for the observed large dispersion of the $L_{\rm X}/L_{\rm B}$ ratios. After an AGN phase galaxies climb up slowly (in ~ 1 Gyr or more) starting from $L_{\rm X} \simeq L_{\rm discr}$, until the transition to the inflow takes place at $L_{\rm X} \simeq L_{\rm in}$ and the AGN is activated, then precipitously falling back to $L_{\rm X} \simeq L_{\rm discr}$ before starting a new cycle.

An attractive feature of this model is that it accounts for the fate of the cooling gas during the inflow phases, an aspect that remains under the rug in other models. Just $\lesssim$1% of the stellar ejecta eventually terminate inside the central BH, thus releasing enough energy to eject the other $\sim$ 99% and enriching in metals the ICM.

Central BH-gas flow interactions may also take place at a much smaller scale. At the deep bottom of the potential well of ellipticals, stellar and gas density, hence radiative losses, can be very high, and *mini-inflows* inside $\sim$ 200 pc are common while the rest of the galaxy is in a wind or outflow regime (Ciotti *et al.* 1991). If central massive BHs are ubiquitous among ellipticals, mini-inflows with typical mass deposition rates $\sim 10^{-3} - 10^{-2} M_\odot$ yr^{-1} may feed a low-level, sporadic AGN activity, that may account for the observed $L_{\rm RADIO} - L_{\rm X}$ correlation, and produce occasional optical flares at the very center of the galaxy. A $\sim 10^6 L_\odot$ ultraviolet flare that might have been generated in this way has been recentlty detected with HST at the center of the elliptical galaxy NGC 4552 (Renzini *et al.* 1995), one of the X-ray brightest galaxies in Fig. 1. Future HST observations – especially in the ultraviolet – should tell us how common these flare events are, and whether their occurrence correlates with other optical and X-ray properties of the host galaxies.

It should be said that probably fueling central BHs from a hot ISM is not the most common way to activate QSOs. The observed peak of the QSO number density at $z = 2$ and beyond must have a very different origin, probably intimately connected to the very formation of galactic spheroids. Central cooling catastrophes in ellipticals have taken place already in probably a minority (likely in the most giant) of them. Hence, the QSO activated in this way would just relate to the tail of the QSO density-redshift distribution.

We may wander if this same process could also work at the much larger scale of galaxy cluster "cooling flows", where a similar embarrassment exists as to where the cooling gas is eventually disposed. May central BHs in cD galaxies affect cluster scale "cooling flows"? Clearly a much energetic QSO is required to compensate for radiative losses in cluster flows ($L_{\rm X} \simeq 10^{45}$ erg s^{-1}), and temporarily revert inflows to outflows near the cluster center. An attempt in this direction has been recently made by Tucker & David (1997).

6 Conclusions

The study of gas flows in ellipticals is a very rich field of research, as such flows connect to the central, active regions of galaxies to the one end, and to the intracluster medium at the other end, thus involving a manifold of astrophysical phenomena: stellar mass loss, supernovae, galactic dark matter, plasma thermodynamics and radiative processes, galactic chemical evolution, galaxy-galaxy and galaxy-cluster interactions, BH and AGNs and their feedback on the ISM, etc. In summary, here stands the issue:

• Gas flows in ellipticals are not in a steady state.

• Outflows and winds appear to prevail over inflows in most ellipticals.

• The evolution of the flows over cosmological times is controlled by the evolution of the rate of Type Ia SNs, that is still insufficiently constrained by direct observations.

• The iron and other element abundances in the ISM of ellipticals provides an indirect measure of the present (recent) SN rate. There still appears to be a factor ~ 2 discrepancy between the expected iron abundance and the abundances derived from X-ray data.

• The total iron contained in clusters of galaxies sets a constraints on the past integral SN rate. Latest estimates favor solar proportions for the elemental abundances in the ICM (i.e., a solar α-elements over iron ratio), which favors a substantial increase of the SNIa rate with lookback time.

• Central BHs can play a main role in the late evolution of the flows, with its interaction with the hot ISM giving rise to intermittencies and potentially to a great variety of observable phenomena.

I am very grateful to Silvia Pellegrini for her permission to use here Fig. 1 and 2, to Luca Ciotti for Fig. 3, and to both of them plus Fabrizio Brighenti and Annibale D'Ercole for stimulating discussions.

References

Anders, E., & Grevesse, N. 1989, Geochimica et Cosmochimica Acta, 53, 197

Arimoto, N., Matsushita, K., Ishimaru, Y., Ohashi, T., & Renzini, A. 1997, ApJ, 477, 128

Arnaud, M., Rothenflug, R., Boulade, O., Vigroux, L., & Vangioni-Flam, E. 1992, A&A, 254, 49

Balbus, S.A. 1991, ApJ, 372, 25

Bertin, G., & Toniazzo, T. 1995, ApJ, 451, 111

Binney, J., & Tabor, G. 1995, MNRAS, 276, 663

Boute, D.A., & Fabian, A.C. 1997, astro-ph/9707117

Brighenti, F., & Mathews, W.G. 1997, astro-ph/9710199

Canizares, C.R., Fabbiano, G., and Trinchieri, G. 1987, ApJ, 312, 503

Cappellaro, E., *et al.* 1997, A&A, 322, 431

Ciotti, L., & Ostriker, J.P. 1997, ApJ, 487, L105

Ciotti, L., D'Ercole, A., Pellegrini, S., & Renzini, A. 1991, ApJ, 376, 380

David, L.P., Forman, W., & Jones, C. 1991, ApJ, 369, 191

D'Ercole, A., Ciotti, L., Pellegrini, S., & Renzini, A. 1990, Windows on Galaxies, ed. G. Fabbiano, J. Gallager, & A. Renzini (Dordrecht: Kluwer), p.

275
Elbaz, D., Arnaud, M., & Vangioni-Flam, E. 1995, A&A, 303, 345
Elvis, M., & Fabbiano, G. 1997, Next Generation X-ray Observatory, ed. M.J.L. Turner (Univ. of Leicester), p. 33
Fabbiano, G. 1989, ARA&A, 27, 87
Fabbiano, G. 1995, Fresh Views of Elliptical galaxies, ASP Conf. Ser. 86, 103
Fabbiano, G., Kim, D.-W., & Trinchieri, G. 1992, ApJS, 80, 531
Faber, S.M., and Gallagher, J.S. 1976, ApJ, 204, 365
Fabian, A.C., Arnaud, K.A., Bautz, M.W., & Tawara, Y. 1994, ApJ, 436, L63
Forman, W., Jones, C., and Tucker, W. 1985, *Ap. J.*, **293**, 102.
Hattori, M., & Habe, A. 1990, MNRAS, 242, 399
Ishimaru, Y., & Arimoto, N. 1997, PASJ, 49, 1
Kritsuk, A.G. 1992, A&A, 261, 78
Leibundgut, B. 1998, Supernovae and Cosmology, ed. B. Binggeli, R. Buser, & L. Labhardt (Basel: University of Basel), in press
Loewenstein, M. 1989, MNRAS, 238, 15
Loewenstein, M., and Mathews, W.G. 1987, 319, 614
Malagoli, A., Rosner, R., & Fryxell, B. 1990, MNRAS, 247, 367
Mathews, W.G., & Backer, J. 1971, ApJ, 170, 241
Matsushita, K., *et al.* 1994, ApJ, 436, L41
Matsushita, K., *et al.* 1997, ApJ, 488, L125
Mushotzky, R. 1994, in Clusters of Galaxies, ed. F. Durret (Gyf sur Yvette: Editions Frontières), p. 167
Mushotsky, R., *et al.* 1996, ApJ, 466, 686
Nulsen, P.E.J., Stewart, G.C., & Fabian, A.C. 1984, MNRAS, 208, 185
Pellegrini, S., & Fabbiano, G. 1994, ApJ, 429, 105
Perlmutter, S., et al. 1997, ApJ, 483, 565
Renzini, A. 1994, in Panchromatic View of Galaxies, ed. G. Hensler, Ch. Theis, & J.S. Gallagher (Gif-sur-Yvette: Editions Frontières), p. 155
Renzini, A. 1997, ApJ, 488, 35
Renzini, A., & Buzzoni, A. 1986, Spectral Evolution of Galaxies, ed. C. Chiosi, A. Renzini (Dordrecht: Reidel), p. 135
Renzini, A., Ciotti, L., D'Ercole, A., & Pellegrini, S. 1993, ApJ, 419, 52
Renzini, A., Greggio, L., di Serego Alighieri, S., Cappellari, M., Burstein, D., & Bertola, F. 1995, Nature, 378, 39
Sarazin, C.L., & White, R.E.III 1987, ApJ, 320, 32
Sarazin, C.L., & Ashe, G.A. 1989, ApJ, 345, 22
Tucker, W., & David, L.P. 1997, ApJ, 484, 602
van den Bergh, S., & Tammann, G. 1991, ARA&A, 29, 363

THE BEGINNINGS OF STELLAR X-RAY ASTRONOMY

R. ROSNER
Enrico Fermi Institute and Dept. of Astronomy & Astrophysics,
The University of Chicago,
5640 South Ellis Avenue,
Chicago IL 60637, USA
E-mail: rrosner@oddjob.uchicago.edu

I review the beginnings of the field of stellar X-ray astronomy, concentrating on the period immediately preceeding, and immediately following, the launch of the *Einstein* Observatory. The wealth of data was such that, within the first two years following launch, major discoveries were made by scientists from both the *Einstein* Observatory PI groups and *Einstein* Observatory Guest Observers which established stellar X-ray astronomy as a new astronomical discipline: Discovery of early and late-type stars, as well as young protostars, as soft X-ray sources; discovery of the "dividing line" separating X-ray emitting and X-ray quiet giant and supergiant stars; and establishment of the "solar-stellar connection" as a paradigm for understanding X-ray emission from late-type stars.

1 Introduction

While X-ray emission from normal stars other than the Sun was observed prior to the launch of the *Einstein* Observatory [1,2,3], the ubiquity of this emission was not realized at that time. Thus, the types of stars seen to that point were either very non-solar (e.g., the RS CVn systems, such as Capella [1,2], whose mean X-ray luminosity is several orders of magnitude larger than that of solar-type stars [4]) or essentially just like the Sun (e.g., the nearby system α Cen A,B [3]) — whether late-type stars, as a general class, were X-ray sources was at that time still unknown; and nothing was known about the X-ray properties of early-type stars. At that time, I was a young assistant professor at Harvard, working closely with G.S. (Pippo) Vaiana, who was then the leader of the solar X-ray astronomy group at the Harvard-Smithsonian Center for Astrophysics (CfA); and I well-recall being shown the initially proposed "strawman" observing list for stars to be observed at the beginning of the *Einstein* Observatory mission. This list had been prepared by scientists working with Riccardo Giacconi's X-ray astronomy group, and had two characteristics which struck me immediately: First, the list was extremely short; second, the list focused mostly on very late-type flare stars, and very nearby stars, presumably based on the assumptions that stellar X-ray emission – if extant at all – was most likely to be correlated with the presumed activity related to stellar flaring, and if present, would be relatively puny in luminosity. These perceptions were bolstered by two distinct arguments, the first of which was based on theoretical views then still in vogue, holding that stellar coronae were largely a consequence of the damping (and consequent plasma heating) of sound waves generated at the photospheric level by turbulent surface convective motions[5]. As a consequence, it was thought that the peak of stellar coronal activity would occur for late F and early G dwarf stars; that the level of X-ray variability for any given spectral type would be small (because the ultimate driver of coronal heating – stellar surface convection – was known to be statistically steady, and relatively uniform in amplitude from star to star for a given spectral type); and that – with the exception of the flare stars – lower-mass stars would have substantially lower coronal X-ray emission levels. The second argument was based on estimates of the stellar contribution to the diffuse soft X-ray background[6], which suggested that this contribution was negligible (a suggestion which later turned out to be largely incorrect[7]), and by implication, that the level of stellar X-ray emission was at solar levels or lower. It is therefore not surprising that the initial *Einstein* "strawman" stellar observing list was exploratory, and aimed at succeeding in those cases for which then-existing arguments pointed to likely success.

The story I will tell in the following recounts how the vigorous field of stellar X-ray astronomy sprung from these inauspicious beginnings. I will focus on the period just before the launch of the *Einstein* Observatory, as well as the period immediately following (i.e., from the summer of 1978 to the summer of 1980), cutting off all original references after 1981; I will therefore make no attempt to survey the current state of affairs of this field (cf. Serio[8], in this volume).

2 Creating the Initial Einstein Observatory Stellar Survey

At the time we first looked at what *Einstein* was to observe during its initial observational period, Vaiana was convinced that the "received wisdom" was simply wrong, and that most late-type stars would be recognized as X-ray sources. He based his opinion on his work in solar X-ray astronomy, which had convinced him that the Sun's magnetic field was largely responsible for the observed solar X-ray emission; since the Sun was not likely to be unusual as far as its magnetic properties were concerned, Vaiana argued that X-ray emission should also not be regarded as an unusual attribute of the Sun. It therefore seemed inescapable that we should be much more ambitious in attempting to detect X-ray emission from stars. The next step was therefore to convince Riccardo Giacconi that the *Einstein* survey should include many more pointings at stars; if he agreed, we were in the position to design an entirely new stellar program for *Einstein.*

I am unfortunately not familiar with the specific arguments that Vaiana marshalled for Giacconi, but I can well-imagine the nature of the forceful presentation that Pippo surely made. But no matter what the manner of this presentation, we do know now that it was successful, as Giacconi gave the go-ahead for a modestly-scaled (but still much larger than originally-planned) retooling of the initial CfA stellar observing plan.

We were of course not the only scientists in the *Einstein* collaboration who wanted to observe stars: The group centered at Columbia University also sought to establish an exploratory stellar program as part of Columbia's share of the Principal Investigator (PI) groups' time allocation; and the NASA/ Goddard Space Flight Center group similarly targeted stars as part of its Solid State Spectrometer (SSS) observing plan. The question was thus how to devise a strategy that minimized conflict within the PI group. The strategy developed by Vaiana was, first, to concentrate on a well-defined set of the nearest examples of each stellar type (avoiding those already chosen by the Columbia group); second, to establish collaborations with *Einstein* Observatory Guest Observers who had themselves identified stars of this sort; third, to establish collaborations with other CfA scientists whose fields were likely to yield serendipitous stellar X-ray sources; and fourth, to leave the initial surveys of open stellar clusters to the Columbia group and Guest Observers. This approach eventually led to the construction of the *Einstein* Observatory Stellar Survey catalog[4]; in the short run, we were able to put together a much more carefully-constructed initial stellar observing plan, which (together with Columbia's and Goddard's stellar observing plans) served for the *Einstein* Observatory's initial observing period.

3 The First Results

The first indications that X-ray observations of stars were going to be as productive and exciting as we hoped came from an entirely serendipitous observation, and for stars which few of us expected to be X-ray sources (especially at the levels that were indeed observed). More precisely, the observation in question was actually a series of observations of the region immediately surrounding the strong galactic binary X-ray source Cyg X-3, which were part of a boresight observational sequence carried out at the very beginning of the mission. This field was expected to be empty of sources other than Cyg X-3, but the very first field showed this expectation to be entirely wrong: The initial Imaging Proportional Counter (IPC) image showed a quartet of sources located roughly 0.4° from Cyg X-3, and further study (which included merging the various IPC images, as well as using observations from the High Resolution Imager [HRI]) shortly thereafter showed a total of six definite point sources, which were readily identified with OB stars in the Cyg OB2 association [9]. I still recall the thrill of looking at a Palomar Sky Survey image of this region, which showed nothing at the position of the powerful Cyg X-3 X-ray source, and immediately recognizing the obvious trapezium of stars in the Cyg OB2 association which were in fact coincident with the four X-ray sources seen initially. Thus, OB stars were confirmed as an entirely new class of stellar X-ray sources.

It needs to be immediately added, however, that what seems obvious today was not at all obvious at the time of these observations: While we were sure that we had detected emission from outer atmospheres (i.e., the winds) of OB stars, this was not a view universally believed by other astronomers and astrophysicists. In particular, there were a number of alternative suggestions, and their proponents quickly made themselves heard. Particularly interesting were the possibilities that we had actually detected emission from the shocked material resulting from the interaction of the powerful winds flowing from OB stars with the surrounding interstellar matter [10,11]; and the alternative possibility that we had detected emission from close binary systems in the predicted "quiet" phase [12] preceeding strong mass exchange associated with high-luminosity compact binary X-ray sources [13,14]. These alternative possibilities were not so readily disposed of, but disposed of they were: The observed emission turned out to be significantly harder than that expected from shocked wind bubbles; and the remarkable fact that we observed every association member within our detection limits, together with observations taken soon thereafter by others of the η Carinae region [15] and of the Orion nebula [16], which also showed a number of X-ray sources coincident with OB stars in that region, argued strongly against the "quiet" binary model — these observations, together with this model, would have implied that every OB star observed by *Einstein* was in fact a binary system of the sort described by van den Heuvel [12], Lamers *et al.* [13], and Conti [14], and implication that did not satisfy Occam's Razor.

This was the first piece of work carried out by what was shortly to become the group comprising the *Einstein* Stellar Survey team: Pippo Vaiana provided the central direction, the drive, and the scientific ambition; Rick Harnden led the analysis of the X-ray data (with the help of Charles Maxson); and I was given the task of making theoretical sense of what we saw.

The initial observational period of the *Einstein* mission led to a number of other major discoveries in this new field of stellar X-ray astronomy: Thus, Ku and Chanan [16] showed that T Tauri stars were luminous X-ray sources, giving impetus to new observational programs focusing on star-formation regions; and Holt *et al.*[17] were able to use the Solid State Spectrometer to obtain the first X-ray line spectrum from a star other than the Sun. (Emission lines from Mg, Si, S, and Fe were detected.) All of these initial results were published in a single volume of the Astrophysical Journal Letters, devoted entirely to results from the *Einstein* Observatory.

It should not be surprising that our group – realizing the terrific opportunity that we were facing – threw itself into the work of implementing the revised stellar observing plan with a great deal of enthusiam. The amount of work to be done was clearly enormous, and we were able to attract two additional scientists who from the beginning became essential members of the stellar X-ray group, Leon Golub and Roberto Pallavicini. A number of students also joined the hunt, including Jay Bookbinder, Jürgen Schmitt, and Ken Topka. In short order, we were well on our way to completing this initial survey, finding the remarkable result (for that time!) that virtually all stars that we had placed on our observing list turned out to be an X-ray source. These initial results were presented at a landmark summer school held at Erice (Sicily, Italy) in the summer of 1979[18], and led to a re-evaluation of the role that stellar observations were to play in the CfA's observing program for *Einstein*: With the strong support of Riccardo Giacconi, Pippo Vaiana led the construction of a new observing plan (aided by Leon Golub, Rick Harnden, Charles Maxson, and me), which eventually formed the core of the *Einstein* Stellar Survey[4]. This represented a significant re-allocation of observing time for the CfA X-ray astronomy group, whose interests had been largely focused on extragalactic astronomy, and would not have occured if it not had been for Pippo Vaiana's vision (which recognized the rich future lying ahead for stellar X-ray astronomy) and Riccardo Giacconi's support (which grew from his faith in the scientific judgement of his junior colleagues).

4 Epilogue

By mid-1979, it had become obvious to the stellar community that *Einstein* Observatory observations were providing an entirely new perspective on the issue of stellar activity; and in short order, the Guest Observer program was swamped by requests for stellar observations – stars became one of the major programmatic targets for the *Einstein* Observatory Guest Observer program until the mission's demise. During these early days of the Guest Observer program, stellar astronomers outside the CfA collaboration also made many major contributions, including surveys of OB stars[19,20,21], of low mass stars[22,23], and of the Hyades open cluster[24]; the discovery of the so-called "coronal dividing line" separating giant stars with and without X-ray emission [25,26,27]; extensive surveys of active evolved stars in binary systems [28,29,30] (viz., RS CVn stars); and more extensive studies of X-ray emission line spectra from "normal" X-ray emitting stars [31].

This period also saw the first steps made in establishing the now well-known "solar-stellar connection": Roberto Pallavicini and collaborators were able to show [32] that the stellar X-ray luminosity for late-type stars with a well-developed surface convection zone correlated very well with the rotation rate of the underlying star. The rest, as they say, is history, and is recorded in the first report of the *Einstein* Observatory Stellar Survey group [4], published in 1981, as well as in the first review of stellar X-ray astronomy [33], published in 1985.

Acknowledgments

The work originally carried out by the stellar X-ray astronomy group at Harvard-Smithsonian Center for Astrophysics, led by G.S. Vaiana, was supported in large part by NASA grants. As is apparent from this article, my work in stellar X-ray astronomy was largely motivated by my interactions with Pippo Vaiana, to whom I will always be grateful, and whose memory I am very pleased to honor. I would also like to acknowledge the wonderful collaborative spirit of our group, and especially acknowledge the contributions of my colleagues Leon Golub, Rick Harnden, and Charles Maxson, and my students Jay Bookbinder and Jürgen Schmitt. I would also like to acknowledge the stellar astronomy group at the Osservatorio Astronomico di Palermo and at the University of Palermo, who continue to vigorously develop the field of study founded by Pippo Vaiana.

References

1. R.C. Catura, L.W. Acton and H.M. Johnson, *ApJL* **196**, L47 (1975).
2. R. Mewe *et al.*, *ApJL* **202**, L67 (1975).
3. J. Nugent and G. Garmire, *ApJL* **226**, L38 (1978).
4. G.S. Vaiana *et al.*, *ApJ* **245**, 163 (1981).
5. R. Mewe, *Space Sci. Rev.* **24**, 101 (1979).
6. P. Gorenstein and W.H. Tucker, *ARA&A* **14**, 373 (1976).
7. R. Rosner *et al.*, *ApJL* **249**, L5 (1981).
8. S. Serio, in this volume.
9. F.R. Harnden *et al.*, *ApJL* **234**, L51 (1979).
10. J.I. Castor, R. McCray and R. Weaver, *ApJL* **200**, L107 (1975).
11. R. Weaver *et al.*, *ApJ* **218**, 377 (1977).

12. E.P.J. van den Heuvel, in *Structure and Evolution of Close Binary Systems*, ed. P. Eggleton (Reidel, Dordrecht, 1976).
13. H.J.G.L.M. Lamers, E.P.J. van den Heuvel and J.A. Pettersen, *A&A* **49**, 327 (1976).
14. P.S. Conti, *A&A* **63**, 225 (1978).
15. F.D. Seward *et al.*, *ApJL* **234**, L55 (1979).
16. W.H.-M. Ku and G.A. Chanan, *ApJL* **234**, L59 (1979).
17. S.S. Holt *et al.*, *ApJL* **234**, L65 (1979).
18. R. Rosner and G.S. Vaiana, in *X-ray Astronomy*, ed. R. Giacconi and G. Setti (Reidel, Dordrecht, 1980).
19. J.P. Cassinelli and G.L. Olsen, *ApJ* **229**, 304 (1979).
20. K. Long and R.L. White, *ApJL* **239**, L65 (1980).
21. G.C. Steward and A.C. Fabian, *MNRAS* **197**, 713 (1981).
22. H.M. Johnson, *ApJ* **243**, 234 (1981).
23. B.M. Haisch, J. Linsky, F.R. Harnden, Jr., R. Rosner, F. Seward, and G.S. Vaiana, *ApJL* **242**, L99 (1980).
24. R.A. Stern, M.C. Zolcinski, S.K. Antiochos and J.H. Underwood, *ApJ* **249**, 647 (1981).
25. T.R. Ayres and J.L. Linsky, *ApJ* **235**, 76 (1980).
26. T.R. Ayres, J.L. Linsky, G.S. Vaiana, L. Golub and R. Rosner, *ApJ* **250**, 293 (1981).
27. J.L. Linsky and B.M. Haisch, *ApJL* **229**, L27 (1979).
28. F.M. Walter and S. Bowyer, *ApJ* **245**, 671 (1981).
29. F.M. Walter and L.V. Kuhi, *ApJ* **250**, 254 (1981).
30. F.M. Walter, W. Cash, P.A. Charles, and S. Bowyer, *ApJ* **236**, 212 (1980).
31. J.H. Swank, N.E. White, S.S. Holt and R.H. Becker, *ApJ* **246**, 208 (1981).
32. R.P. Pallavicini, L. Golub, R. Rosner, G.S. Vaiana, T. Ayres and J.L. Linsky, *ApJ* **248**, 279 (1981).
33. R. Rosner, L. Golub and G.S. Vaiana, *ARA&A* **23**, 413 (1985).

STELLAR CORONAL STRUCTURES

S. SERIO
Istituto e Osservatorio Astronomico, Piazza del Parlamento 1, 90134 Palermo (Italy)
E-mail: serio@oapa.astropa.unipa.it

The discovery of X-ray emission from normal stars throughout most of the HR diagram was among the momentous findings of the *Einstein* Observatory. The mechanism responsible for X-ray activity in late type stars is believed to be essentially similar to the one operating in the Sun, i.e. dissipation of magnetic energy in an optically thin corona. Coronal magnetic fields should thus induce structuring of the coronal plasma in loop-like structures, as observed in the Sun. I shall discuss on the structure of stellar coronae, essentially along the lines of a program initiated by the late G.S. Vaiana after the early observations of the solar corona by *Skylab*, and continued in Palermo. I shall show how spectrally resolved X-ray observations can evidence the existence of stellar coronal loops, both in the static case and during flares.

1 Introduction

Already in 1968, early X-ray images of the Sun obtained during the program for the development of X-ray optics initiated by R. Giacconi at American Science and Engineering showed that the solar corona is highly inhomogeneous, the hot plasma being confined in structures associated with photospheric magnetic fields [1]. That program culminated with the launch of *Skylab* in 1973, whose S-054 telescope produced thousands of X-ray images of the solar corona, showing the widespread confinement of the coronal plasma in structures or "loops", on scales ranging from the few 10^8 *cm* of the coronal Bright Points, to the several 10^9 *cm* of the Active Regions, up to those comparable to the solar radius of Large Scale Structures [2].

The observation of X-ray emission from late type stars throughout most of the HR diagram since the early results of the *Einstein* mission [3] has been interpreted as evidence that stellar coronae are, similarly, made up of loop-like structures.

Under the inspiration of the late G.S. Vaiana, our group in Palermo has dedicated some effort to the study of magnetically confined coronal loops, of their properties, and of the related diagnostic. In the following, after a brief outline of the physics of magnetically confined coronal plasma (sect. 2), I shall highlight some recent developments in the study of stellar coronal loops, both static (sect. 3), and flaring (sect. 4), especially with the aim to show that we have now tools for the diagnostic of the structure of stellar coronae, which will become very effective with the forthcoming CCD spectra from *AXAF* and *XMM*.

2 Basic Physics of Coronal Loops

The basic physics of coronal loops is simple. Magnetic confinement, when the plasma pressure is smaller than the magnetic pressure, does in fact force the plasma to move essentially along the field lines. The problem is, therefore, one dimensional, and governed by compressible hydrodynamic equations; in the case of loops of constant cross section, they assume the form [4]:

$$\frac{dn}{dt} = -n\frac{\partial v}{\partial s} \tag{1}$$

$$nm_H\frac{dv}{dt} = -\frac{\partial p}{\partial s} - nm_Hg_s + \frac{\partial}{\partial s}\left(\mu\frac{\partial v}{\partial s}\right) \tag{2}$$

$$\frac{d\epsilon}{dt} + (p+\epsilon)\frac{\partial v}{\partial s} = E_H(s,t) - n^2\beta P(T) + \mu\left(\frac{\partial v}{\partial s}\right)^2 - \frac{\partial}{\partial s}\left(\kappa_o T^{5/2}\frac{\partial T}{\partial s}\right) \tag{3}$$

$$p = (1+\beta)nkT \tag{4}$$

$$\epsilon = \frac{3}{2}p + n\beta\chi \quad . \tag{5}$$

In the above equations, n, T and p are, respectively, the plasma density, temperature and pressure, while ϵ is the internal energy density, $E_H(s,t)$ is a source term for coronal heating, which can be both stationary and time varying, and $P(T)$ is the optically thin radiative loss functions from a plasma with appropriate abundances; s is the coordinate along the length of the tube and v the plasma velocity, g_s is the component of the solar gravity along the loop, β the ionization fraction, μ the viscosity coefficient, m_H the hydrogen mass, k the Boltzman constant, χ the hydrogen ionization potential, and κ_o Spitzer's coefficient of thermal conductivity.

2.1 Static Coronal Loops

Whenever the source term E_H in (3) is constant and the plasma velocity $v = 0$, then the loop is hydrostatic, and the relevant equations reduce to:

$$\frac{dp}{ds} = -nm_H g_s \tag{6}$$

$$E_H(s) = n^2 \beta P(T) + \frac{d}{ds}\left(\kappa_o T^{5/2} \frac{dT}{ds}\right) \tag{7}$$

A typical description of a loop is that of having its legs anchored in the photosphere, being symmetrical about its apex and semicircular in shape, for convenience. When the radius of such a loop is much smaller than the pressure scale height $s_p = 2kT/gm_H$, Rosner, Tucker and Vaiana (RTV[5]) scaling laws apply:

$$E_H = \frac{\kappa_o T_{max}{}^{7/2}}{L^2} = 10^5 p^{7/6} L^{-5/6} \quad , \tag{8}$$

$$T_{max} = 1.4 \cdot 10^3 (pL)^{1/3} \quad . \tag{9}$$

where L is the semi-length of the loop, and T_{max} is the temperature at its top.

Using the RTV scaling laws, and $x = s/L$, it is easy to see that (7) becomes:

$$\frac{T_{max}{}^{1/2}}{3.1 \cdot 10^{19} \kappa_o K^2} P(T) - \frac{1}{T_{max}{}^{7/2}} \frac{d}{dx}(T^{5/2} \frac{dT}{dx}) = 1 \quad , \tag{10}$$

i.e., the distribution of temperature along the loop is independent on its length and on plasma pressure within. This result is very important, since it shows a fundamental limitation of static coronal models: although one may fit stellar coronal observations with loop models, in the hypothesis that X-ray emission is dominated by a system of identical single loops much smaller than s_p, we can only determine T_{max} or, equivalently, the product of pressure and length, but *not a definite value for the length of the loops.*

In this case, therefore, loop models are essentially equivalent to isothermal models, often used to fit coronal spectra. In fact, we can express the X-ray luminosity for an isothermal atmosphere as $L_x \propto p^2 P(T)/T^2\ V$, where V is the volume of the emitting plasma. If we assume that a system of identical loops covers a fraction f (filling factor) of the stellar surface S_*, the *effective* coronal temperature will be related through (10) and the spectral response of the instrument to the maximum temperature. Then, using the scaling law in

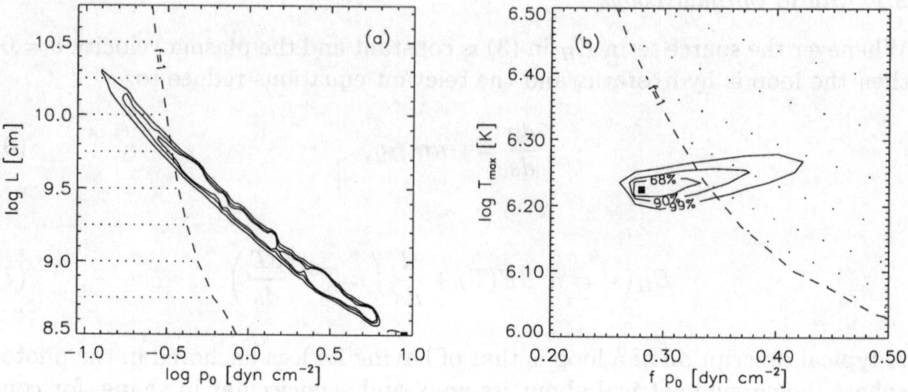

Figure 1: Analysis of the *ROSAT* PSPC spectrum of Procyon with hydrostatic loop models. The lines show confidence regions at the 68%, 90% and 99% levels in the p, L (panel a), and in the fp, T (panel b) loop parameter space. The dashed line is the locus of surface filling factor $f = 1$, and the unphysical region where $f > 1$ is shaded. The closing of the contours at high p/low L is due to insufficient sampling of the parameter space.

(9), we can write $L_x \propto p^2P(T)/T^2V \propto TP(T)pfS_*$. Therefore, a description of a degenerate identical-loops corona in terms of its maximum (or effective) temperature and of the average pressure (i.e. the product of filling factor and plasma pressure), is essentially equivalent to a one temperature model.

The left panel of Fig. 1 shows a typical "banana" shaped fit of a loop model to *ROSAT* data in the p, L plane, dramatizing the uncertainty in the determination of the parameters of the fitting model [6]. Similar results are obtained in the L, T and in the p, T planes. In contrast, the right panel shows that the average coronal pressure fp and the maximum loop temperature T_{max} are reasonably well determined.

It is worth noting, however, that the spectra of isothermal model coronae differ from the spectra of one loop coronae, since these last are not isothermal. While, with good statistics and with the improved spectral resolution of the *ROSAT* PSPC with respect to the *Einstein* IPC, one temperature fits are often unacceptable, fits with loop models perform better, even if they often cannot determine anything else than the loop's maximum temperature and the average pressure [7 8].

Of course, the degeneracy of the static loop models is broken if the height of the loops is comparable to the pressure scale height. In this case the scaling law (9) becomes[9]:

$$T_{max} = 1.4 \cdot 10^3 (p_o L)^{1/3} \; exp[-L/2s_p] \quad , \tag{11}$$

where p_o is the pressure at the base of the corona. The effect of gravity is indeed responsible for closing the contours of confidence intervals at high L and low p in Fig. 1a. Our ability to "image" static stellar coronae, i.e. to obtain information about the dimensions of dominant static loops, rests therefore with the possibility that they are comparable to the pressure scale height. We are somewhat comforted that, in the case of the solar corona, where $s_p \approx 6 \cdot 10^3 T(K)\ cm$, the dimensions of active region loops are, although generally smaller, essentially comparable to s_p. If this remains the case in stellar coronae, then it will be possible to obtain some information about their structure. Moreover, also independent density estimates, possible with density sensitive line ratios, together with X-ray spectra can be of help.

Generally, fits of loop models to pulse height spectra obtained with the *Einstein* IPC have not been satisfactory, since they have given poor constraints on the length of the loops, or required very large loops. Attempts to overcome the problem of degeneracy considering loops with a variable cross section have also been tried[10 11]; this is possible, however, at the expense of adding an extra parameter to the fit, the ratio of top to bottom loop cross section. Moreover, the values fitted for this parameter are either near to 1[10], or in obvious contrast[11] with observations of the solar corona, which show coronal loops with generally little, if at all, variations in their cross section.

2.2 Flaring Loops

The study of variability often offers a way of estimating the physical size of an astrophysical source. This is particularly true for a coronal plasma, whose radiative cooling time can be easily related to the volume of the cooling region.

The typical application here is that of building a flare model, using numerical solutions of equations (1) to (5), for a loop during the decay of a flare. Reale *et al*[12] for example, have analyzed *Einstein* IPC data for a flare in Prox. Cen, to derive a loop's half length of $7 \cdot 10^9\ cm$, slightly smaller than the stellar radius, but with aspect ratio similar to solar coronal loops. They have also considered the influence on the X-ray spectral distribution, during the flare decay, of possible deposition of further energy, and determined that the data were compatible with such deposition acting only for a short fraction of the decay.

Basically, one models the source term E_H in (3) as a function of time. The flare is caused by its sudden increase above the steady state level, and the decay starts as soon as it starts to decrease. Disregarding viscosity, assuming complete ionization ($\beta = 1$) and that the plasma at the flare peak is essentially static ($v = 0$), it is easily seen that (3) can be reduced to [13]:

$$\frac{dS}{dt} = \frac{kE_H}{p} \quad , \tag{12}$$

where $S = \ k\ ln(T^{5/2}/p)$ is the entropy per particle at the top of the loop, and E_H is the peak value of the energy deposition term. In the hypothesis of a sudden decrease of E_H, using RTV scaling laws, one obtains a new scaling law relating the flare peak temperature T_o and the half length of the loop to the entropy decay time τ:

$$\tau = \frac{3.7 \cdot 10^{-4} L}{\sqrt{T_o}} \approx 120 \frac{L_9}{\sqrt{T_7}} \quad , \tag{13}$$

where L_9 and T_7 are, respectively, the loop semi-length and the flare peak temperature in units of 10^9 *cm* and $10^7 K$.

Extensive simulations of flaring loop models have also shown that the decay of peak temperature and density are exponential, although with different time scales, and that a relationship of the type

$$T \propto n^{\xi} \tag{14}$$

holds during most of the decay phase. The simulations also show that $\xi = 2$ when the switching off of the energy term in (3) is abrupt, while it can get as low as 1/2 (the value also given by the RTV scaling law) when it is more gradual [13]. These results allow a powerful diagnostic tool for flare decay, since they can be used to ascertain the amount of heat deposited during the decay phase, if any, together with the dimensions of the loops, simply by a comparison with a grid of numerical flare models.

3 Fitting Stellar X-ray Data with Static Loop Models

The improved energy resolution of the *ROSAT* PSPC with respect to the *Einstein* IPC has allowed new insight on the properties of stellar coronal plasma and methods are being developed to fit stellar coronal pulse height spectra with static coronal models, whose extension to the *ASCA* CCD spectra and to those that will be obtained from *AXAF* and *XMM* appears very promising. The number of independent channels in these spectra does in fact help to profit on the dependence on gravity, and the simulations show that best model fits reproduce well the parameters of the models, with an indetermination of the loop length less severe than with proportional counter spectra [8].

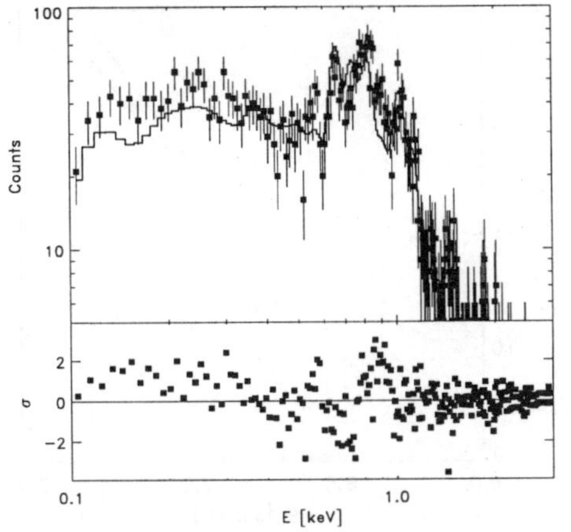

Figure 2: A two temperature fit (solid line) to a simulation of an *XMM* EPIC 5000 counts spectrum, in the hypothesis that it is originated by a solar-like star, with a single dominanting loop with maximum temperature $t = 2.3 \cdot 10^7$ K and $L = 10^9$ cm. The large residuals imply that it will easily be possible to distinguish between single loop and two temperature model coronae.

A second, often used, model to fit stellar data when the isothermal model does not work well, has been a two-temperature model. A detailed comparison of loop model fits iwith one and two-temperature fits has been conducted by Ciaravella, Maggio and Peres [14]. While coronal loop *ROSAT* PSPC and *ASCA* SIS spectra with sufficient statistics ($\approx 10^3$ counts or more) are not fitted by isothermal models, two temperature models provide a good fit. This indeed implies that the two temperatures components often found in standard procedures of stellar X-ray data analysis do not necessarily mean the presence of two different components in the stellar coronae. The two components come out naturally as a result of the differential emission measure distribution along the loop and of the two temperature model adopted for the fit.

With increased count statistics and spectral resolving power, loops with high plasma pressure and comparable in length to the pressure scale height are not fitted even by two temperature models. Figure 2 shows a two temperature fit of a simulated loop model with the *XMM* EPIC camera. The fit is clearly unacceptable, showing that with the forthcoming CCD cameras we should be able to make further progress in understanding the structure of stellar coronae.

One of the characteristics of loop models that have made them somewhat unpopular, is the fact that often the formal fit to real data give very high value for the loop lengths. For example, Fig. 3 shows a single loop fit to a ≈ 2000

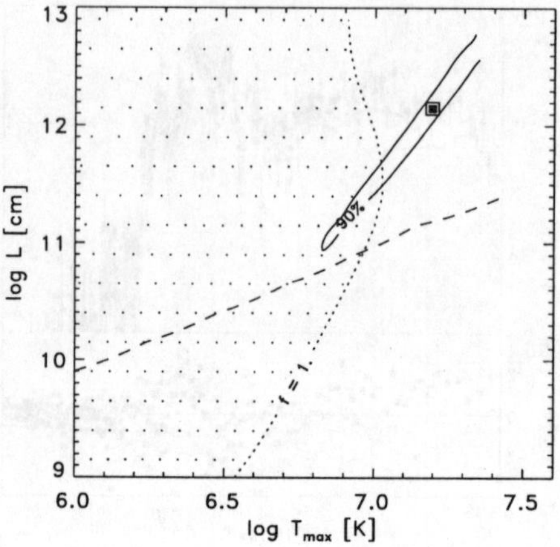

Figure 3: 90% confidence level contour plots of the fit of a one loop model to a ≈ 2000 counts *ROSAT* PSPC observation of a solar-like Hyades star (VB 77). The square corresponds to the formal value of the best fit, the dashed line to the pressure scale height, and the unphysical region, where $f < 1$ is shaded.

counts *ROSAT* PSPC pulse height spectrum of VB 77, a solar-like Hyades star. Of course the occurrence of extremely high loop lengths has to be taken with caution, since typically even stringent confidence levels of the fit span several orders of magnitude in the length of the structures, because of the degeneracy of the loop models. Even a small perturbation, for example the presence of an additional component in the corona, could introduce a large discrepancy between input model data and fitted values. One example is given in Fig. 4 showing how a fit of the same data shown in Fig. 3 with a two loop model [15] improves sensibly the physical meaning of the fit. While the formal value of the length given from the one loop fit was $\approx 10^{12}$ *cm*, i.e. one order of magnitude larger that the stellar radius, the loop giving most of the contribution is now well determined, and small.

In the case of *ROSAT*, two loop fits are probably applicable only in a few cases, because of the requirement of high signal to noice spectra. The increase in sensitivity and in energy resolving power foreseen with the *AXAF* and the *XMM* CCDs will certainly allow to study the contribution of different structures to the same stellar corona.

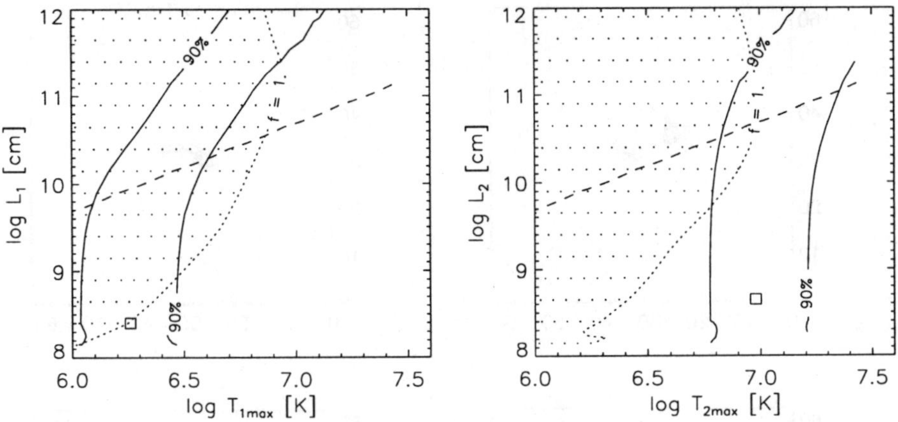

Figure 4: Same as in Fig. 3, for a two loop model fit to the same data. The length of the dominant loop (right panel) is now constrained to a small value.

4 Stellar Coronal Loop Lengths from Flare Decays

To extend the results for Prox Cen mentioned in sect.2.2 above to other flares would require time consuming case by case numerical simulations. Reale *et al.*[16] have developed a method to estimate the size of flaring loops, parametrizing the results of an extensive set of numerical simulation of stellar flares. Equations (13) and (14) are given, in fact, in terms of temperature and density at the top of the loop, while actual stellar observations can be rather expressed in terms of some kind of spectral property or "color", that can be related to a global *effective temperature*, and of the total count rate, that can be related to the emission measure. This method, therefore, has to be adapted and specifically calibrated for each specific instrument, folding simulated light curves and spectral distributions with the instrumental response.

To ascertain the capabilities of the method, a first application has been made to a number of solar flares observed by the *Yohkoh* Soft X-ray Telescope. In this case, temperature can be estimated by pixel to pixel filter ratios, density (or, better, emission measure) from pixel counts, and the length eventually obtained can be compared to that actually observed on the X-ray images. The results of the simulations can be interpolated with a simple relationship among the ratio of the observed decay time of the flare light curve to τ in (13) and the slope in the temperature-emission measure plane.

The method can be easily applied in most cases, when the slope in the

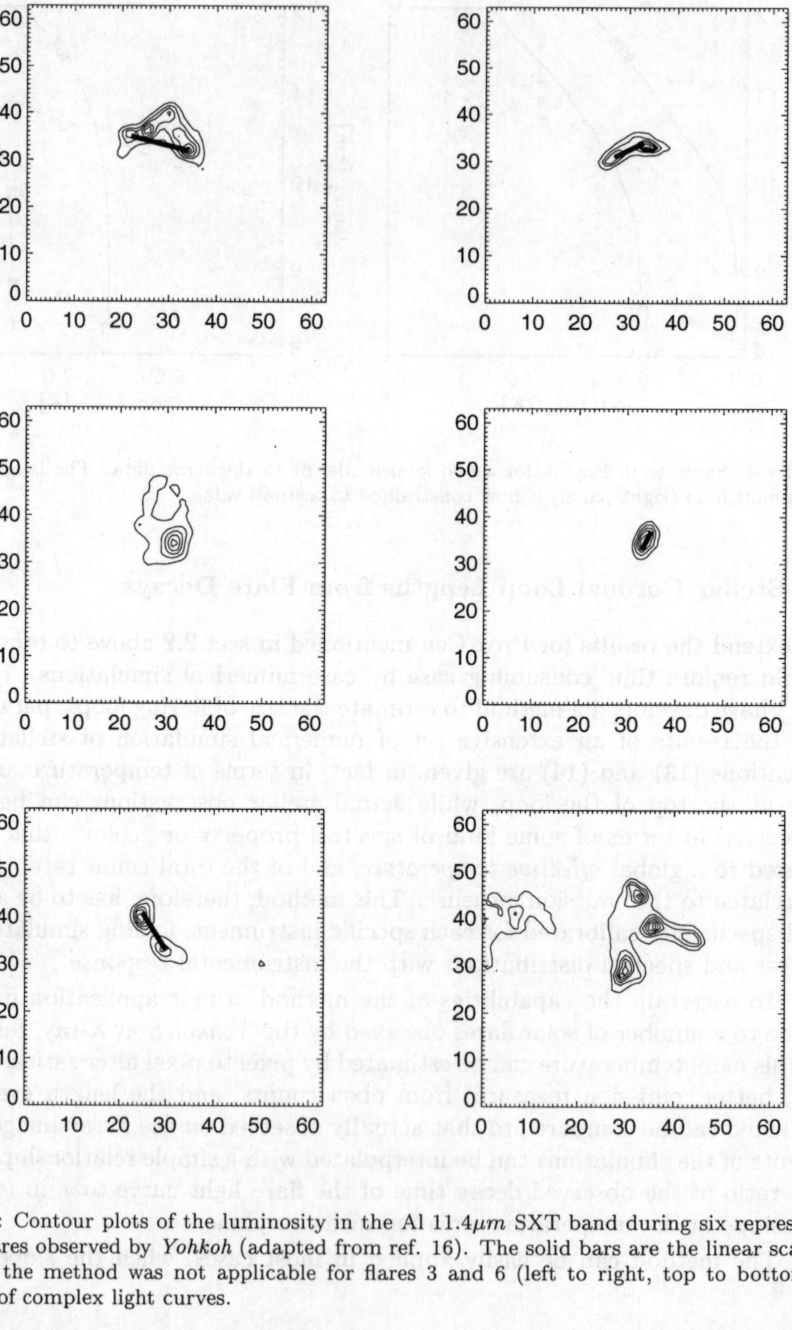

Figure 5: Contour plots of the luminosity in the Al 11.4μm SXT band during six representative flares observed by *Yohkoh* (adapted from ref. 16). The solid bars are the linear scales derived; the method was not applicable for flares 3 and 6 (left to right, top to bottom), because of complex light curves.

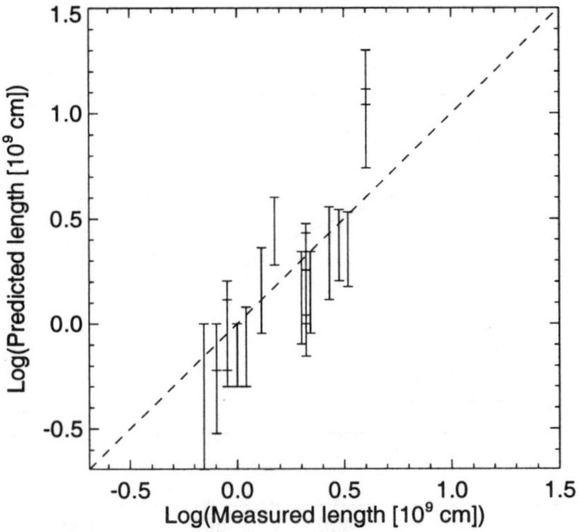

Figure 6: Ranges of estimated linear dimensions as a function of observed dimensions for 17 flares observed by *Yohkoh* (adapted from ref. 16).

Temperature-Emission measure plane is well defined and larger than .3, and when the light curve shows a clear exponential decay. When these conditions are met, it gives results in good agreement with those directly derived from the *Yohkoh* images. Fig. 5 shows a sample of the results, for 6 *Yohkoh* flares. In two cases the method could not be applied, the third and the sixth (left to right, top to bottom), because these flares showed complex evolving morphology and lighcurves, while in all other cases the agreement between the derived lengths and the dimensions of the flaring loops is excellent. The results of the comparison of the lengths derived directly from the images and those derived using the method above are shown in Fig. 6

The obvious success of this first attempt has induced to extend the method to observations of stellar coronae. One immediate problem stays in the fact that the observations of the *ROSAT* PSPC are generally not well fitted with one temperature. Moreover, if the best fit temperature is used anyhow, simulations show lerge systematic errors in the inferred lengths. The cure here is to substitute to the temperature a *Spectral Shape Index* (SSI), obtained as a linear combination of PSPC channel counts optimized for the analysis of loop models [17], and to develop the appropriate relationship between slopes and decay times.

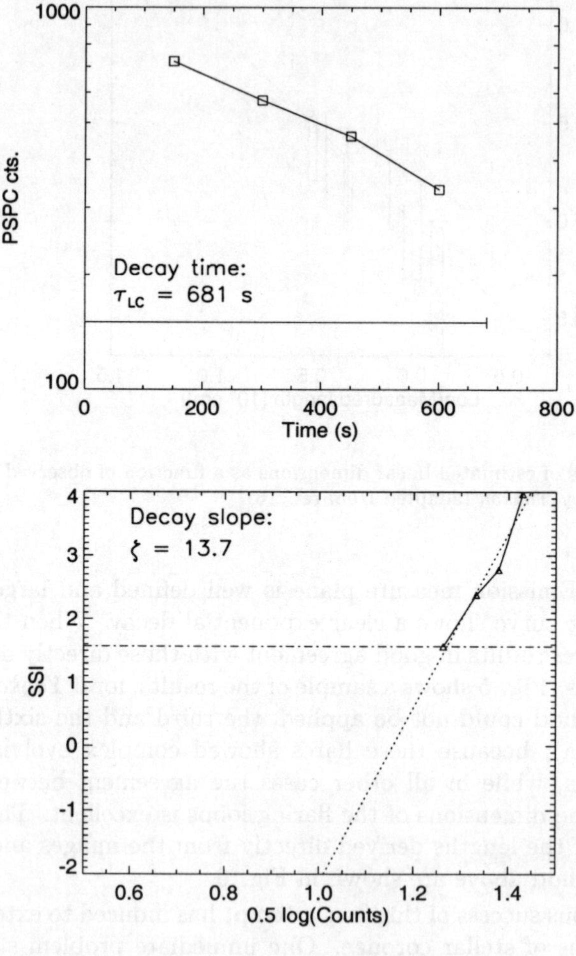

Figure 7: *ROSAT* PSPC light curve (upper panel), and trajectory in the Count Rate-Spectral Shape Index plane (lower panel) for the decay of a flare observed on the dMe star AD Leo (from ref.??16) The decay time and the slope determine that the flare occurred in a loop $4.1 \cdot 10^9$ *cm* high, with little heat deposition during the decay.

An example of the application of the method to a Rosat observation of a flare on the dM3.5e star AD Leo is shown in Fig. 7 [18]. The decay of the light curve is well described by an exponential, and the trajectory in the SSI-Log(Count rate) plane is essentially linear. One can derive that very little heating is deposited in the loop during the decay, and that the length of the loop is $4.1 \cdot 10^9$ *cm*.

Although in this case we cannot perform a direct comparison, as in the case of solar loops, it appears that the method is very powerful, based on simple assumptions, and easy to apply. Indeed, we plan to apply it extensively to stellar flares observed by *ROSAT*, and to develop its extension to CCD spectra from the forthcoming X-ray missions.

5 Conclusions

The standard view of the coronae of late type stars is that they are somewhat similar to the solar corona, where the plasma is obviously confined by magnetic fields in loop-like structures.

Our ability to confirm this view with the observation of static coronae rests in the capability of energy resolved observations. I have shown that the temperature distribution of static coronal loops small with respect to the coronal pressure scale is independent on their length. If the emission of a stellar corona is dominated by an ensemble of such loops with the same maximum temperature, information about their length can only be obtained by means of line ratio density diagnostics. When the length of the loops is comparable to the pressure scale height, then is this degeneracy removed. Spectral resolved observations of sufficient statistics, such as those that will be obtained with the large X-ray observatories to be launched before the end of the century will be able to give this information, even in the case of more complex models involving the contributions of loops of different sizes.

On the other hand, in the case of stellar flares, we have already the capability to estimate the dimensions of the structures were the flares occurr. The reassuring picture that is emerging is that these structures are similar to those present in the Sun.

Acknowledgements

I am grateful to my colleagues in Palermo, and in particular to A. Maggio, G. Micela, G. Peres and F. Reale, for extensive discussions and for allowing to cite unpublished results. This work has been supported by Ministero dell'Università e della Ricerca Scientifica e Tecnologica.

References

1. G.S. Vaiana, W.P. Reidy, T. Zehnpfenning, L. Van Speybroeck, and R. Giacconi, 1968, *Science* **161**, 564 (1968)
2. G.S. Vaiana, A.S. Krieger, and A.F. Timothy, *Sol. Phys.* **32**, 81 (1973)
3. G.S. Vaiana *et al.*, *Ap. J.* **245**, 163 (1981)
4. G. Peres, R. Rosner, S. Serio, and G.S. Vaiana, *Ap. J.* **252**, 791 (1982)
5. R. Rosner, W.H. Tucker, and G.S. Vaiana, *Ap. J.* **220**, 643 (1978)
6. S. Serio, in *Magnetodynamic Phenomena in the Solar Atmosphere-Prototypes of Stellar Magnetic Activity*, ed. Y. Uchida *et al.* (Kluwer, Dordrecht, 1996).
7. A. Ciaravella, G. Peres, A. Maggio, and S. Serio, *Astron. Astrophys.* **306**, 563 (1996)
8. A. Maggio and G. Peres, *Astron. Astrophys.* **306**, 553 (1996)
9. S. Serio, G. Peres, G.S. Vaiana, L. Golub, and R. Rosner, *Ap. J.* **243**, 288 (1981)
10. R.A. Stern, S.K. Antiochos, and F.R. Harnden Jr., *Ap. J.* **305**, 417 (1986)
11. J.R. Lemen, R. Mewe, C.J. Schrijver, and A. Fludra, *Ap. J.* **341**, 474 (1989)
12. F. Reale, G. Peres, S. serio, R. Rosner, and J.H.M.M. Schmitt, *Ap. J.* **328**, 256 (1988)
13. S. Serio, F. Reale, J. Jakimiec, B. Sylwester, and J. Sylwester, *Astron. Astrophys.* **241**, 197 (1991)
14. A. Ciaravella, A. Maggio, G. Peres, *Astron. Astrophys.* **320**, 945 (1997)
15. A. Maggio and G. Peres, *Astron. Astrophys.*, in press.
16. F.Reale, R. Betta, G. Peres, S. Serio, and J. McTiernan, *Astron. Astrophys.*, in press.
17. G. Micela, *et al.*, in preparation.
18. G. Reale, G. Micela, G. Peres, R. Betta, S. Serio, *Mem. SAIt*, in press.

UNDERSTANDING X-RAY STARS: The Discovery of Binary X-ray Sources

E. J. Schreier
Space Telescope Science Institute, Baltimore, MD 21218, USA
E-mail: schreier@stsci.edu

H. Tananbaum
Harvard-Smithsonian Center for Astrophysics, Cambridge, MA 02138, USA
E-mail: ht@cfa.harvard.edu

The discovery of binary X-ray sources with UHURU introduced many new concepts to astronomy. It provided the canonical model which explained X-ray emission from a large class of galactic X-ray sources: it confirmed the existence of collapsed objects as the source of intense X-ray emission; showed that such collapsed objects existed in binary systems, with mass accretion as the energy source for the X-ray emission; and provided compelling evidence for the existence of black holes. This model also provided the basis for explaining the power source of AGNs and QSOs. The process of discovery and interpretation also established X-ray astronomy as an essential sub-discipline of astronomy, beginning its incorporation into the mainstream of astronomy.

1 Introduction

During X-ray astronomy's pioneering decade, the 1960's, a few tens of X-ray sources were discovered via instruments flown on rockets and balloons. Several of these sources were identified via positional coincidence with known astronomical objects, and a number of physical mechanisms were hypothesized to explain the X-ray emission – thermal bremsstrahlung, inverse Compton scattering, synchrotron emission, line emission, and black body radiation. The early rocket flights and subsequent optical and radio observations indicated that some of these mechanisms were indeed present, e.g. synchrotron emission from the Crab nebula, but the basic mystery persisted - how could objects, especially compact ones, emit 1000 times more energy in X-rays than in the optical or radio bands. The sources were probably too small to be explained by thermal bremsstrahlung, synchrotron, or inverse Compton emission from an extended region; and too luminous to be explained by black body radiation from a neutron star. The mystery persisted until after the launch of UHURU on December 12, 1970.

In the course of an incomparably exciting one-year period starting in January 1971, more than one hundred new X-ray sources were detected. Many of these were found to vary. As discovery followed discovery, a picture of a new class of objects gradually developed, remarkable for the variability of their X-ray emission on a broad range of timescales. A model emerged which could explain the origin and variability of the vast amounts of energy radiated in the X-ray band. Accretion of matter onto a collapsed object in a binary stellar system became the canonical model to explain these luminous, variable galactic X-ray sources.

The type of collapsed object in the binary system varies – white dwarf, neutron star, black hole – as does the nature of the companion star, from supermassive early O and B stars to low-mass late-type stars. The details of the compact object, companion, and overall system geometry determine the great range of characteristics observed for the X-ray sources. But the basic mechanism of matter transferred from a star in a binary system to a compact companion, accelerated by gravity and often forming an accretion disk, and emitting energy in the form of X-rays as it falls onto the compact object remains the fundamental model for a large set of galactic X-ray sources. In fact over the past three decades, accretion of matter onto collapsed objects (albeit somewhat larger, by up to 9 orders of magnitude) also has become the model of choice to explain the energy source in active galaxies and QSOs.

2 The History

2.1 Cygnus X-1

The first variable galactic X-ray source studied in detail with UHURU was Cygnus X-1. Although Cygnus X-1 had been reported to vary in intensity between different balloon and rocket flights in the late 1960s, UHURU revealed variability on time scales shorter than 1 second. The initial discovery resulted from the observation of rapid variability superimposed on the expected triangular shape of the transfer function, as UHURU's collimators scanned past the direction of Cygnus X-1. Dr. Minoru Oda, visiting at American Science and Engineering and working on the new data with Riccardo and the UHURU group, noted that the pulsations looked at times periodic, but varied from scan to scan. We hypothesized that this could be caused by an underlying rapid periodicity, beating against the 0.096 second telemetry sampling rate. An intensive search for an underlying periodicity led to the regime of milliseconds, too short to be explained by anything other than a collapsed object – a neutron star or black hole[1,2].

Even in the absence of regular pulsations, as later concluded, the extreme and rapid variability of the source required the presence of a compact object. Figure 1 shows the irregular variability of Cygnus X-1 on 4 different timescales (0.096s, 0.48s, 4.8s, and 14.4s) with the triangular response of the UHURU

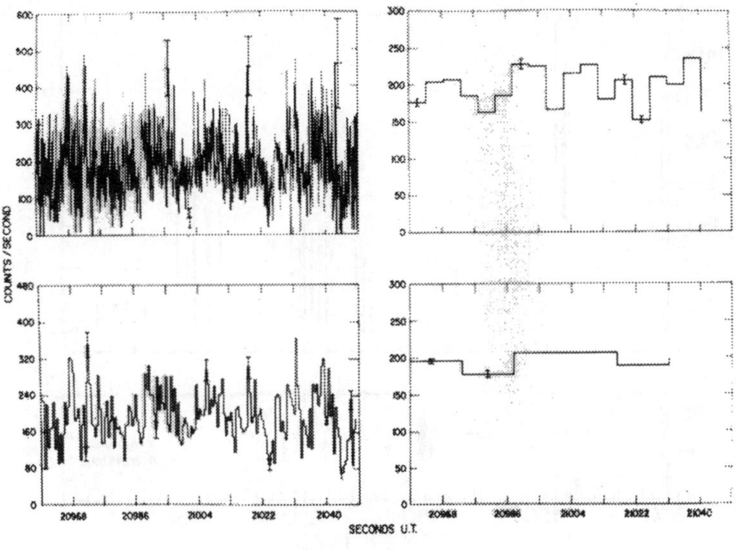

Figure 1: Cyg X-1 short time-scale variability

collimator removed.

The focus of study turned to the detailed nature of the pulsations and to the optical and/or radio identification of the X-ray source. A rotating modulation collimator experiment on a rocket flight was planned by the MIT X-ray group to better determine the location of the source. Given the UHURU discovery of rapid time variability, the MIT group optimized half of the rocket experiment to study the intensity fluctuations. The accurate location of Cygnus X-1 made possible by UHURU and by the MIT rocket flight [3] led to the discovery of a radio source [4,5] and the subsequent optical identification with a 9th magnitude B0 supergiant, a 5.6 day spectroscopic binary system [6,7]. The mass function of the system led to a mass estimate for the X-ray source well in excess of 3 solar masses. There was, and still is, no known way to have a neutron star or white dwarf in excess of 3 solar masses, leading to the likelihood that the compact object we were observing was a black hole.

Since the idea of a black hole actually being observed was revolutionary, it was necessary to work through all of the arguments in the chain with great care. The identification of the radio and optical source with the X-ray source was based on a positional coincidence of order $1'$. So this was considered to

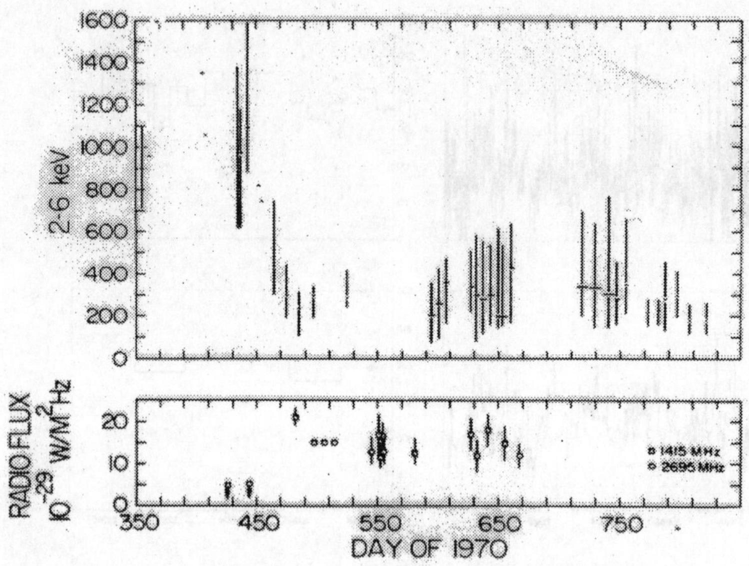

Figure 2: X-ray and radio transition in Cyg X-1

be the weak link by supporters and skeptics alike. Fortunately, as shown in Figure 2, a remarkable transition with a decrease of a factor of 4 in the average 2-6 keV X-ray intensity from March to April 1971 was noted to occur at the same time as the weak radio source, coincident with the 9th magnitude BO supergiant, first appeared [8]. This discovery convinced the UHURU team (and much of the community) of the validity of the identification and the likelihood that Cygnus X-1 was a black hole. Subsequent correlated X-ray and radio transitions as well as an arc sec X-ray position from the Einstein Observatory have confirmed the identification as a certainty.

Further intense observational work and refinements in theory followed the initial Cygnus X-1 discoveries. We took the dramatic step of slowing down the rotation rate of UHURU to create longer continuous data streams to better monitor the X-ray variability. Our observations over the next months, as well as those with other experiments [9,10] revealed no constant underlying period, but rather pulse trains of varying quasi-periodicities ranging down to milliseconds. Refinements in the mass estimate of the primary star and in the mass function led to the lower limits for the compact source as low as 3 solar masses. But essentially, the hypothesis that Cygnus X-1 was a compact object too massive to be a neutron star, accreting matter from a massive companion, withstood the test of time. An accretion disk around a several-solar-mass black hole, with a modest accretion rate as low as 10^{-9} solar masses per year, can produce the X-ray luminosity, with irregularities in the inner regions of the disk producing the quasi-periodicities. More important than the detailed parameters, the presence of a relatively massive collapsed object in a binary system showed one way to detect the presence of a black hole through the effects of its gravitational field. The impact of this discovery, providing compelling evidence for the existence of black holes, is certainly one of the most important and exciting breakthroughs from X-ray astronomy.

2.2 Centaurus X-3

After the discovery of the rapid variability of Cygnus X-1, we began to search for comparable behavior from other bright sources with UHURU. The best method for discovery, given the unknown nature of the phenomena we were studying, was the thorough inspection of the computer generated "strip charts" of our scans! All blips, large and small, were worthy of careful checking for reappearance on other scans and for other appropriate analyses. The long, slow scans of the galactic plane invoked to study Cygnus X-1 quickly revealed the presence of intensity fluctuations in other galactic X-ray sources.

The first additional variable source found during this intensive near-real-time monitoring was signalled by the notation of several "???" on the computer printout by one of the authors of this paper (EJS) during a late night session. The variability in the signal from a very strong source in the general direction of Centaurus was even more dramatic than had been seen with Cygnus X-1 – several pulses with nearly 100% modulation. The data suggested either a hardware problem or the modulation of the instrument response function by a source whose intensity varied with a period of approximately 5 seconds (see Figure 3). The next morning the other author arrived to find Riccardo poring over the printouts with other scans of the same source showing the same five second pulsations!

With Centaurus X-3, unlike Cygnus X-1, the same pulsation period was always seen. There was no need to hypothesize a hidden underlying period. We did have to write new computer programs to measure and refine the period by fitting trial periods to the pulses, and by comparing pulse phase from one observation to the next. We learned to correct for the orbital motion of the earth – we discovered what radio astronomers had been doing for a few years – and then to search for changes in the period. The high Q of the pulsations, and the similarity to radio pulsars, led to the interpretation of the compact object

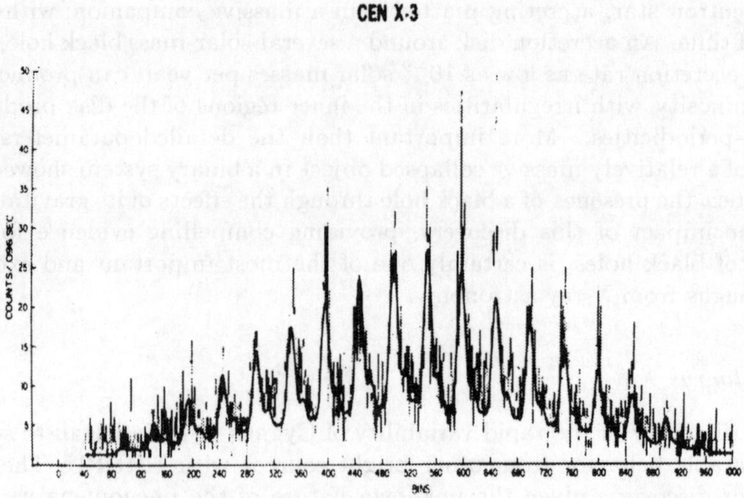

Figure 3: Periodic 4.8s pulsations from Cen X-3

in Cen X-3 as a rotating neutron star [11]. A pulsating white dwarf model was also hypothesized and remained an alternative interpretation for some time, based in part on the relative long period of the pulsations, although it pushed the limits of existing stellar models. Later observations confirmed the long-term persistence of the underlying period, other examples of faster pulsing X-ray sources were found (see 2.3 below), and detailed measurements showed the speed-up of the period - these factors combined to rule out the white dwarf possibility, leaving the rotating neutron star model as compelling.

The evolution of our understanding of Cen X-3 proceeded unevenly over the next few months. As the earlier data were studied in more detail, small changes in the pulsation period were seen within a day, as shown in Figure 4, where we fit constant periods to short intervals of pulse arrival times. The data also revealed large changes in the average intensity, with transitions occurring within the order of an hour, as seen in Figure 5.

As we accumulated more data and the changes in pulsation period were plotted along with the intensity data, as shown in the original hand-drawn Figure 6, a correlation became obvious. The intensity transitions, in both directions, were eventually shown to be periodic. The repeating step function

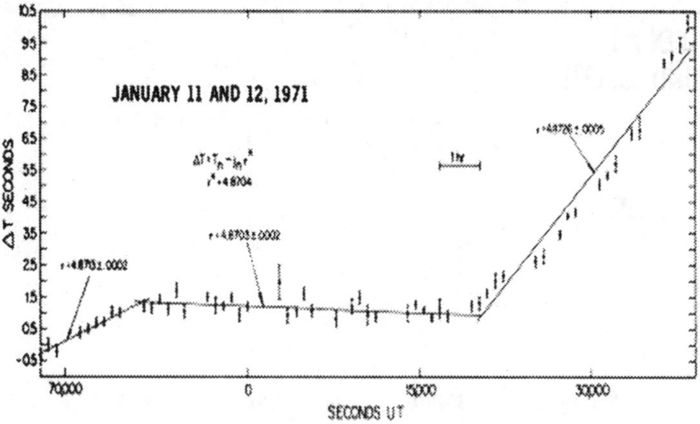

Figure 4: Small changes in the Cen X-3 pulsation period

in intensity – high for 1.5 days, low for half a day – was then quickly interpreted as an occultation of the pulsating source by a companion object, suggesting that the X-ray source was in orbit around a large star.

If the pulsating x- ray source were itself moving in an orbit of high inclination, consistent with occultations, then Doppler variations (delays and advances in pulse arrival times) would be expected; the magnitude of the pulsation period changes were indeed consistent with this interpretation. The final epiphany was the recognition that the period measurements could all be fitted with a sine curve. The sinusoidal variation in the 4.85 second X-ray pulse period, correlated with the two-day periodic intensity variations as shown in Figure 7, provided the unambiguous signature of a compact object in an occulting binary star system [12] (see Figure 8).

The compact object-binary companion model for Cen X-3 allowed estimates of masses and orbit sizes from the X-ray data alone! In our excitement, we re-discovered the binary mass function (with attendant mistakes pointed out by our optical astronomy colleagues), not realizing that this could be found in textbooks. Eventually, a 13th magnitude B0 supergiant was identified with the system, consistent with the predictions of a massive early-type star made

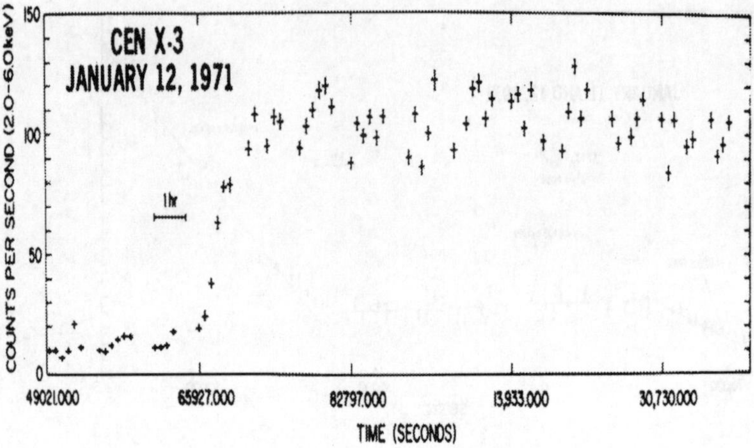

Figure 5: Large changes in Cen X-3 average intensity

from the X-ray data. The strong stellar wind associated with the massive star likely provides the source for the accreting material.

We should note that Riccardo Giacconi and his two primary collaborators on this work (the authors of this contribution) were trained as physicists, which no doubt facilitated the building of the UHURU instrumentation and the development of the software to analyze the data. The intellectual path from large intensity changes to regular eclipses and from erratic changes in the pulse period to a sinusoidal variation was, however, a somewhat tortuous one, which illustrates the movement, described by Tucker and Tananbaum in their contribution to this volume, of X-ray astronomy into the larger domain of traditional astronomy (where observers and theorists alike were already quite familiar with binary systems and their characteristics). Of course, during this transition, the X-ray astronomers brought along the black holes and other interesting toys to entertain their new-found colleagues.

We proceeded to monitor the underlying pulse and orbital periods for secular changes. The pulsation period was found to undergo small fluctuations in both directions, but on the average, it decreased over the course of several years of observations. The speed up in rotation was consistent with transfer of

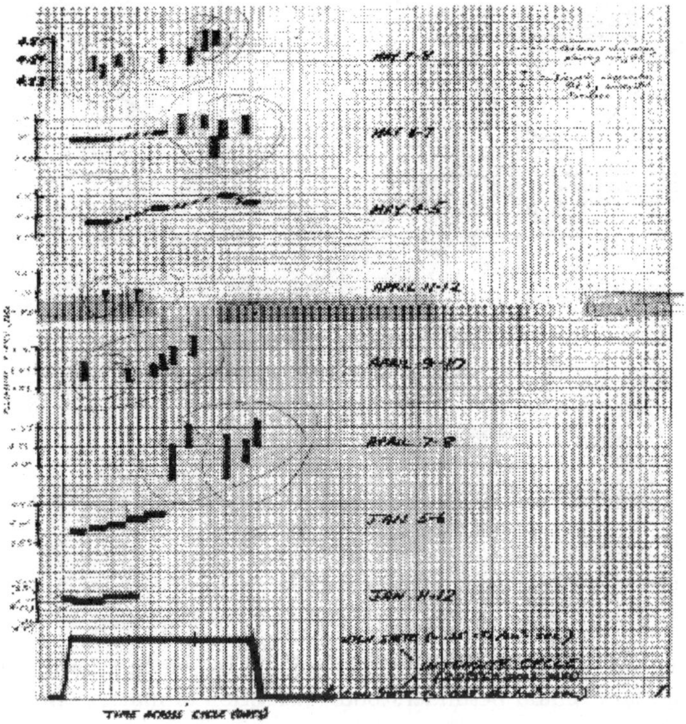

Figure 6: Correlation of period changes with intensity for Cen X-3

angular momentum to the neutron star via the accreting material; it provided evidence that there was an accretion disk even in the presence of a strong wind from the primary star, as shown in Figure 9.

The presence of fluctuations, and the fact that the speed up was less than predicted, suggested a balance between a speed-up torque and dissipative slow-down torques, interpreted as occurring naturally via an equilibrium of the accreting material at the magnetopause, with physically reasonable values for the Alfven and co-rotation radii. Variations in intensity were observed, consistent with this hypothesis. Irregular variations in the orbital period were also observed through the years, suggesting more complicated mass loss and mass exchange effects.

The discovery, first of the regular pulsations, and then of the binary nature

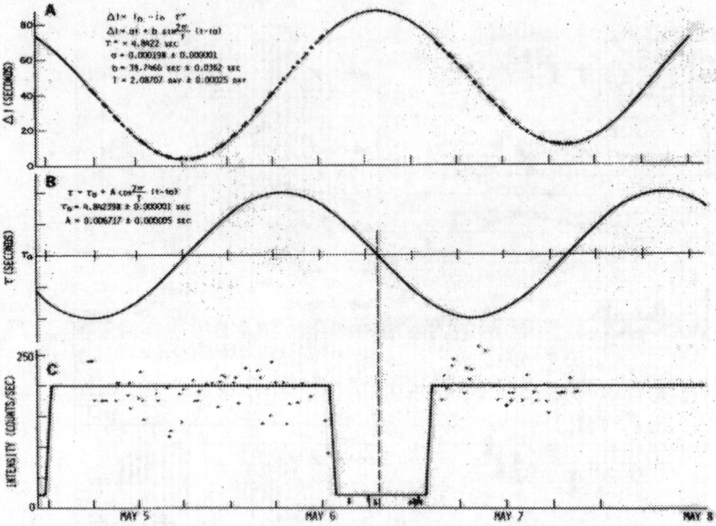

Figure 7: Signature of compact object in binary system Cen X-3

of Centaurus X-3, provided the canonical model by which a large class of "X-ray emitting stars" could be understood - accretion of matter onto a compact object in a binary system.

2.3 Hercules X-1

The discovery of Hercules X-1 followed the initial discoveries of variability in Cygnus X-1 and periodic pulsations in Cen X-3 by several months. That Her X-1 was even discovered to be "interesting" is a somewhat amazing story in itself. With a location nearly 40° above the galactic plane, Her X-1 does not lie in a region scanned very often by UHURU. In fact, if Giacconi and his team had not applied the philosophy of "soft failure" in the design and approach to UHURU (see contribution by Tucker and Tananbaum), it is interesting to speculate what, if anything, we might know today about Her X-1. Digressing briefly, we note that a few near-disasters befell UHURU. In the first six weeks after launch, the transmitter overheated and the satellite nearly fell silent, before being recovered; then the on-board tape recorder failed, setting off a scramble to establish a set of ground stations, providing $\sim 50\%$ coverage for data transmitted in real-time. Later the transmitter again became intermittent

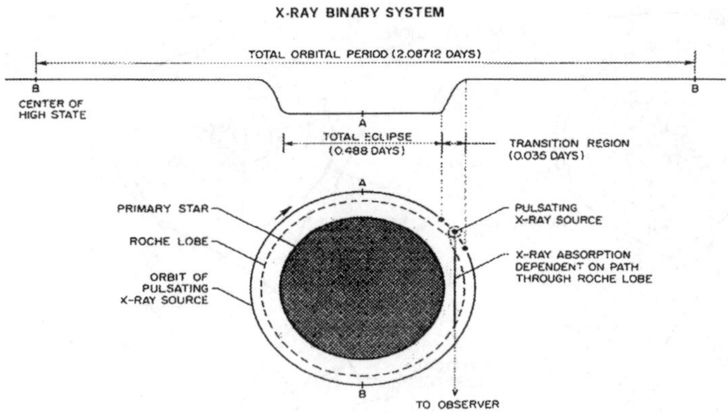

Figure 8: Schematic for compact object in occulting binary star system

leading to several months with large data gaps and lack of satellite control, which in turn led to the loss of the star trackers when they crossed near enough to the Sun to destroy their N-shaped reticle pattern.

Due to Sun angle constraints, related to the transmitter temperature problems, scans covering our primary targets of interest in November 1971 could not go along the Galactic plane, but instead included the region where we would soon discover X-ray pulsations from Hercules X-1. Moreover, to study its pulsations and changes in average intensity, the $\sim 1°$ attitude control and determination provided by the Sun sensors and magnetometers was sufficient. Thus, a major discovery was made with UHURU after many of its primary capabilities were no longer operational (illustrating the value of the "soft failure" philosophy).

A slowed down scan across the constellation Hercules produced the data shown in Figure 10. A period of 1.2368 ± 0.0006 seconds could be fit to this initial observation of duration ~ 100 seconds. Note that we had utilized another feature of UHURU and placed the higher time resolution electronics (0.096 sec) on the broader (5° FWHM) collimator. The band of sky being scanned had been chosen to include Sco X-1 and Cyg X-1, which allowed us to validate an

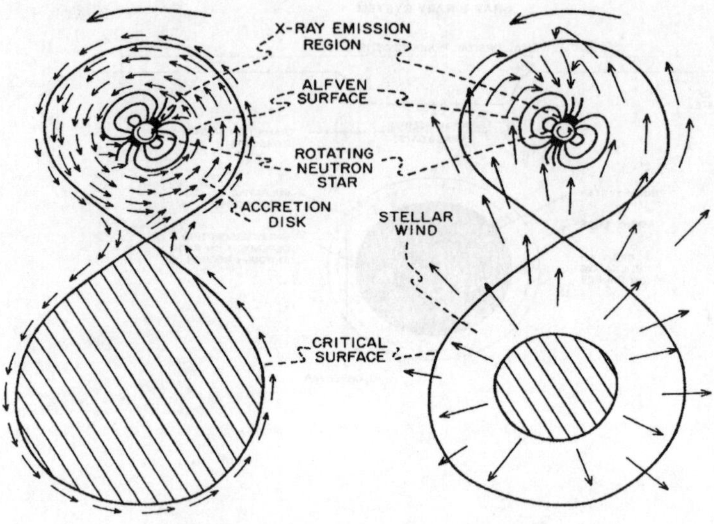

Figure 9: Mass transfer in binary star system

approximate location for the source. The pulsing source was consistent with the position for a relatively weak source in the 2nd UHURU Catalog. So it is fair to say that there would have been a weak source and a crude position for Her X-1, but it is unlikely to have been a target for follow-up observations without the serendipitous discovery of its bright, pulsing state by this scan in November 1971. The intensity also showed large changes on a time scale of days, and extensive coverage for the next 5 months, revealed a binary eclipse with a period of 1.7 days, as well as a new phenomenon - the 35-day cycle[13].

For ~ 10 days, Her X-1 was observed as intense and pulsing, with regular 1.7 day eclipses. It then remained weak or off for ~ 25 days. This surprise led us to monitor its intensity (or at least look for turn-ons and turn-offs) for most of the next year, until the effective end of "normal" UHURU operations with the battery failure in April 1973. While understanding the pulsations and eclipses for Her X-1 was relatively straightforward given the progress with Cen X-3, the UHURU team and the community now had a new mystery to work on - the 35-day cycle of Her X-1. Was this cycle truly periodic (no, but apparently it is quite regular) and how did we explain it (perhaps the most accepted explanation today is precession of a tilted or warped accretion disk).

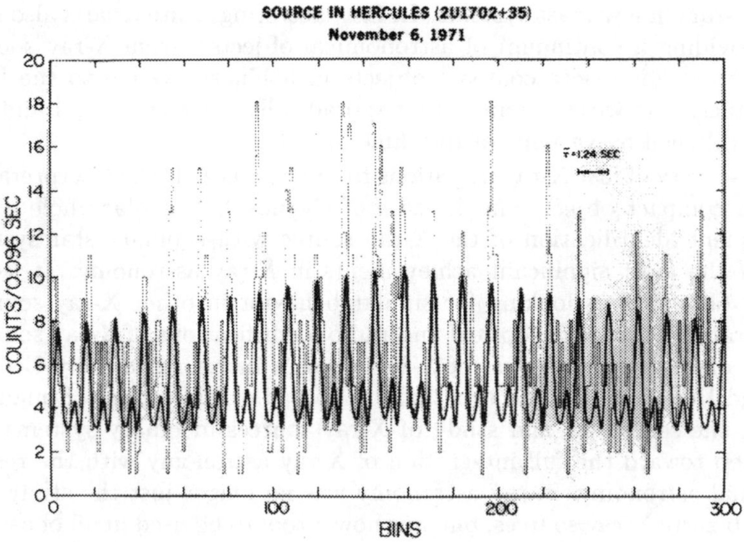

Figure 10: Her X-1 Pulsations

Hercules X-1 was seen to have a Doppler velocity $\sim$ 40% that of Cen X-3 and thereby a mass function $\sim$ 20 times smaller. An optical identification, replete with 1.7 day intensity variations due to heating of one face of the companion star by the X-ray source, was soon discovered by Bahcall & Bahcall [14] and Liller [15]. Mass estimates for the companion star were of order 1.5-2.5 M_o, so Her X-1 provided a system different from Cyg X-1 and Cen X-3 with their massive, early-type companion stars.

The richness of the Her X-1 system (we should also mention X-ray intensity dips possibly related to the crossing of our line of sight by a stream of infalling matter feeding the accretion disk) led to its becoming the most widely studied compact X-ray source (and possibly one of the most widely studied astronomical objects) for an extended time in the 70's.

3 Implications

Currently there are hundreds of X-ray sources detected in our galaxy and in other galaxies which can be explained by compact objects in binary star systems. Of these, there are nearly 20 black holes or black hole candidates, with most of the remainder being neutron stars orbiting high mass primaries, or neutron stars in low mass systems. In fact, accreting white dwarfs also emit X-rays, providing a continuum of astronomical objects - from X-ray sources powered by accretion onto compact objects in a binary system to the large class of cataclysmic variables and other sporadically variable optical and UV sources - explained by variants of the same model.

The discovery of fast X-ray pulsations from Cygnus X-1, the interpretation requiring a compact object – in the case of Cygnus X-1, a black hole – and the subsequent identification of the X-ray source with a binary star system, was one of the most significant achievements in X-ray astronomy. It led to systematic observations looking for similar behavior in other X-ray sources, to concentrated searches for optical and radio identifications of X-ray sources, and to the discovery of the class of X-ray binaries including initially Centaurus X-3 and Hercules X-1, and eventually a host of other systems. Equally as important, the search for and study of X-ray sources in binary systems was a major step toward the full integration of X-ray astronomy with the rest of observational astronomy. X-ray astronomy was no longer just the study of a handful of bizarre X-ray sources, but was now a tool to be used in all of astronomy. The combined force of optical, X-ray and in some cases radio astronomy was required to determine the nature of the objects we discovered. In the case of Cygnus X-1, for example, X-rays determined the size scale of the compact object, radio provided the identification, and optical measured the mass. Meanwhile, the presence of evolved objects in binary systems contributed to a better understanding of supernovae and to the entire picture of stellar evolution, including mass exchange binary systems. One example of change was the recognition (again the data demanded it) that a supernova explosion need not totally disrupt a binary system. Through mass exchange the more evolved star might be the less massive in the system by the time it underwent a supernova explosion. Although the field of X-ray astronomy had to wait until the Einstein Observatory near the end of the '70s for true imaging capabilities, and the resultant ability to attract "guest observers" from outside the ranks of X-ray astronomers, through UHURU it was already moving rapidly towards joining the mainstream of astronomy.

4 Coda (a word we learned from Riccardo)

The experience of working closely with Riccardo and the rest of the small and very dedicated UHURU group during this formative period was unique and tremendously rewarding. The opportunity to see and interpret so many new and exciting facets of the universe, and to participate intimately in the process of discovery was something very special to all of us. We are tremendously grateful to Riccardo for sharing this opportunity with us.

We acknowledge the contributions of the entire UHURU group at American Science and Engineering, Inc. and subsequently at the Harvard-Smithsonian Center for Astrophysics, and the other X-ray astronomers, radio and optical observers who contributed to this work. In the spirit of this retrospective as opposed to formal review, we have included only a sampling of source references for the work presented here. Key participants in the analysis of the early UHURU X-ray observations of Cygnus X-1, Centaurus X-3 and Hercules X-1 included: G. Fabbiano, R. Giacconi, P. Gorenstein, H. Gursky, C. Jones, E. Kellogg, R. Levinson, (who played a particularly important role in uncovering the binary nature of Cen X-3), M. Oda, E. Schreier, and H. Tananbaum. Relevant references for this early work in the field of binary X-ray sources are contained in many reviews, including relevant chapters in two books:

"The Galactic X-ray Sources", by H. Gursky and E. Schreier (1975) in *Neutron Stars, Black Holes and Binary X-Ray Sources*, (H. Gursky and R. Ruffini, ed.), Reidel Publishing Co, Dordrecht.

"Compact X-Ray Sources", by H. Tananbaum and W. Tucker (1974) in *X-Ray Astronomy*, (R. Giacconi and H. Gursky, ed.), Reidel Publishing Co, Dordrecht.

References

1. M. Oda *et al*, Astrophys. J. Lett **166**, L1 (1971).
2. E. Schreier *et al*, Astrophys. J. Lett **170**, L21 (1971)
3. S. Rappaport *et al*, Astrophys. J. Lett **168**, L17 (1971)
4. L. Braes & G. Miley, *Nature* **232**, 246 (1971)
5. R. Hjellming & C. Wade, Astrophys. J. Lett **168**, L21 (1971)
6. B. Webster & P. Murdin, *Nature* **235**, 37 (1972)
7. C. Bolton, *Nature* **235**, 271 (1972)
8. H. Tananbaum *et al*, Astrophys. J. Lett **177**, L5 (1972)
9. S. Holt *et al*, Astrophys. J. Lett **166**, L65 (1971)
10. S. Shulman *et al*, Astrophys. J. Lett **168**, L49 (1971)
11. R. Giacconi *et al*, Astrophys. J. Lett **167**, L67 (1971)
12. E. Schreier *et al*, Astrophys. J. Lett **172**, L79 (1972)
13. H. Tananbaum *et al*, Astrophys. J. Lett **174**, L143 (1972)
14. J. Bahcall & N. Bahcall, Astrophys. J. Lett **178**, L1 (1972)
15. W. Liller, *IAU Circ. 2427* , (1972)

X-RAY EMISSION FROM NEUTRON STARS
- Some personal reflections and recent developments -

Joachim Trümper
Max-Planck-Institut für Extraterrestrische Physik,
85740 Garching, Germany
E-mail: jtrumper@mpe-garching.mpg.de

After a few remarks about the early history of the subject we present a short review of the present observational situation regarding the X-ray emission from isolated neutron stars. In total 32 objects have been detected with ROSAT, the majority of which are radio pulsars showing non-thermal (magnetospheric) emission. For three radio pulsars and three point sources in SNRs thermal emission has been seen which probably comes from the photospheric of the cooling neutron star. A third class comprising the objects represents neutron stars according matter from the interstellar medium.

1 Introduction

Neutron Stars have been a subject of theoretical considerations since the 1930's. The first to consider stars made of nuclear matter was Lew Landau. Rosenfeld[1] has recalled "that when the news of the neutron's discovery reached Copenhagen, there was a lively discussion on the same evening about the prospects opened by this discovery. In the course of it Landau improvised the conception of neutron stars "Unheimliche Sterne", weird stars which would be invisible and unknown to us unless by colliding with visible stars they would originate explosions, which might be supernovae." That was in 1932 during a visit of Landau to Niels Bohr's institute. Two years later and independently, Baade and Zwicky made their remarkable prediction of the existence of neutron stars, and their formation in a stellar collapse accompanied by a supernova explosion [2]. The first quantitative theory of neutron stars published by Oppenheimer and Volkoff in 1939 treated a degenerate neutron liquid in Schwarzschild's metric and determined its Chandrasekhar mass to be $\sim 0.7\ M_\odot$ [3].

2 Historical remarks

The first observational evidence of the existence of objects as compact as neutron stars or black holes came only in 1962 when Riccardo Giacconi and his team discovered the bright X-ray source Scorpius X-1 [4]. This source actually outshined the sun in hard X-rays and it soon became clear that the only way to explain such energetic emission was the accretion of matter onto compact objects, either neutron stars or black holes. Today we know that Sco X-1 is a neutron star in a low mass binary system while Cyg X-1, discovered a little later, turned out to be very likely a black hole in a high mass binary system. A detailed study of these and other compact objects became possible when Uhuru was launched in 1971, Riccardo's second milestone contribution to astrophysics. Uhuru discovered the X-ray pulsations in Cen X-3 and, little later in Her X-1 which showed that these objects were spinning magnetized neutron stars in close binary systems. At present we know a few hundred X-ray binaries with neutron stars and some of them with black holes. They are located in the Milky way and in nearby galaxies, and a wealth of observational and theoretical work has been done on these exotic objects and the extreme physical conditions prevailing in them.

In 1971 we started a program in X-ray astronomy at the University of Tübingen (AIT) which we continued after 1975 at the Max-Planck-Institut in Garching in collaboration with AIT. Our main goal was to observe X-ray binaries by means of spectrometers flown on balloons. With Uhuru already up we came late; but in the hard X-ray band accessible from balloon altitudes we did not face satellite competition yet. One of the early results was the discovery of cyclotron lines in Her X-1 which allowed the first spectroscopic determination of a neutron star magnetic field. The balloon flight was in May 1976, and it took us until December to analyse the data and convince ourselves of the reality of the effect. The final reassurance came on 13 December, and one day later I phoned Riccardo to ask him whether I could give a late talk at the Texas symposium which was taking place during that week in Boston. After I had told him about the new effect, he asked me to buy a ticket. Two days later I gave a short plenary talk arranged by him [5].

At that time Riccardo and his team were already deeply involved in the preparation for HEAO-2 which after its launch was named Einstein observatory. This project became another stunning success and completed Riccardo's hattrick. Using the new possibility to explore the faint X-ray sky Riccardo himself concentrated on the study of clusters of galaxies and on finding the "holy grail of X-ray astronomy", the origin of the X-ray background. Although the Einstein deep surveys allowed to make important steps, they resolved only 15-20% of the background into discrete sources. Obviously one had to dig much deeper. In the mid 1980's Riccardo asked me whether we were interested in a joined ROSAT deep survey project in collaboration with Maarten Schmidt. I invited a young member of our team, who had profound experience in the calibration of the PSPC and computer simulations of ROSAT observations, to participate. It was GFCnther Hasinger who during the following years became our team leader of the ROSAT deep survey project. The results of his highly successful undertaking are described by G. Hasinger and G. Zamorani in this book.

Before coming back to neutron stars after this excursion a few remarks on the early history of ROSAT may be appropriate. We had started to work with Carl Zeiss Company in 1972/73 on X-ray mirrors. Our aim was to reduce the surface microroughness from the 2nm reached at that time to much lower levels and to use Zerodur instead of metal or fused silica. At the end, the ROSAT Zerodur mirrors had a microroughness $\sigma = 0.25$ nm, resulting in an extremely low scattering halo (scaling with σ^2) and in a high contrast of the telescope. In 1975 we made a proposal to our ministry for a Roentgen satellite. One of our main objectives was a very sensitive all sky survey which would be followed by detailed observations of selected sources. During my visits to Cambridge I discussed our ideas with Riccardo. His reaction was very positive and he suggested to include a copy of the Einstein HRI provided by SAO to be used for high resolution pointed observations. At the same time he was somewhat sceptical about the scientific validity of the all sky survey, because a lot of telescope time would be wasted looking at empty regions. Nevertheless, Riccardo supported ROSAT a lot, and that helped to get American astrophysicists and NASA interested in the mission. He also participated in the international Spitzingsee meeting which was organized by MPE in 1978 to discuss the future of High Energy Astrophysics in general and our satellite plans in particular [6].

3 X-ray emitting single neutron stars

Already at the beginning of X-ray astronomy, in the early 1960's single neutron stars had been proposed to be X-ray sources because of their expected high surface temperatures and relatively long ($\sim 10^5 - 10^6$ yr) cooling times [7]. One of the possible candidates was the strong X-ray source located in the region of the Crab nebula. Using the lunar occultation technique the famous NRL rocket experiment showed that the X-ray source was extended and located right at the position of the nebula [8]. Friedman and his collaborators had hoped to see a point source, the neutron star, and with this experiment they came very close to discovering it. Actually, with a somewhat larger collecting area and better photon statistics they would have seen the point source on top of the nebula emission. Thus the epochal discovery of neutron stars was left to radio astronomers. This event had an enormous impact on astrophysics and astrophysicists. Many people got attracted by the new field including myself.

Pulsed X-ray emission from the Crab pulsar had been discovered by a NRL rocket experiment soon after its radio discovery. But the Einstein observatory was the first telescope sensitive enough to detect the much weaker

X-ray emission from other pulsars. Pulsations were found for the LMC-pulsar (PSR 0540-69) and PSR 1509-58, while PSR 0656+14, PSR 1929+20 and PSR 0950+08 could just be detected. In addition point sources in the supernova remnants RCW 103 and G269.5+10 were found which showed X-ray fluxes consistent with cooling neutron star emission. Another pulsar, PSR 1055-52, was detected with EXOSAT a few years later.

While it was obvious that the young and fast pulsars in the Crab and in the LMC emit nonthermal radiation probably connected with the acceleration of highly relativistic particles the case was not so clear for the other sources. But the fluxes were roughly consistent with those predicted for neutron stars by standard cooling theories and this remained the preferred interpretation for quite a while.

In comparison with its predecessors ROSAT provided a much larger sensitivity, in particular at low X-ray energies (< 0.28 keV). This and the relatively good spectral resolution of the PSPC facilitated spectral studies and the distinction between thermal and nonthermal spectra. The improved angular resolution combined with the extremely low particle background of the PSPC (~ 1 count/pixel/day!) made it a very powerful instrument for studies of very faint sources. At the same time the small HRI error boxes helped to make identification and to fight source confusion. Last not least neutron star studies profited from the long pointing times of ROSAT and the long mission duration. More recently extremely important complementary information has come from ASCA providing a better spectral resolution and going to higher X-ray energies, up to 10 keV which facilitated the measurement of power law spectra. In the following we shall review briefly the current observational situation.

4 Nonthermal X-rays from radio pulsars

So far 27 radio pulsars have been detected in X-rays, 11 of which show X-ray pulsations at the radio pulse period [10]. This list includes Geminga, the first radio-quiet pulsar discovered in X-rays [9], which may have been seen in the radio band recently [20]. The compilation also includes 7 millisecond pulsars which all were found with ROSAT, RX J0437-47 being the first and most prominent example. A few of these sources which we call the three musketeers (PSR 0656-14, Geminga and PSR 1055-52) show a dichotomic behaviour [11]: a soft thermal component seen at $E < 0.28$ keV and a hard power law component detected at $E \geq 0.5$ keV which are phase shifted with respect to each other by $\sim 100^o$.

Using power law fits for all pulsars (after substracting the thermal components in the cases of the three musketeers) and using standard distances derived from the radio dispersion measure one can derive the 4s-luminosities. It turns out that these luminosities measured in the ROSAT band (0.1–2.4 keV) show a surprisingly narrow linear correlation with the spin down power $\dot{E}$: $L_x = 10^{-3}\dot{E}$ as shown in Figure 1 [10]. A similiar correlation had been found earlier by other authors [11] [12] on the basis of much smaller samples, more limited spectral resolution, and much poorer photon statistics. The most striking feature of the new data is that the correlation holds in the same way for normal pulsars and for the much older millisecond pulsars. The scatter of the distribution is impressively small in view of the effects which are expected to widen it, such as distance errors and differences in the aspect angles under which the pulsar beams are seen. The narrow intrinsic correlation between X-ray luminosity and rotational energy strongly suggests that the X-ray luminosity is drawn mainly from the rotational energy and not from the heat content. This implies a nonthermal origin of the X-rays which is supported by ROSAT/ASCA spectra measured for a couple of sources showing power laws at energies ≥ 0.5 keV.

5 Thermal radiation from Cooling Neutron Stars

Neutron stars are born hot. They cool very rapidly in the beginning due to neutrino emission which dominates until ages of $\sim 10^4$ yr, after which photon cooling from the surface becomes dominant.

Fitting blackbody spectra to the soft thermal components of the three musketeers mentioned above yields blackbody luminosities which roughly agree with the predictions of standard cooling models, as shown in Fig. 2. Other data points shown in this figure were obtained for point-like X-ray sources in the supernova remnants RCW103, Puppis A, and G 269.5+10, using the estimated ages and distances of the supernova remnants. None of these three sources in SNR shows pulsations, neither in radio nor in X-ray. But their spectra are quite soft, roughly compatible with blackbody spectra of cooling neutron stars. We conclude with all reservation that in these six objects we are seeing predominantly photospheric radiation from the hot neutron star.

It is well-known that blackbody spectra are bad approximations in the case of normal stars or white dwarfs. Realistic photospheric spectra have been calculated for neutron stars by several authors (e.g. [13] [14] [13]) taking into account the opacities of dense hydrogen and helium (both magnetic and unmagnetic) or iron atmospheres. They generally show that hydrogen and helium models give better fits to the observed X-ray spectra than blackbody or pure iron models. Compared with blackbody fits hydrogen helium photospheric models yield lower temperatures and larger radii. Perhaps the most convincing spec-

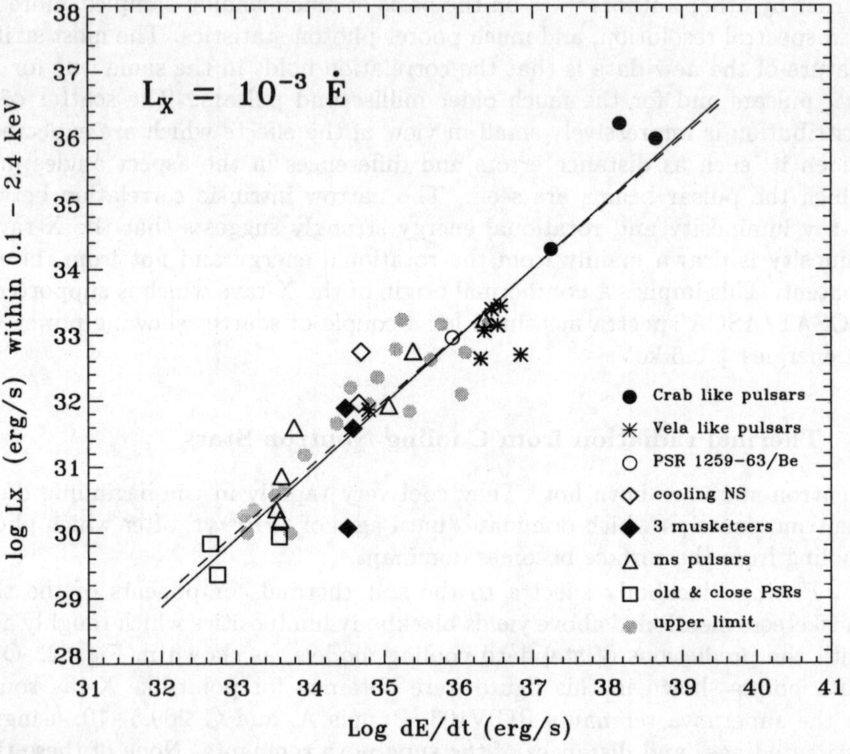

Figure 1: X-ray luminosity vs. spin-down energy loss for all rotation-powered pulsars detected by ROSAT as of October 1996. We note that for the three musketeers Geminga, PSR 0656+14 and PSR 1055-52 the low energy thermal component has been subtracted from the data. The solid line represents $L_x \sim \dot{E}^{1.03}$, the dashed line $L_x = 10^{-3} \times \dot{E}$. 1σ upper limits for the X-ray luminosity of all pulsars which have a $\dot{E}/4\pi d^2 \geq 1.5 \times 10^{-11}$ erg s^{-1}cm^{-2} and have been observed but were not detected in pointed and serendipitous ROSAT observations are indicated.

trospheric study has been made recently for 1E 1207.4-5209 in G 269.5+10 for which both ROSAT PSPC and ASCA spectra are available [16]. Furthermore the interstellar hydrogen column can be determined using the SNR spectrum taken by ROSAT and since the SNR is a shell source of relatively large diameter the "SNR contamination" of the ASCA point source spectrum is negligible. A blackbody fit to the spectral data yields an unrealistically small radius of the neutron star of $1.2(d/2\text{kpc})$ km and a unredshifted temperature $T_{bb} = 3{\times}10^6$ K. For a hydrogen photospheric model the fit gives a much more realistic radius of $(6-13)(d/2kpc)$ km and a $T_{eff} = (1.4-1.9) \times 10^6$ K.

Finally we mention that for the three musketeers, optical emissions have been discovered with HST. In the case of Geminga the optical flux falls right on top of the Raleigh-Jeans part of the photospheric spectrum fitting the ROSAT data [17]. A narrow spectral feature which appears in the optical spectrum of Geminga has been interpreted in terms of the proton cyclotron resonance in a magnetic field of 1012 G which roughly coincides with that derived from magnetic braking [18]. In the two other cases the optical fluxes are about a factor of 10 and 4 above the Raleigh-Jeans curve [18] for PSR 0656+14 and PSR 1055-52, respectively.

6 Candidates Neutron stars accreting from the Interstellar Medium

Pulsar birth rates and supernova rates suggest that up to 10^9 neutron stars exist in our Galaxy, a million times more than have been actually observed yet. After $\sim 10^7$ years when the radio emission has faded away and the surface cooled below 10^5 K isolated neutron stars become invisible unless they accrete matter from the interstellar medium [19]. The correponding accretion rate and luminosities can be calculated using the Bondi-formula which predicts that the luminosity is proportional to the density of the ISM and the inverse cube of the neutron star velocity. The number of old isolated neutron stars expected to show up in the ROSAT all sky survey has been estimated to range from 2.000 to 10.000 [19].

In the light of these high expectations the outcome of intensive searches in the ROSAT data bases have been disappointing. They yielded so far only two good candidates and a few possible ones. The criteria of these searches have been a soft X-ray spectrum and a sufficiently faint counterpart in the optical/UV in order to discriminate against single white dwarfs, supersoft sources etc.

The first of the candidate RX J1856.5-3754 is a bright ROSAT source (3.3 cts/s), having a very soft spectrum ($kT = 57$ eV, $N_H = 1.4 \times 10^{20}\text{cm}^{-2}$ which has been identified with a 26th magnitude blue stellar object [22 23]. Its

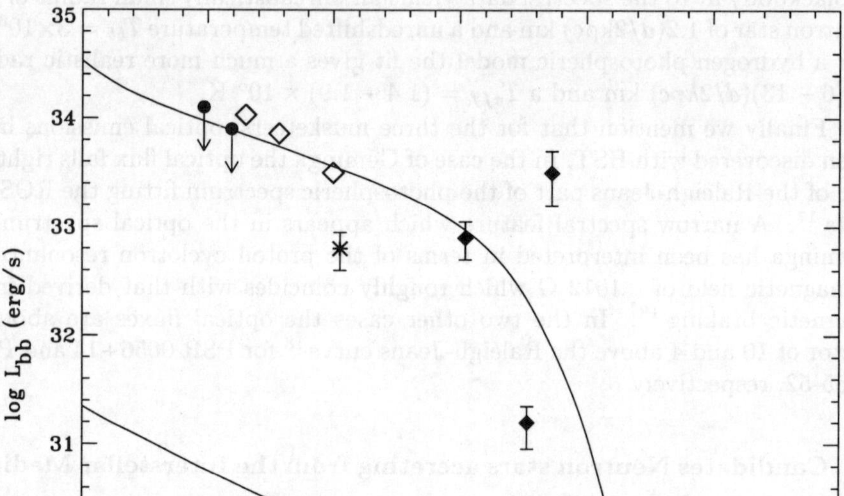

Figure 2: Bolometric blackbody luminosities L_{bol}^{∞} for 8 neutron stars of increasing age: Crab pulsar (upper limit); point sources in RCW 103, Puppis A, G 269.5+10; pulsars PSR 0656+14, Geminga and PSR 1055-52. For comparison the curves for standard and accelerated neutron star cooling (FP model) are shown (Umeda et al. 1993). The error bars are obtained from spectral fits, but do not include errors in the distances.

optical flux roughly fits the Rayleigh-Jeans extrapolation of the X-ray spectrum supporting the identification with a hot neutron star. The second source RX J0720.4-3125 was discovered in the ROSAT all sky survey. ROSAT pointed observations showed a constant count rate of 1.64 PSPC cts/s [24]. The very soft spectrum can be fitted by a blackbody of $kT = 79$ eV, $N_H = 1.3 \times 10^{20}$ cm^{-2}. No optical counterpart could be found so far, and the $f_x/f_o >$ 10.000 is consistent with a single neutron star. The most interesting point is that this source shows a very significant coherent pulsation at 8.39115 seconds indicating that the neutron star is rotating and possesses a magnetic field. From the condition that the Alfven radius should be smaller than the Kepler rotation radius one can infer a magnetic field strength of less than 10^{10} G. On the other hand it must be of that order to produce the strong pulsation.

Why is the number of such sources so much lower than the predictions ? One main point is that the search cannot be extended down to the ROSAT survey limit because at the faint end the photon statistics is too poor to establish the soft nature of the spectrum. Another reason is that at fainter X-ray fluxes even the most powerful optical telescopes available are not sensitive enough to constrain the f_x/f_o or even to detect the optical emission.

One has to better understand these systematics and increase the source statistics before one can draw conclusions about space densities of old neutron stars. A very important aspect is that these studies should constrain the velocity distribution of neutron stars at small velocities where the radio pulsar methods do not work.

7 Concluding remarks

After 35 years of X-ray astronomy this new discipline has become an integral and indispensible part of astrophysics. Riccardo has led the way with the first rocket experiment, the Uhuru satellite and the Einstein observatory. Others have continued. Today X-ray astrophysics encompasses objects ranging from the moon and comets out to the most distant quasars, from the extremely compact neutron stars to the clusters of galaxies, the largest physical formations in our universe.

After 30 years of neutron star astrophysics we are now seeing the faint photospheric emission from these tiny stars. The low fluxes present a challenge both for X-ray and optical telescopes. But Landau's unheimliche Sterne have become visible in the proper sense. In a few years detailed spectroscopic studies with future instruments will give a wealth of new information about their physical properties: elements abundances, gravitational redshifts, magnetic fields, masses and radii and the behaviour of matter at densities exceeding those of atomic nuclei.

References

1. L. Rosenfeld: Proceedings of the Sixteenth Solvay Conference on Physics, Editions de l'Université de Bruxelles, p. 174 (1974)
2. W. Baade and F. Zwicky, Phys. Rev. 46, 76 (1934)
3. J.R. Oppenheimer and G.M. Volkoff, Phys. Rev. 55, 374 (1939)
4. R. Giacconi et al., Phys. Rev. Letters 9, 439 (1962)
5. J. Trúmper et al., Proc. 16. Texas Symposium on Relativistic Astrophysics, Annals of the New York Academy of Sciences Vol 302, p 538, (1977)
6. X-ray and Gamma-ray Astronomy in the 1980's, Eds. B. Aschenbach, K. Pinkau, European Science Foundation report (1978)
7. S. Tsuruta, Physics Reports Vol. 290, (1997) in press
8. S. Bowyer et al., Science 146, 912 (1964)
9. J.P. Halpern and S.S. Holt, Nature 357, 222 (1992)
10. W. Becker and J. Trümper, Astron. Astrophys. (1997) in press
11. H. Oegelman, in "The Lives of the Neutron Stars", Eds. A. Alpar, G. Kiziloglu and J. van Paradijs, Kluwer Academic Publisher, (1995)
12. F.D. Seward and Z.R. Wang, ApJ 332, 199 (1988)
13. R. Romani, ApJ 313, 718 (1987)
14. V.E. Zavlin et al., Astron.Astrophys. 215, 141 (1996)
15. G. Pavlov et al., Astron.Astrophys. 289, 837 (1994)
16. V.E.. Zavlin, (1997), private communication
17. R.D. Meyer et al., ApJ. 433, 265 (1994)
18. G.F. Bignami et al., ApJ. Letters 456, L111 (1996)
19. J.P. Ostriker et al., ApJ. Letters 6, 179 (1970)
20. A.D. Kuzmin and B. Ya. Losowskii, Astronomy Lett. 23, 295 (1997)
21. O. Blaes and P. Madau, The Astrophys. Journal 403, 690-705, (1993)
22. F. Walter et al., Nature 379, 233 (1996)
23. R. Neuhäuser, Astron.Astrophys. in press
24. F. Haberl et al., Astron.Astrophys. in press

X-RAY IMAGING OF CROWDED FIELDS AND GRAIN-SCATTERED HALOS

GEORGE W. CLARK
Center for Space Research and Department of Physics, Massachusetts Institute of Technology, Cambridge, MA 02139

JONATHAN W. WOO
Harvard-Smithsonian Center for Astrophysics, Cambridge, MA 02138

Abstract

The advantages of grazing incidence X-ray telescopy, pioneered by Riccardo Giaconni and his associates, are mentioned, and the prospects for studies of the extra-galactic X-ray background and grain-scattered halos of eclipsing X-ray binaries by *AXAF* are described.

1 X-ray Telescopes

The technical foundation for much of present and most of future X-ray astronomy was laid in the seminal papers on grazing incidence X-ray concentrators by Giacconi & Rossi (1960) and on Wolter-style X-ray telescopes by Giacconi et al. (1965). Giacconi and his associates pressed the technology and politics of X-ray telescopes toward larger areas and higher resolution from the earliest days of extra-solar X-ray astronomy. They produced the *Apollo* telescope for solar X-ray imagery, and then the *Einstein* observatory. Soon *AXAF*, the *Advanced X-Ray Astrophysics Facility*, will be launched, a goal established by Giacconi toward which he and many others have worked for three decades. It will provide better than 0.5 arcsec angular resolution, non-dispersive CCD imaging spectroscopy to 10 keV, and transmission grating spectroscopy with a resolution of 1000.

Mechanically collimated proportional counters are the essential instruments for studies of milli- and micro-second variability of X-ray sources like those based on data now coming from the *Rossi X-Ray Timing Explorer*. They provide the required large collecting areas at much lower cost than imaging systems. But, the practical limit of mechanical collimation is about 1° FWHM, which can result in confusion even among bright point sources. And only an image can reveal the details of extended objects.

2 A New Transient Be X-Ray Binary in a Crowded Field

Figure 1 displays three images of a small patch of the Small Magellanic Cloud recorded by the grazing incidence telescope and PSPC image detector on the *ROSAT* observatory in October 1991, October 1992, and June 1993. All three show the binary pulsar SMC X-1 (which was

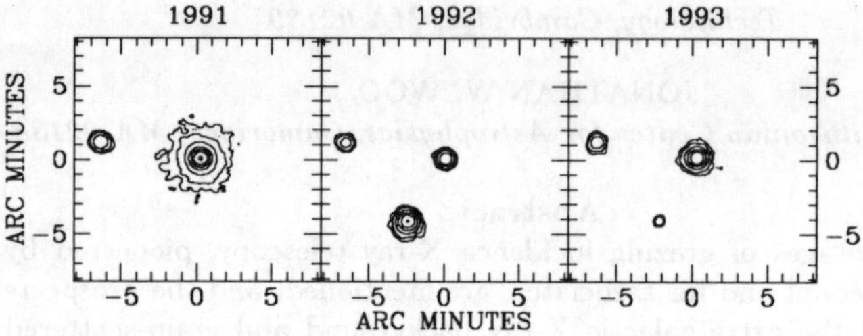

Figure 1: Three *ROSAT* images of SMC X-1 and its near neighbors in the Small Magellanic Cloud. The persistent source to the left is the foreground star HD 8191. The source seen only in the 1992 and 1993 images is the Be-type transient, RX J0117.6-7330.

mostly in eclipse during the 1992 and 1993 observations), and the much fainter foreground star HD 8191. The 1992 and 1993 images show, in addition, a bright non-pulsing, Be-type transient, RX J0117.6-7330 (Clark, Remillard & Woo 1997). The three sources, separated by $\sim$ 5 arcmin, would have been recorded as one by a mechanically collimated detector. The derived light curve would have been dominated by the eclipse transitions of SMC X-1, and would have shown a pecularly high flux in the depth of the eclipse due to the transient and HD8191.

In the 1993 image the transient is faded by a factor of $\sim$ 246, but is clearly detected because imaging limited the area from which background is counted with the source.

These images illustrate the advantages of spatial resolution and reduced background afforded by even the modest angular resolution of the *Einstein* and *ROSAT* observatories. It is exciting to anticipate the huge advantage that *AXAF* will provide for the analysis of the extra-galactic X-ray background where the ultimate contributors may be a million times fainter than SMC X-1 and crowded on a scale of arc seconds.

3 Extended Images of Dust-Scattered Halos

Among the studies of extended objects that *AXAF* will greatly enhance is the analysis of grain-scattered X-ray halos. A scattered photon suffers a travel-time delay that depends on the angle of scatter and the distances of the source and the scattering grain. The halos of eclipsing binary X-ray stars are particularly interesting because the shutter action of an eclipse cuts off the primary beam from the neutron star. This initiates a variation in the form and spectrum of the halo that contains information about the distance of the source and about the properties and distribution of the grains along the line of sight.

Studies of the halos of 4U 1538-52 and Cen X-3, based on data from *Ginga* and *ASCA*, indicate that the average density of interstellar grains is substantially less than that of solid "astronomical silicate", which supports the idea that interstellar grains are fluffy aggregates of smaller solid particles (Clark, Woo, & Nagase 1994; Woo et al. 1994).

Figure 2 shows sequences of simulated AXAF images of 4U 1538-52 and its halo at two energies from orbital phase -0.2 to +0.2. The eclipse occurs between phases -0.1 and +0.1. The source was set at a distance from earth of $D = 5.5$ kpc, and dust was distributed evenly from $0.3D$ to $1.0D$. At the center of each image is the one bright (saturated) pixel struck by photons coming directly from the source when it is out of eclipse, or by photons scattered by circumstellar matter when the source is in eclipse. All the surrounding events are due to the grain-scattered photons. During the eclipse the low-energy image is gradually hollowed out, leaving only the more widely scattered and more delayed photons. After the eclipse, the reverse process occurs as the halo recovers from the inside out. The same sequence occurs in the high-energy image, but on a shorter time scale, and with much weaker intensity because the scattering efficiency decreases with the square of the energy. When real images of halos come available from *AXAF*, the challenge will be to find models of grain composition and spatial distribution which yield computed halos that match data like these.

4 The Future

AXAF will be about as big as the *Large X-Ray Telescope* envisioned in the original 1969 proposal of the consortium which ultimately produced the *Einstein Observatory*. Though only half the size of the *LXT*, it achieved remarkable results in its two-year life in low orbit, as have the *ROSAT* and *ASCA* observatories in their many years of operation, but also in low orbit. With the great advances that have since been made in X-ray mirror technology, X-ray CCDs, and X-ray diffraction gratings, and with a long life in high orbit, *AXAF* will provide access to a vast new realm of X-ray spectroscopy and observational cosmology.

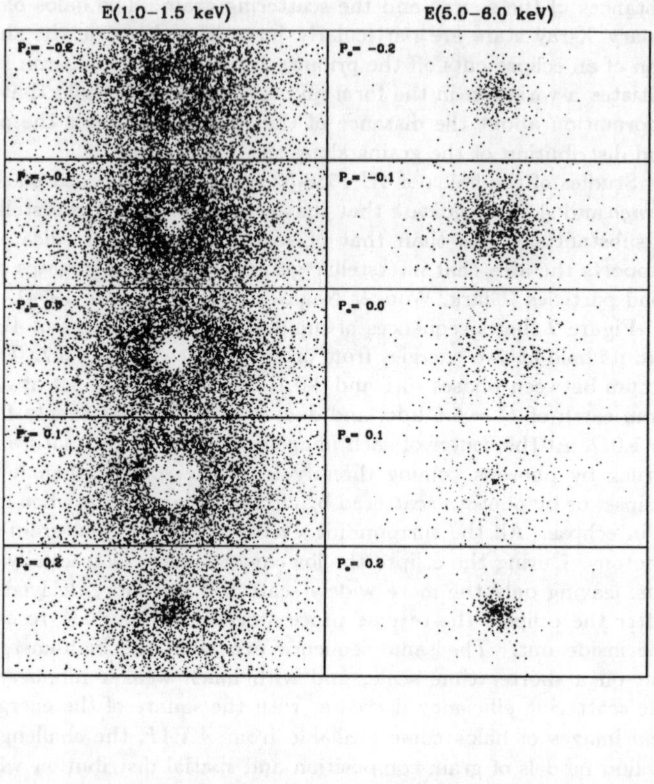

Figure 2: Simulated 2-ksec exposure *AXAF* images of 4U 1538-52 and its grain-scattered halo before, during, and after the eclipse. All photons coming directly from the source are recorded in one saturated pixel at the focal point; all the other events are due to halo photons.

References

1. Giacconi, R. & Rossi, B. 1960, J. Geophys. Res., 65, 773
2. Giacconi, R., Harmon, N. F., Lacey, R. F., & Szilagyi, Z. 1965, J. Opt. Soc. Am., 55, 345
3. Clark, G. W., Woo, J., & Nagase, F. 1994, Ap. J., 422, 336
4. Woo, J. W., Clark, G. W., Day, C. S. R., Nagase, F., & Takeshima, T. 1994, Ap.J., 436, L5
5. Clark, G. W., Remillard, R. A., & Woo, J. W. 1997, Ap. J., 474, L111

Iron K_α-Line Emission from Neutral Interstellar Gas

Rashid Sunyaev
Max-Planck-Institut für Astrophysik
Karl-Schwarzschild-Str. 1, 85748 Garching, Germany
E-mail: sunyaev@mpa-garching.mpg.de
Space Research Institute
Moscow, Russia
E-mail: rs@hea.iki.rssi.ru

1 Introduction

Every atom of iron in the interstellar space of our Galaxy is illuminated by X-Ray background radiation. Iron atoms are absorbing from time to time photons with energy above the K-edge and emit fluorescent photons with an energy of 6.4 keV.

In 1962 Riccardo Giacconi and his collaborators discovered X-ray Background Radiation of the Universe. This was well before the discovery of the Cosmic Microwave Background in 1964 and this was the first real Background Radiation of the Universe coming to us from cosmological distances. Nobody in the 60's was able to measure the optical or infrared radiation energy density in space between galaxies. Even in the long radiowaves spectral band the foreground synchrotron radiation of the Galaxy was much brighter than the sum of emission from weak distant radiogalaxies.

Using the EINSTEIN Observatory Riccardo Giacconi made an attempt to resolve the X-Ray background and to identify the main sources contributing to its intensity. And only now — 35 years after the discovery of the X-Ray Background Radiation of the Universe – the ROSAT spacecraft and ASCA have clarified that observed sources (mainly very distant and weak QSO's and AGNs) contribute 75% of its observed intensity. Riccardo Giacconi was working on the interpretation of ROSAT and Einstein data with an excellent team of coworkers. The next step would be to use AXAF and to find: 1. the nature of the remaining 25% of the sources contributing to the background and; 2. the reason for a difference of source counts in soft and standard X-Ray bands.

There are many ways the X-Ray background might become of great importance for cosmology - we could discuss the dipole anisotropy in the future ABRIXAS maps of the sky, small scale angular fluctuations, the dependence of X-ray background intensity on redshift connected both with the expansion of the Universe and cosmological evolution of the sources of this background.

However in this paper I wish to show that X-Ray background radiation opens a new way to investigate the neutral interstellar gas in our Galaxy and other galaxies. The upcoming missions AXAF, XMM, SPECTRUM-X, ABRIXAS, ASTRO-E are using grazing incidence optics with high flux sensitivity up to 10 keV, good energy resolution of the detectors at 6 keV due to the use of X-Ray CCDs (150 eV), bolometers (10 eV), Bragg spectrometer (5 eV) and gratings. This will create the possibility of investigating Fe and other abundant elements, K-alpha lines from interstellar gas activated by X-Ray background photons (see original discussion in Vainshtein, Sunyaev, 1980). The great X-ray missions of the next Millenium beginning with the CONSTELLATION and XEUS have such high energy resolution and high surface area that without any doubt the way proposed below to investigate interstellar matter might become as common as the role played today by the 21 cm line of neutral hydrogen and the CO 2.6 mm line.

It is well known that we could investigate the K-edge absorption jumps in the spectra of bright X-Ray sources due to neutral or weakly ionised matter on the line of sight towards the source (Brown and Gould, 1970). Unfortunately it is usually difficult to separate absorption in the binary system or in the matter around the extragalactic source and in the interstellar gas. In addition it is necessary to mention that K-edge absorption has the spectral width of the order of E/3. In the case of a fluorecsent line the spectral width of interstellar fluorescent line is close to its natural width of 1-3 eV. Therefore with devices able to give very good energy resolution it is much better to investigate the narrow K-alpha emission lines in the spectrum of X-Ray background radiation.

Although the Galaxy is practically transparent at these energies, a certain proportion of photons will be absorbed by atoms and ions of Fe, S, and Si. This absorption will be accompanied by emission of the K_α lines of these elements, with the energies E_0 given in Table I. By measuring the equivalent widths of the K_α lines in the background emission spectrum one should be able to determine the total abundance of iron, sulfur, and silicon atoms in any given direction. Alternatively, if the abundance of these elements is (considered known) accepted, one can estimate the total mass of interstellar matter – gas and dust. Since dust grains are transparent to X-Rays with $h\nu > 1keV$, the atoms bound in grains will absorb X-Rays just as though they were free atoms.

2 Equivalent width of a K_α line in the background spectrum

Let us evaluate the total luminosity of the Galaxy in the K_α line of an element with the atomic number Z. We shall write the cross section for absorption of an X-Ray photon having an energy $E > E_1$ (where E_1 is the photoionization limit of the K shell) in the form

$$\sigma = \sigma_0 (E_1/E)^{2 \cdot 8}.$$

The values of E_1 are given in Table I, along with Veigele's (1973) values for σ_0. After absorption has taken place, the vacancy in the K shell can be filled either by emission of a K_α photon or by the Auger effect. The quantum yield (the relative probability of the emission of a K_α photon) is given by

$$K = \frac{A_0 Z^{\frac{1}{4}}}{A_0 Z^{\frac{1}{4}} + \Gamma} = \frac{1}{1 + (C/Z)^4}; \quad C = (\frac{\Gamma}{A_0})^{\frac{1}{4}} \approx 30.7,$$

where $A_0 Z^4$ and Γ (which does not depend on Z) denote the probabilities of a radiative transition and of Auger Decay. Column 7 of the table gives the quantum yield K for Fe, S, Si.

If we take the background intensity in the 1-40 keV energy range to be (Gursky and Schwartz, 1977)

$$I_d(E) = I_0 (\frac{1keV}{E})^\gamma \ ; \qquad (I_0 = 8.5 keV/cm^2 \cdot sec \cdot keV \cdot sr, \gamma = 0.4),$$

we will obtain for the probability j_0 that in unit time a K_α photon will be emitted by a given atom of an element of atomic number Z:

$$j_0 = 4\pi K \int_{E_1}^{\infty} \sigma(E) I_d(E) \frac{dE}{E} = \frac{4\pi I_o \sigma_0 K}{2.8 + \gamma} (\frac{1keV}{E_1})^\gamma.$$

Column 6 of the table shows that for Fe, S, Si atoms emission of this type will typically occur once every 170 billion years.

Let M denote the combined mass of interstellar gas and dust; then their total luminosity in the K_α line will be

$$L_0 = \alpha \frac{M}{m_p} j_0 E_0 = \frac{4\pi K_\alpha \sigma_0 I_0}{2.8 + \gamma} \frac{M}{m_p} E_0 (\frac{1keV}{E_1})^\gamma,$$

where α denotes the abundance of the element and m_p is the mass of a proton. For the K_α line of iron, which has $\alpha \approx 4 \cdot 10^{-5}$, we have

$$j_0 = 1.8 \cdot 10^{-19} sec^{-1}, L_0 = 9.10^{35} (\frac{M}{10^{10} M_\odot}) erg/sec.$$

The Galaxy would be transparent to such radiation. The value (5) amounts to about 0.3% of the total luminosity of our Galaxy in the 2-10 keV range, with the bulk of the radiation coming from X-Ray binary sources like Sco X-1, Cyg X-1, and the galactic bulge LMXBs. But X-Ray binaries will themselves radiate in the K_α line, because their X-Rays will be reprocessed at the surface of the normal star, in the accretion disk, and in the cool gas surrounding the source (see discussion in Basko et al., 1974). One should therefore select the radiation field so as not to include any bright galactic sources.

Table I

Elem.	Abund. 10^{-5}	E_0 (keV)	E_1 (keV)	σ_0 10^{-20} cm^2	j_0, 10^{-20} s^{-1}	K	L_0, 10^{35} erg/s	W eV
Fe	4.0	6.40	7.11	3.53	1.83	0.34	9	360
S	1.6	2.31	2.47	11.6	1.86	0.069	1.3	38
Si	3.3	1.74	1.84	16.0	1.75	0.042	1.9	45

The K_α emission line of iron is a doublet, with its components separated by 12 eV.

Let $N(R, \theta, \varphi)$ represent the number density of atoms of interstellar matter in the direction (θ, φ), at distance R; then the flux in the K_α line will be expressed by

$$F(E_0) = \frac{\alpha j_o E_0}{4\pi} \int N(R, \theta, \varphi) sin\theta d\theta d\varphi dR$$

and the equivalent width of the line will be

$$W = \frac{F(E_0)}{I_d(E_0)\Omega_0} = \frac{\alpha \sigma_0 K E_0}{2.8 + \gamma} (\frac{E_0}{E_1})^\gamma \int N(R, \theta, \varphi) \frac{sin\theta d\theta d\varphi}{\Omega_0} dR,$$

where Ω_0 denotes the solid angle of the detector, and the integral extends over a volume limited by the size of the Galaxy and by the instrument aperture.

To obtain a rough estimate let us adopt a very simple model Galaxy: a homogeneous gaseous disk of radius $R_0 = 10kpc$ and thickness $2b_0 = 0.2kpc$. The solar system would be located at the edge of the disk. If the instrument aperture does not extend beyond the disk, that is, if $\theta < b_0/2R_0 \approx 0^o.6, \varphi << 2\pi$, then it extends in the direction toward the galactic center

$$W = \frac{2\alpha\sigma_0 K}{2.8+\gamma}(\frac{E_0}{E_1})^\gamma E_0 N R_0 = \frac{\alpha\sigma_0 E_0 K}{(2.8+\gamma)\pi R_0 b_0}\frac{M}{m_p}(\frac{E_0}{E_1})^\gamma. \quad (1)$$

For the K_α line of Fe we obtain $W \approx 360M/(10^{10}M_\odot)eV$. Despite this large equivalent width the K_α line should be narrow: each component of the doublet will be broadened by galactic rotation to several electron volts, that is, to values only slightly in excess of the natural width of the line.

The total number of atoms with nuclear charge Z along the line of sight is related to the observed equivalent width of the K_α line by

$$\int N_z dR = \frac{W}{E_0}\frac{1}{\sigma_0 K}(\frac{E_1}{E_0})^\gamma(2.8+\gamma) = 10^{20}aW,$$

where $a = 0.43, 1.78, 2.8cm^{-2}/keV$ for Fe, S, Si, respectively and $\gamma = 0.4$.

In deriving Eq. (1) we have assumed that the detector is so oriented that the major axis of its aperture is exactly in the galactic plane.

Equation (1) is also applicable to the case where a line in a distant galaxy is being observed, provided the galaxy is turned edgewise to the observer. The solid angle Ω_0 should then be small enough so as not to extend beyond the galaxy. If the galaxy has its plane facing the observer, W will be smaller by a factor of R_0/b_0.

The estimates given above for the luminosity and equivalent width are valid only if the Galaxy is transparent to radiation with $E > E_0$. In the plane of the disk the optical depth is quite large, reaching $\tau = 0.4$ at the edge of K_α absorption by Fe atoms. For S and Si the optical depth is still larger; but at an angle to the disk τ will be far smaller.

In addition to the mean emission of the Galaxy as a whole, it would be interesting to search for K_α line radiation in prominent gaseous complexes and giant molecular clouds.

As Z decreases the equivalent width of the K_α line becomes smaller, but the amplitude of the K-discontinuity at the photoabsorption limit E_1 increases. Accordingly, for carbon, oxygen, and neon it would be preferable to observe K-discontinuities in the diffuse background spectrum. The method of investigating the interstellar medium by observing K-discontinuities in the absorption spectra of X-Ray sources and the diffuse background has been discussed in literature on several occasions (see for example, Brown and Gould, 1970).

3 The giant molecular clouds in central 100 parsec of the Galaxy

Another way to produce bright iron K_α line due to fluorescence is to illuminate a giant molecular cloud by a strong flux of X-Rays with a continous spectrum above 7.1 keV.

Sunyaev, Markevitch and Pavlinsky (1993) have noted that a strong concentration of molecular hydrogen in the giant clouds in the Galactic Center region should lead to the appearance of the diffuse emission, related to the scattering of the radiation from compact X-ray sources and Sgr A* source by the molecular hydrogen in these clouds. It was also noted, that detection of a strong fluorescent line of neutral iron ($h\nu = 6.4$ keV) and a characteristic spectrum of scattered components may provide unique information on the activity of this region during the last several hundreds years. Moreover, scattering of the emission from the nucleus of the Galaxy by the neutral and molecular cloud of the whole Galaxy should lead to the appearance of the diffuse component, which contains the information of the activity of the nucleus during past tens of thousands of years. Recent ASCA observations of the diffuse emission of the Galactic Center region gave new information in support of this hypothesis. Along with emission lines from the hydrogen- and helium-like ions of heavy elements, the powerful iron fluorescent line (K_α) was detected, which surface brightness distribution roughly follows the distribution of giant molecular clouds. The Sgr B2 complex was found to be especially bright in the 6.4 keV line, at the level of 1.7 10^{-4} $photon\ s^{-1}cm^{-2}$ (Koyama et al., 1996, see Fig.1). The X-Ray spectrum of Sgr B2 is given in Fig.2. Such a high flux in the fluorescent line requires a powerful source of primary X-ray emission which gives rise to the reprocessed radiation. Given the distance from the Sgr B2 complex to the GC ($\sim 40'$ or about 100 pc in projection) Sunyaev, Markevitch & Pavlinsky (1993) and Koyama et al. (1996) estimated the luminosity of the putative GC source a few hundred years ago in excess of 10^{39} $erg\ s^{-1}$.

The linear dimension of the Sgr B2 cloud exceeds 40pc (Lis and Goldsmith, 1989). If the flare of Sgr A* in X-Rays was short (one or several years) the illumination front would produce K_α-photons only in restricted regions of the cloud. It is very interesting that the illumination front is moving with a velocity exceeding the speed of light (see fig. 3 and discussion in Sunyaev, Churazov, 1998)

4 Scattering of X-Ray lines on neutral gas

The 6.4 keV K_α line flux from Sgr B2 cloud and the column density of the cloud $N_H L \gtrsim 2\ 10^{23}\ cm^{-2}$ are so high that future missions like CONSTELLATION and XEUS have a good chance to observe the weak spectral features connected

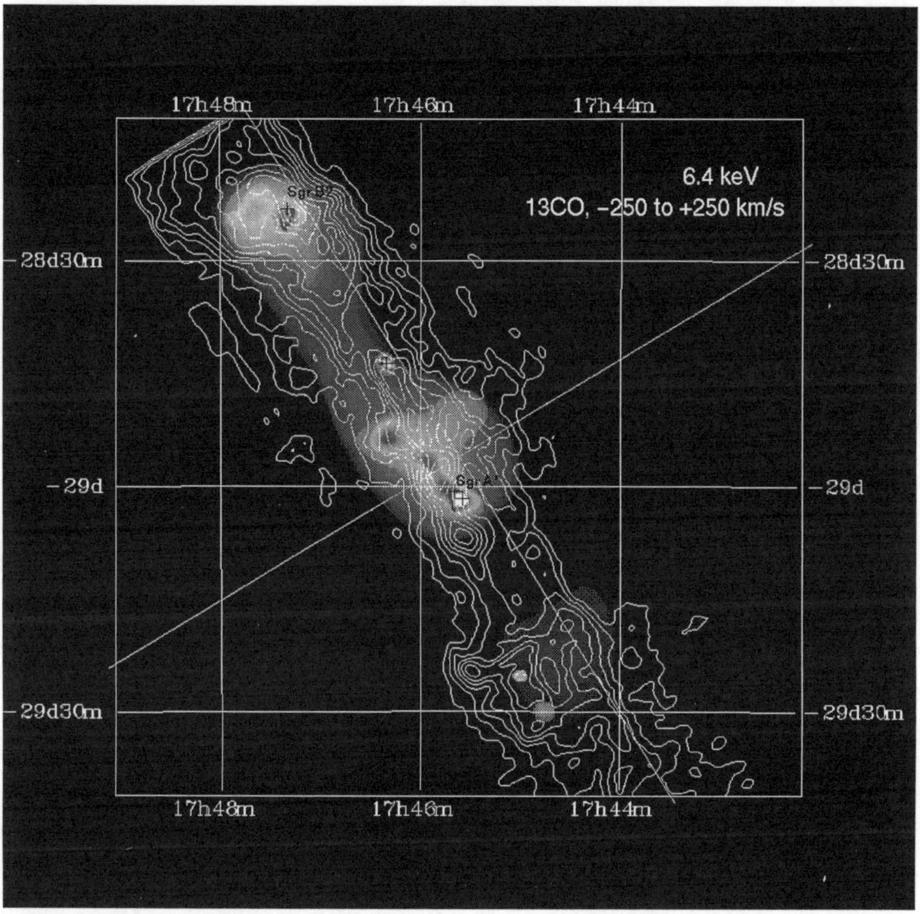

Figure 1: Wavelet filtered maps of the Galactic Center region in neutral iron K_α line obtained by ASCA. (Gilfanov, Sunyaev, 1999) The contours show velocity integrated intensity distribution of the ^{13}CO emission. The line stretching from the upper-left to the lower right marks position of the Galactic plane

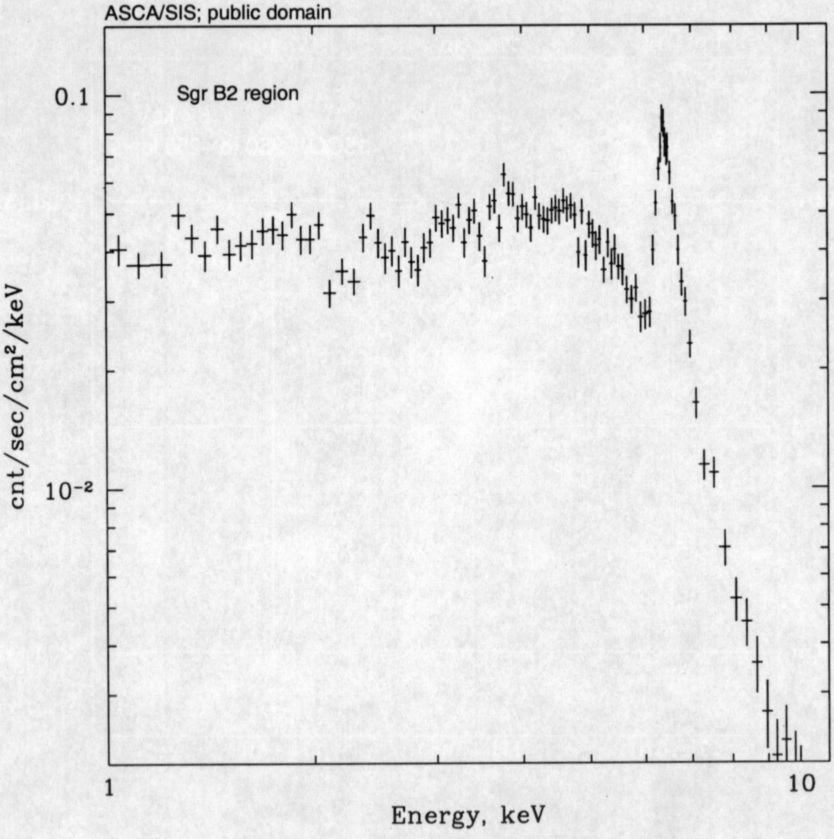

Figure 2: The overall spectrum of Sgr B2 region observed by ASCA/SIS in 1993 (Gilfanov, Sunyaev, 1999). Prominent iron K_α line dominates the spectrum.

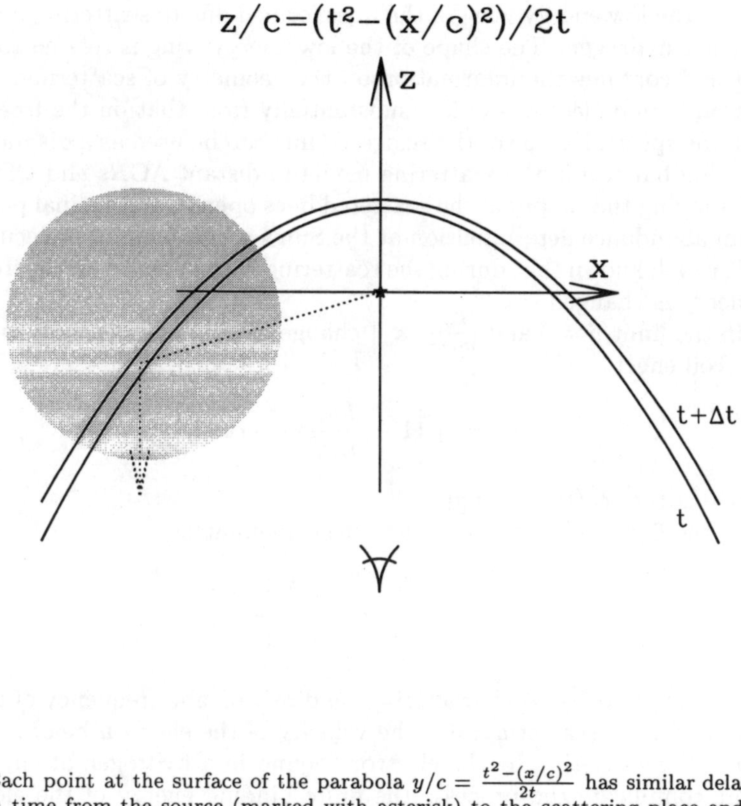

Figure 3: Each point at the surface of the parabola $y/c = \frac{t^2-(x/c)^2}{2t}$ has similar delay t in propagation time from the source (marked with asterisk) to the scattering place and then to the observer. In the case of a short flare the fluorescent photons which are observed at a given moment of time are produced in the neutral matter located at the surface of the parabola. For a steady source which faded at a given moment of time the exterior of the parabola is filled with primary continuum photons.

with the scattering of the X-Ray line photons on the molecular hydrogen and helium atoms in the cloud.

Bolometric observations with high energy resolution (better than 12 eV) of ASTRO-E, Bragg spectrometer (resolution $\sim$ 5eV) of Spectrum-X-Gamma and gratings of XMM and AXAF can verify another prediction – the specific shape of the low energy wing of the lines formed due to scattering by atomic or molecular hydrogen. The shape of the low energy wing is related to the recoil effect and contains the information on the geometry of scattering. The recoil effect on bound electrons differs substantially from that on the free electrons. Thus, the spectral shape of the scattered line can be used as a diagnostic of the ionization balance in the scattering media in distant AGNs and QSOs. Moreover studying the shapes of the scattered lines opens the principal possibility of helium abundance determination at the Sun's surface and in molecular clouds.

It is well known that during the scattering of the photon on the free electron its energy is changed.

In the limit $\vec{v} = 0$ and $\frac{h\nu_1}{m_e c^2} \ll 1$ change of the photon energy is related to the recoil effect:

$$\nu_2 = \nu_1 \left[1 - \frac{h\nu_1}{m_e c^2}(1 - cos\theta) \right] \tag{1}$$

where θ is the scattering angle.

In the limit $\frac{h\nu_1}{m_e c^2} \ll \frac{v}{c}$ Doppler effect dominates:

$$\nu_2 = \nu_1 \frac{1 - \frac{\vec{v}}{c}\vec{\Omega}_1}{1 - \frac{\vec{v}}{c}\vec{\Omega}_2} \tag{2}$$

where $\vec{\Omega}_1, \nu_1$ and $\vec{\Omega}_2, \nu_2$ characterize the direction and frequency of the photon before and after scattering, $\vec{v}$ is the velocity of the electron before scattering.

For the scattering by the electron, bound in a hydrogen atom, additional factors complicate the process: the finite binding energy of the electron and the motion of the electron in the atom. Because of the discrete energy levels of the electron in the exited state not all values of the photon energy change are possible. Because of the "random" motion of the electron within the atom change of the photon energy is no longer an unambiguous function of the geometry. For free electrons Pozdnyakov et al. (1979) noted, that even at rather low electron temperatures the $kT_e \sim 1$ eV profile of a scattered line apparently broadens due to the Doppler effect. Note that in this example electron velocity was $v \approx 600$ km/s, the value is not unusual for accretion disks in X-ray binary systems and galactic nuclei. The characteristic velocity of the electron in a hydrogen atom is $v \sim \alpha c \sim 2000$ km/s (α is the fine structure constant; $1/\alpha = \hbar c/e^2 = 137$) and motion with such velocity should significantly affect the amount of energy transferred to the photon during the scattering. This ambiguity of energy transfer does not violate conservation laws, since heavy nucleus can carry the necessary momentum, having negligible kinetic energy.

Depending on the final state of the electron, the scattering process can be divided into three branches:

> **A)** Rayleigh (coherent) scattering: $\gamma_1 + H = \gamma_2 + H$. The energy of the photon remains the same, only direction changes. The recoil effect is $\sim m_p/m_e$ times lower than in the case of scattering by a free electron.
>
> **B)** Raman scattering: $\gamma_1 + H = \gamma_2 + H(n,l)$, where $H(n,l)$ denotes one of the excited states of the hydrogen atom. The energy of the photon decreases by the value of excitation energy: $h\nu_2 = h\nu_1 - E_{n,l}$. Raman satellites of the line appear.
>
> **C)** Compton scattering: $\gamma_1 + H = \gamma_2 + e^- + p$, accompanied by ionization of the atom. The energy of the photon decreases by the value of ionization potential and kinetic energy of the electron after scattering: $h\nu_2 = h\nu_1 - 13.6eV - E_e$.

It is necessary to mention that in a non-relativistic limit the sum, over differential crossections of these three branches, is equal to the Thomson differential crossection.

For the scattering by a helium atom along with the increase of Rayleigh scattering (by a factor of 2) the structure of the lines, corresponding to Raman scattering, changes strongly. In particular the gap between the ground and the first excited state in the helium atom is ~20 eV. Note, that for ~ 6 keV photons the wavelength $\lambda \sim 2\mathring{A}$ is comparable with the size of the atom and the parity selection rule is not strict. Due to the stronger binding of the electron in the helium atom, the electron momentum distribution is substantially wider than that in atomic or molecular hydrogen. A comparison of the spectra scattered by hydrogen and helium atoms is given in Fig.5.

The energy gap is twice as large as that for the hydrogen atom and significant differences in the recoil profile allow one to hope that studying the recoil spectrum might become a method for helium abundance determination. Note that even after multiple scatterings the photon does not appear in this energy gap.

The difference between scattering by molecular and atomic hydrogen is most prominent for scattering by small angles (i.e. for Rayleigh scattering). Firstly, coherent (Rayleigh) scattering by molecule will be “enhanced” by a

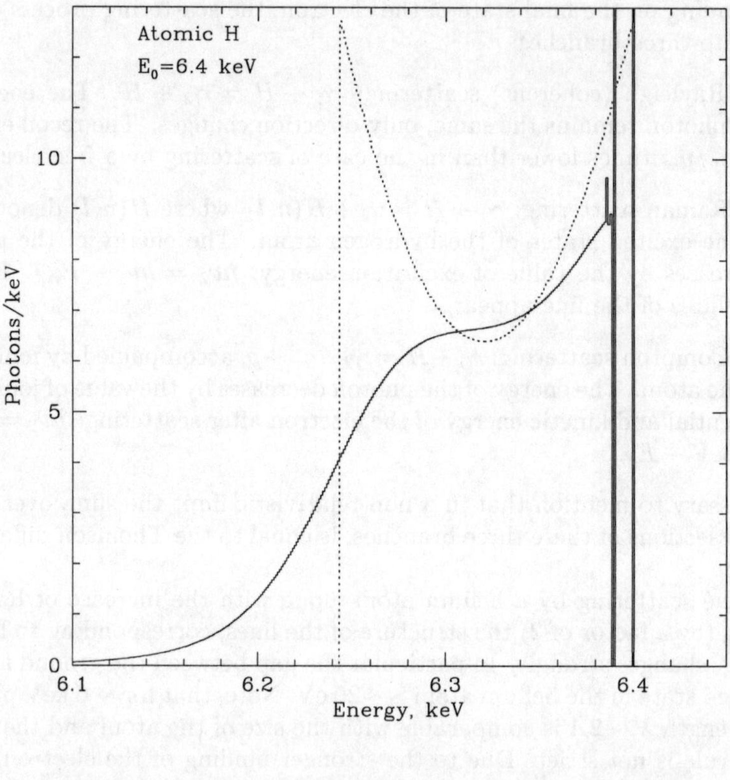

Figure 4: Neutral hydrogen: Spectrum of the scattered 6.4 keV photons averaged over all scattering angles (solid line). About 13.5 % of photons undergo elastic scattering, 2.3 % of the photons change energy by 10.2 eV – Raman satellite of the line appears, corresponding to the excitation of the second level in the hydrogen atom. The same spectrum calculated for scattering by free cold electrons is shown by the dotted line.

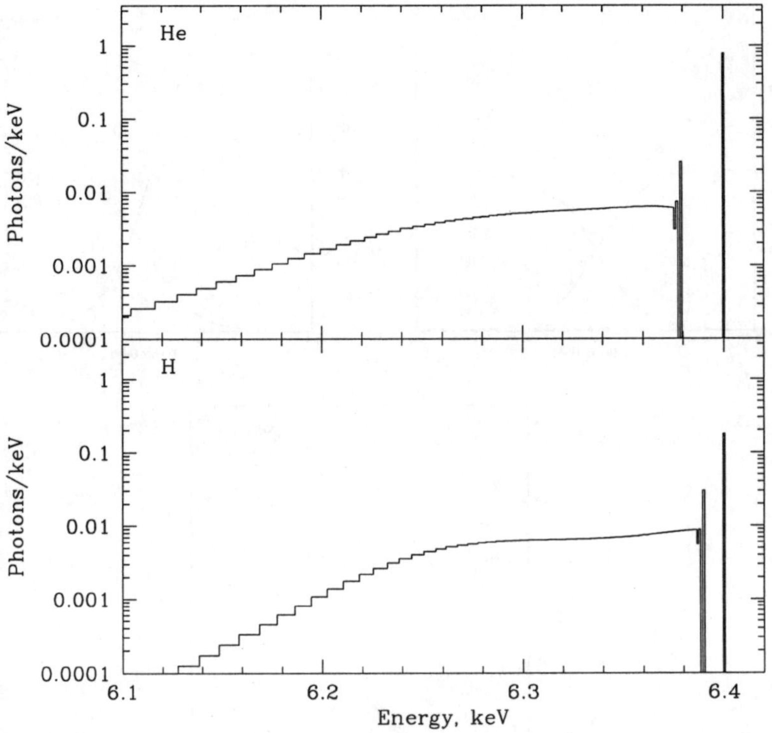

Figure 5: Spectra of monochromatic 6.4 keV line, scattered by hydrogen (lower figure) and helium (upper figure) atoms and averaged over all angles. One can expect to see such spectra if the source of the monochromatic line is located at the center of an optically thin cloud.

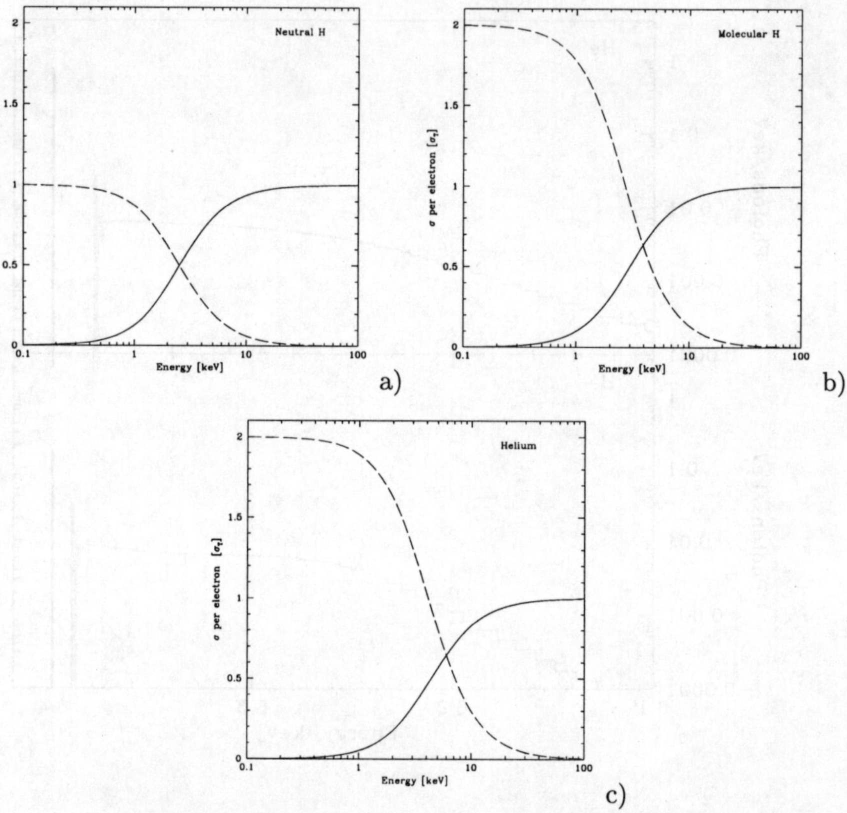

Figure 6: a) The contributions of coherent (Rayleigh – dashed line) and incoherent (Raman and Compton – solid line) processes to the total crossection of photon scattering by a hydrogen atom as the function of photon energy. b) the same as in a) but for molecular hydrogen. c) the same as in a) for helium (using tabulated data of Hubbell et al., 1975)

factor of Z i.e. a factor of 2 per electron (Fig.6b). Secondly the structure of electron terms in the molecule differs somewhat from that in the hydrogen atom. As a result the gap between the unshifted line (due to Rayleigh scattering) and the line corresponding to the Raman scattering with excitation of the first electron term in the molecule will be ~ 11 eV compared to 10.2 eV in atomic hydrogen. Compton scattering by large angles will be very similar to that for atomic hydrogen. The analysis of scattered line spectra may provide information about helium abundance in the molecular clouds in the central 100 parsec of the Galaxy and in the gas concentrations around QSO's and AGN's.

The major effects are the *smearing of back scattering peak and the appearance of the energy gap* – lack of photons in the interval from initial photon energy $h\nu_0$ down to $h\nu_0 - E$, where E is the excitation energy of the first levels above ground level ($E = 10.2$ eV for atomic hydrogen, $E \approx 20$ eV for neutral helium, $E = 40.8$ eV for He II ion, $E \approx 11$ eV for molecular hydrogen). The possibility of the detection of Raman satellites of X-ray emission lines is very attractive. The energies of these lines are shifted by the excitation energy of various levels of the atoms of hydrogen and helium. Of course with the current X-ray CCD and cooled solid state detectors with an energy resolution of about 100 eV one can expect detection of effects only in the zone with $\Delta h\nu \sim 100$ eV.

References

1. Basko M., Sunyaev R., Titarchuk L., A&A **31**, 249 (1974).
2. Brown R.L. & Gould R.J., Phys.Rev.D. **1**, 2252 (1970).
3. Gilfanov M., Sunyaev R., Astron.Letts , in press (1999).
4. Gursky M & Schwartz D.A., Ann.Rev.Astron.Astrophys. **15**, 541 (1977).
5. Hubbell J. *et al*, J. Phys.Chem.Ref.Data **4**, 471 (1975).
6. Koyama *et al*, Publ.Astron. Soc. Japan **48**, 249 (1996).
7. Lis D.C. and Goldsmith P.F., ApJ **337**, 704 (1989).
8. Pozdnyakov L.A., Sobol I.M., Sunyaev R.A.,A&A **75**, 214 (1979).
9. Sunyaev R.A., Markevitch M., Pavlinsky M., ApJ **407**, 606 (1993).
10. Sunyaev.R., Churazov E., MNRAS **297**, 1279 (1998).
11. Sunyaev R., Churazov E., Astronomy Letters **22**, 648 (1996).
12. Vainshtein L. & Sunyaev R., Soviet Ast. Letters **6**, 673 (1980).
13. Veigele W.J., Atomic Data **5**, 51-111 (1973).

factor of Z (the Z-factor of X-ray emission, Fig. 6b). Secondly, the structure of electron terms in the molecule differs somewhat from that in the hydrogen atom. As a result, the gap between the unshifted line (Thomson or Rayleigh scattering) and the line corresponding to the Raman scattering with excitation of the first electron level in the molecule will be ≈ 11 eV, compared to 10.2 eV in atomic hydrogen. Compton scattering by large angles will be very similar to that for atomic hydrogen. The analysis of scattered line spectra may bring information about helium abundance in the molecular clouds in the central 100 parsec of the Galaxy, and in the gas concentrations around QSOs and AGNs.

The major effects are the appearance of back scattering peaks and the appearance of the energy gap - lack of photons in the interval from the initial photon energy $h\nu_0$ down to $h\nu_0 - E$, where E is the excitation energy of the first levels above ground state ($E = 10.2$ eV for atomic hydrogen, $E \approx 20$ eV for neutral helium and $E \approx 11$ eV for molecular hydrogen). The possibility of the detection of Raman satellites of X-ray emission lines is very attractive. The energies of these lines are shifted by the excitation energy of various levels of the atoms of hydrogen and helium. Of course with the current X-ray CCD and cooled solid state detectors with an energy resolution of about 100 eV one can expect detection of effects only in the zone with $\Delta h\nu \geq 100$ eV.

References

1. Basko M., Sunyaev R., Titarchuk L., A&A 31, 249 (1974).
2. Brown R.L. & Gould R.J., Phys Rev D 1, 2252 (1970).
3. Gilfanov M., Sunyaev R., Astron Letters, in press (1999).
4. [illegible] M. & Edwards D.A., Ann Rev Astron Astrophys 16, 341 (1978).
5. [illegible] L. et al., J. Phys. Chem. Ref. Data 4, 471 (1975).
6. Koyama K. et al. Publ Astron Soc Japan 48, 249 (1996).
7. Lis D.C. and Goldsmith P.F., ApJ 337, 704 (1989).
8. Pozdnyakov L.A., Sobol I.M., Sunyaev R.A., A&A 75, 214 (1979).
9. Sunyaev R.A., Markevitch M., Pavlinsky M., ApJ 407, 606 (1993).
10. Sunyaev R.A., Churazov E., MNRAS 297, 1279 (1998).
11. Sunyaev R.A., Churazov E., Astronomy Letters 22, 648 (1996).
12. Vainshtein L. & Sunyaev R., Soviet Astr. Letters 6, 673 (1980).
13. Wiese W.L., Atomic Data 5, 71-141 (1973).

GENERAL RELATIVITY WITH QUASI-PERIODIC OSCILLATIONS FROM LOW MASS X-RAY BINARIES

L. STELLA

Osservatorio Astronomico di Roma, Via Frascati 33, I-00040 Monteporzio Catone (Roma), Italy, e-mail stella@coma.mporzio.astro.it; affiliated to I.C.R.A.

The relativistic precession model for quasi periodic oscillations, QPOs, in low mass X-ray binaries is reviewed. The behaviour of three simultaneous types of QPOs is well matched in terms of the fundamental frequencies for geodesic motion in the gravitational field of the accreting compact object, for reasonable star masses and spin frequencies. The model ascribes the higher frequency kHz QPOs, the lower frequency kHz QPOs and the horizontal branch oscillations to the Keplerian, periastron precession and nodal precession frequencies of matter inhomogeneities orbiting close to the inner edge of the accretion disk. The remarkable correlation between the centroid frequency of QPOs in both neutron star and black hole candidate low mass X-ray binaries is very well fit by the model. Some testable predictions are described. QPOs from low mass X-ray binaries might provide an unprecedented laboratory to test general relativity in the strong field regime.

1 Introduction

Old accreting neutron stars, NSs, in low mass X-ray binaries, LMXRBs, display a complex variety of quasi-periodic oscillation, QPO, modes in their X-ray flux. The *low frequency* QPOs ($\sim 1 - 100$ Hz) that were discovered and studied from high luminosity Z-sources in the eighties are further classified into horizontal, normal and flaring branch oscillations (HBOs, NBOs and FBOs, respectively), depending on the simultaneous position occupied by a source in the X-ray colour-colour diagram (for a review see van der Klis 1995). The kHz QPOs (range of ~ 0.2 to ~ 1.3 kHz) that were revealed and investigated with RXTE in a number of NS LMXRBs (see van der Klis 1998, 1999, 2000 and references therein) involve timescales similar to the dynamical timescales close to the NS. A common phenomenon is the presence of a pair of kHz QPOs (centroid frequencies of ν_1 and ν_2) which drift in frequency while mantaining their frequency difference $\Delta\nu \equiv \nu_2 - \nu_1 \approx 250 - 360$ Hz approximately constant. kHz QPOs show remarkably similar properties across NS LMXRBs of the Z and Atoll groups, the luminosity of which differs by a factor of ~ 10 on average. During type I bursts from six Atoll sources, a nearly coherent signal at a frequency of $\nu_{burst} \sim 290 - 580$ Hz has also been detected. In a few cases ν_{burst} is consistent, to within the errors, with the frequency separation of the kHz QPO pair, $\Delta\nu$, or twice its value, $2\Delta\nu$.

The presence of HBOs has been firmly established in both Z-type and

Atoll LMXRBs; their frequency, ν_{HBO}, ranges from $\sim$ 15 to $\sim$ 60 Hz. The frequency changes of the kHz QPOs and HBOs are positively correlated with the instantaneous accretion rate. In a few Atoll sources some evidence has been found for an equivalent of the NBOs and FBOs of Z-sources (Wijnands et al. 1999; Psaltis, Belloni & van der Klis 1999).

The origin of these QPOs is still debated. According to beat frequency models, BFMs, disk inhomogeneities at the magnetospheric boundary and the sonic point radius are accreted at the beat frequency between the local Keplerian frequency, ν_ϕ, and the NS spin frequency, ν_s, giving rise to the HBOs and the lower frequency kHz QPOs, at ν_1, respectively (Alpar & Shaham 1985; Lamb et al. 1985; Miller et al. 1998). The higher frequency kHz QPOs are attributed to the Keplerian motion ($\nu_2 = \nu_\phi$) at the sonic point radius. The frequency separation of the QPO pair therefore, yields the NS spin frequency ($\nu_s = \Delta\nu$). The inferred ν_s possess a narrow distribution ($\sim 250 - 360$ Hz), far from the mass shedding limit of any NS model and considerably less extended than that of fast millisecond radio pulsars ($\sim$ 600 Hz). The NSs of LMXRBs should therefore be equilibrium rotators with a narrow frequency range despite their different average mass transfer rates, magnetic field strenghts and evolutionary histories. This implies that angular momentum is transferred outward at the same rate at which is received through accretion. Possibilities include gravitational radiation losses (Bildsten 1998; Andersson et al. 1999) or magnetic torques (White & Zhang 1997; Psaltis et al. 1999). Moreover, if LMXRBs are the progenitors of millisecond radio pulsars, then the BFM would require a different evolutionary path leading to the formation of radio pulsars with $\nu_s > 400$ Hz. Recent findings undermine the applicability of simple BFMs to a growing number of LMXRBs with kHz QPOs. First, the separation $\Delta\nu$ decreases by up to $\sim$ 100 Hz as ν_2 increases in Sco X-1 (van der klis et al. 1997), 4U1608-52 (Mendez et al. 1998a,b), 4U1735-44 (Ford et al. 1998) and 4U1728-34 (Mendez & van der Klis 1999). Owing to poor statistics, a similar variation of $\Delta\nu$ in other sources would have remained undetected (Psaltis et al. 1998). Second, there are currently two sources (4U1636-53, Mendez et al. 1999, and 4U1728-34, Mendez & van der Klis 1999) for which ν_{burst} is significantly different from $\Delta\nu$ and its harmonics. BFMs are not applicable to black hole candadates, BHCs, since the *no hair theorem* excludes the possibility that an offset magnetic field or radiation beam can be stably anchored to the black hole, as required to produce the beating with the disk Keplerian frequency.

A variety of alternative QPO models has been proposed. The basic features of the relativistic precession model, RPM, are reviewed here (Stella & Vietri 1998a, 1999; Stella, Vietri & Morsink 1999). In the RPM the QPO signals arise from the fundamental frequencies of the motion of blobs in the vicinity of the NS. The orbits of these blobs are supposed to be slightly eccentric and tilted. As in BFMs, the higher frequency kHz QPOs at ν_2 are produced by the ϕ-motion (*i.e.* the Keplerian motion) of inhomogeneities orbiting the inner disk boundary, while the lower frequency QPO signal at ν_1 originates from their periastron precession, which is primarily determined by strong-field effects. The HBOs are due to the nodal precession of the same inhomogeneities, an effect which is dominated by frame-dragging around fast-rotating collapsed stars. The RPM can be applied to BHCs as well. Within this model both the quadratic dependence of the HBO frequency on ν_2 and the decreasing separation $\Delta\nu$ for increasing ν_2, observed in several Z and Atoll sources are naturally explained. Moreover the RPM fits very well the remarkable correlation between the centroid frequencies of QPOs (or peaked noise components) from low mass X-ray binaries,that was discovered by Psaltis, Belloni & van der Klis (1999). This correlation extends over nearly 3 decades in frequency and encompasses both NS and BHC systems. If this interpretation were confirmed, kHz QPOs and HBOs would provide an unprecedented test for strong–field general relativity, GR.

2 Periastron Precession and kHz QPOs

We consider here only infinitesimally eccentric and tilted blob orbits, under the assumption that the blobs' motion is dictated by the star's gravity alone. In the case of a circular geodesic in the equatorial plane ($\theta = \pi/2$) of a Kerr black hole of mass M and specific angular momentum a, the coordinate frequency measured by a static observer at infinity is (Bardeen et al. 1972)

$$\nu_\phi = \pm M^{1/2} r^{-3/2} [2\pi(1 \pm a M^{1/2} r^{-3/2})]^{-1} \tag{1}$$

(we use units such that $G = c = 1$). ν_ϕ deviates from its Keplerian value at small radii. The upper sign refers to prograde orbits. If we slightly perturb a circular orbit introducing velocity components in the r and θ directions, the coordinate frequencies of the small amplitude oscillations within the plane (the epicyclic frequency ν_r) and in the perpendicular direction (the vertical frequency ν_θ) are given by (Okazaki, Kato & Fukue 1987; Kato 1990; de Felice & Usseglio-Tomasset 1996; Perez et al. 1997)

$$\nu_r^2 = \nu_\phi^2 (1 - 6Mr^{-1} \pm 8aM^{1/2} r^{-3/2} - 3a^2 r^{-2}) \ , \tag{2}$$

$$\nu_\theta^2 = \nu_\phi^2 (1 \mp 4aM^{1/2} r^{-3/2} + 3a^2 r^{-2}) \ . \tag{3}$$

In the Schwarzschild limit ($a = 0$) ν_θ coincides with ν_ϕ, such that the nodal precession frequency $\nu_{nod} \equiv \nu_\phi - \nu_\theta$ is identically zero. ν_r, instead, is always lower than the other two frequencies, reaching a maximum for $r = 8M$ and going to zero at $r_{ms} = 6M$. This qualitative behaviour of ν_r is preserved in the Kerr field ($a \neq 0$). Therefore the periastron precession frequency $\nu_{per} \equiv \nu_\phi - \nu_r$ is dominated by a "Schwarzschild" term over a wide range of parameters (Stella & Vietri 1999).

In the RPM the higher and lower frequency kHz QPOs are identified with $\nu_2 = \nu_\phi$ and $\nu_1 = \nu_{per}$, respectively. Therefore $\Delta\nu \equiv \nu_2 - \nu_1 = \nu_\phi - (\nu_\phi - \nu_r) = \nu_r$. For $a = 0$, Eqs. 1-2 give

$$\nu_r = \nu_\phi(1 - 6M/r)^{1/2} = \nu_\phi[1 - 6(2\pi\nu_\phi M)^{2/3}]^{1/2} \quad . \tag{4}$$

The curves in Fig. 1A show ν_r vs. ν_ϕ for $a = 0$ and selected values of M, the only free parameter in Eq. 4. It is easy to see that ν_r decreases after a maximum of $\nu_r(max) = 358(2M_\odot/M)$ Hz for a ϕ-frequency of $\nu_\phi = 2\nu_r(max) \simeq 0.65\ \nu_\phi(r_{ms})$. The measured $\Delta\nu$ vs. ν_2 for eleven NS LMXRBs is also plotted. It is apparent that for NS masses in the 2 $M_\odot$ range, the simple model outlined above is in qualitative agreement with the measured values, including the decrease of $\Delta\nu$ for increasing ν_2 seen in Sco X-1, 4U1608-52, 4U1735-44 and 4U1728-34. The model above is only an approximation: first, the spacetime around a fast rotating NS is different from a Kerr spacetime (due to the star's oblateness induced by rotation); second, the blobs' orbits might possess a finite (though small) eccentricity. Analytical formulae to partly correct for these effects were derived by Stella & Vietri (1999); Fig. 1B shows the fit to the observed $\Delta\nu$ vs. ν_2 relationship in Sco X-1 that was obtained through them. The orbital eccentricity, in particular, was varied in order to obtain different frequencies, while keeping the periastron distance $r_p = a(1-e)$ fixed. The best model for a non-rotating NS is shown in Fig. 1B. The model reproduces fairly accurately the data with a minimum number of free parameters, the NS mass ($M \sim 1.9$ M$\odot$) and periastron distance ($r_p \simeq 6.2\ M$). The latter value is close to the marginally stable orbit radius. When ν_s is allowed a finite value (say $\nu_s \sim 300 - 600$ Hz), fits of very similar quality are obtained, the parameters of which differ only slightly from those given above. In essence, the effects induced by the NS rotation on ν_r are small, though non-negligible.

The behaviour of the curves in Fig. 1A,B and therefore the ability of the model to match the observations, reflects the properties of the strong field Schwarzschild metric, since lower order expansions fail to reproduce the observed frequencies (see Stella & Vietri 1999). This can be seen through a simple argument. In the RPM interpretation, the epicyclic frequency ν_r is $\ll \nu_\phi$ in kHz QPO sources. This is unusual, the epicyclic frequency in most cases being close to ν_ϕ. However, if one remembers that $4\pi^2\nu_r^2 \equiv d^2V_{eff}/dr^2$, where V_{eff} is the effective potential, one sees that the inequality above is natural for orbits

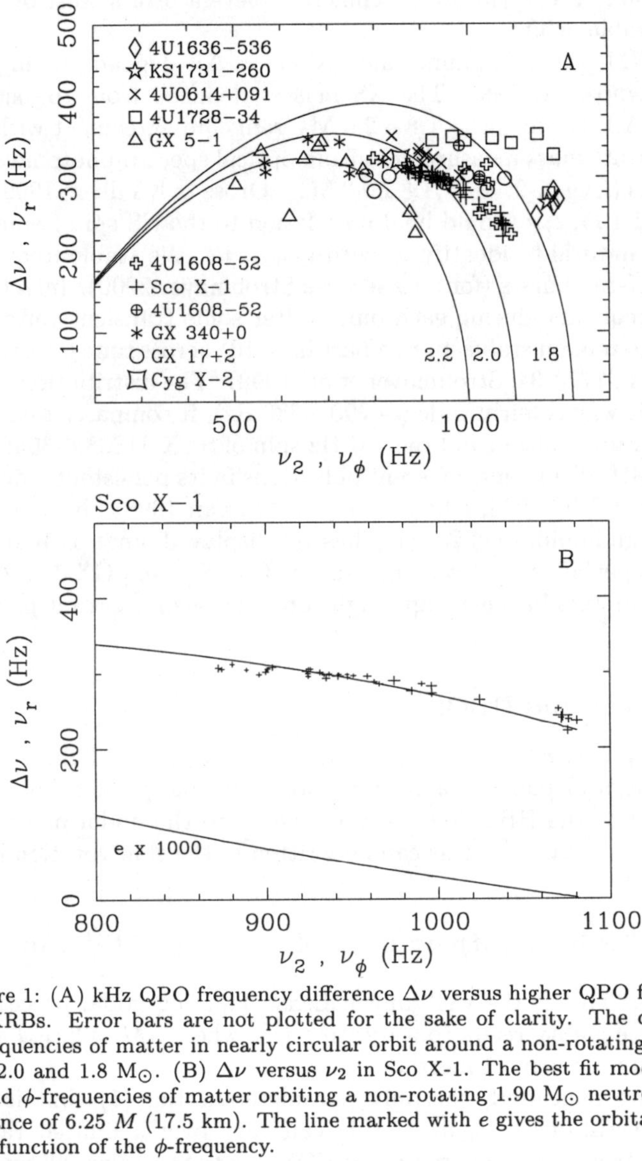

Figure 1: (A) kHz QPO frequency difference $\Delta\nu$ versus higher QPO frequency ν_2 for eleven LMXRBs. Error bars are not plotted for the sake of clarity. The curves give the r- and ϕ-frequencies of matter in nearly circular orbit around a non-rotating neutron star, of mass 2.2, 2.0 and 1.8 $M_\odot$. (B) $\Delta\nu$ versus ν_2 in Sco X-1. The best fit model corresponds to the r- and ϕ-frequencies of matter orbiting a non-rotating 1.90 $M_\odot$ neutron star at a periastron distance of 6.25 M (17.5 km). The line marked with e gives the orbital eccentricity ($\times 1000$) as a function of the ϕ-frequency.

close to the marginally stable one, for which the effective potential notoriously has an inflection point, $d^2V_{eff}/dr^2 = 0$. Thus the relevant orbits must be close to the innermost stable orbit.

Within the RPM, the maximum value of $\nu_r = \Delta\nu$ depends mainly on the mass of the compact object. The NS masses deduced from the simple modelling in Fig. 1A,B are in the $\sim 1.8 - 2.0$ M$_\odot$ range, in agreement with the only relatively accurate mass measurement from optical spectro-photometry in any of these systems (Cyg X-2; $M = 1.8 \pm 0.2$ M$_\odot$; Orosz & Kuulkers 1999). In general within the RPM, $\Delta\nu$ should hold no relation to the NS spin frequency ν_s. Therefore, it is natural to identify ν_s with ν_{burst}, *i.e.* the stable frequency seen during type I X-ray bursts (for a review see Strohmayer 2000). Infact in a given source, the frequency during each burst, after some transient variation, has been observed to remain stable up to a part in $\sim 10^4$ across time intervals of a few years (e.g. in 4U1728-34; Strohmayer et al. 1998). The distribution of NS spins inferred in this way is fairly wide ($\sim 290 - 580$ Hz); it compares well with that of millisecond radio pulsars and the 401 Hz spin of SAX J1808.4-3658, the only bursting LMXRB displaying coherent pulsations in its persistent emission (Wijnands & van der Klis 1998; Chakrabarty & Morgan 1998). None of the LMXRBs of the (high luminosity) Z-group has yet displayed burst oscillations; therefore their spin period is still to be measured. Cyg X-2 and GX 17+2, the only type I X-ray bursters in the group, might provide this important piece of information.

3 Nodal Precession and HBOs

If the blobs' orbits giving rise to the signals at ν_ϕ and ν_{per} are slightly tilted relative to the equatorial plane, nodal precession will take place around the spin axis. In the RPM the HBO frequency is related to the nodal precession frequency. From Eqs. (1) and (3) this can be written in the slow rotation limit $(a/M \ll 1)$

$$\nu_{nod} \simeq 4\pi a\nu_\phi^2 \simeq 6.2 \times 10^{-5}(a/M)m\nu_\phi^2 \text{ Hz} \simeq 4.4 \times 10^{-8}\ I_{45}m^{-1}\nu_\phi^2\nu_s \text{ Hz}\ , \quad (5)$$

where $M = m$ M$_\odot$. This is the well known Lense-Thirring nodal precession formula. The latter equality refers to a rotating NS, where $aM = 2\pi\nu_s I$ with $I = 10^{45} I_{45}$ g cm^2 its moment of inertia.

If ν_ϕ and ν_s are measured, the only parameter in Eq. (5) that is not identified from observations is $I_{45}m^{-1}$; this can vary over a limited range, $0.5 < I_{45}m^{-1} < 2$, for virtually any mass and EOS (see the rotating NS models of Friedman, Ipser & Parker 1986 and Cook, Shapiro & Teukolsky 1994). The stellar oblateness induced by the star's rotation (and thus the quadrupole and higher multipole terms in the gravitational potential), gives rise to correction terms in the nodal precession frequency also (see Morsink 1999 and Morsink & Stella 1999 for a post-Newtonian formula). Their relative importance increases for high values of ν_s and ν_ϕ. Yet the Lense-Thirring term dominates over a wide range of parameters, such that a $\sim \nu_\phi^2$ dependence is expected.

Since the relativistic nodal precession model was proposed, an approximately quadratic dependence of ν_{HBO} on the higher frequency kHz QPOs has been measured in a number of LMXRBs. This dependence was originally suggested on the basis of a few power spectra of the Atoll source 4U1728-34 (Stella & Vietri 1998a). Ford & van der Klis (1998) analysed a large set of power spectra from the same source and determined that the frequency ν_{low} of the $\sim 10 - 50$ Hz QPOs scales as $\nu_2^{2.11\pm0.06}$. Note that the low frequency QPO vs. ν_2 relation of this source appears to be double-branched, with the centroid frequency shifting by a factor of ~ 2 across different observations. This suggests that on occasion the 2nd harmonic of ν_{HBO} is excited instead of the fundamental. Stella & Vietri (1998b) first noticed that the HBO frequency of the Z-source GX 17+2 displays a nearly quadratic dependence on ν_2. Psaltis et al. (1999) carried out a systematic study of Z-sources and determined that the HBO frequency is consistent with a ν_2^2 scaling (Cyg X-2 and Sco X-1 show evidence for a somewhat flatter dependence). for frequencies up to $\nu_2 \sim 900 - 1000$ Hz. A somewhat flatter dependence or even a saturation is observed sometimes at higher frequencies. In essence these results confirmed one of the basic features of the RPM, namely the nearly quadratic dependence of the nodal precession frequency on the ϕ-frequency.

If the NS spin frequency is measured, then for any value of ν_ϕ the model yields a predicted nodal precession frequency which is uncertain only by a factor of a few, mainly due to the allowed range of $I_{45}m^{-1}$ (Stella & Vietri 1998a). Only in the Atoll source 4U1728-34 burst oscillations and simultaneous kHz QPOs and HBOs have been detected unambiguously (Strohmayer et al. 1996; Ford & van der Klis 1998; Mendez & van der Klis 1999) [a]. Therefore the QPO frequencies of 4U1728-34 can be used to test both the ν_{HBO} and the $\Delta\nu$ versus ν_2 relationships predicted by the RPM, when the NS spin derived from burst oscillations is used ($\nu_{burst} \simeq 364$ Hz). In order to take fully into account of all the effects that contribute determining the blobs' geodetic motion in the vicinity of the NS, we adopted a numerical approach and computed the spacetime metric of the star using Stergioulas' (1995) code, an equivalent of

[a] Note that in the case of KS1731-260, for which $\nu_{burst} \simeq 524$ Hz, it is still unclear whether the ~ 26 Hz QPOs, that where were observed together with kHz QPOs at ~ 910 and ~ 1170 Hz, represent HBOs or the equivalent of NBOs in an Atoll source see Smith et al. 1997; Wijnands & van der Klis 1997; Psaltis, Belloni & van der Klis 1999.

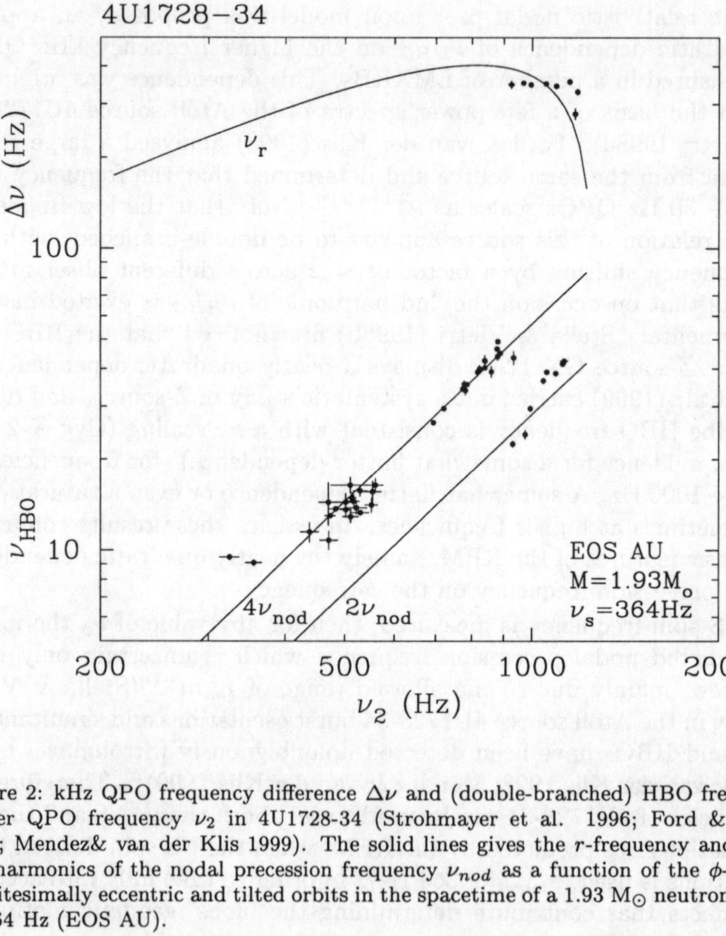

Figure 2: kHz QPO frequency difference $\Delta\nu$ and (double-branched) HBO frequency versus higher QPO frequency ν_2 in 4U1728-34 (Strohmayer et al. 1996; Ford & van der Klis 1998; Mendez& van der Klis 1999). The solid lines gives the r-frequency and the 2nd and 4th harmonics of the nodal precession frequency ν_{nod} as a function of the ϕ-frequency, for infinitesimally eccentric and tilted orbits in the spacetime of a 1.93 $M_\odot$ neutron star spinning at 364 Hz (EOS AU).

that of Cook et al.(1992) (see also Stergioulas and Friedman 1995). ν_{nod} was calculated according to the prescription of Morsink & Stella (1999), whereas ν_r was obtained from the second radial derivative of the effective potential (see Stella, Vietri & Morsink 1999). ν_r and ν_{nod} were derived as a function of ν_ϕ for infinitesimally small tilt angles and eccentricities (Morsink & Stella 1999; Stella, Vietri & Morsink 1999).

Fig. 2 shows the measured values of $\Delta\nu$ and ν_{HBO} versus ν_2 in 4U1728-34. Relatively high NS masses (see also Sect. 2) and stiff EOSs such as AU and UU (Wiringa et al. 1988) are required in this application of the RPM. The solid lines in Fig. 2 are for a 1.93 $M_\odot$ NS with EOS AU and $\nu_s = 364$ Hz. A good agreement is obtained if the HBO frequency, the lower of the two branches seen in 4U1728-34, is identified with the 2nd harmonics of ν_{nod} (*i.e.* $2\nu_{nod}$) (see also Morsink & Stella 1999; Stella & Vietri 1999a). Correspondingly the upper HBO branch is well fit by $4\nu_{nod}$. The geometry of tilted orbits in the innermost disk regions might be such that a stronger signal is produced at the even harmonics of the nodal precession frequency (e.g. Psaltis et al. 1999). The frequency range and trend of the epicyclic frequency ν_r in this model are also in reasonable agreement with the $\Delta\nu$ measurements; a more complex model is clearly required in order to fit these data more accurately. The analogy with the modelling of the Sco X-1 kHz QPO frequencies described in Sect. 2, suggests that introducing a finite (though small) eccentricity in the blobs' orbits would greatly improve the quality of the fit. Detailed modelling based on finite eccentricity geodesics in the spacetime of fast rotating NSs are in progress and will be reported elsewhere.

In summary, the model presented here is capable of reproducing the salient features of both the $\Delta\nu$ versus ν_2 and ν_{HBO} versus ν_2 relationships, with just two free parameters (M and the EOS), the allowed range of which is fairly limited (moreover the EOS cannot even be varied continuously !). Concerning Z-sources and all other Atoll sources for which burst oscillations have not been detected yet, the NS spin can still be regarded as a free parameter. The application of the RPM to the HBOs of these sources can therefore be used to constrain the spin of their NSs. Conversely, in those Atoll source in which ν_{burst} is measured, but HBOs have not been detected yet, the RPM can be used to predict ν_{HBO}. These issues are briefly addressed in the next Section.

4 THE RPM AND THE PBV CORRELATION

Psaltis, Belloni & van der Klis (1999) recently identified two QPOs and peaked noise components the frequency of which follows a tight correlation over nearly three decades. This correlation (hereafter PBV correlation) involves both NS and BHC LMXRBs spanning different classes and a wide range of luminosities (see the points in Fig. 3). In kHz QPO NS systems, these components are the lower frequency kHz QPOs, ν_1, and the low frequency, HBO or HBO-like QPOs, ν_{HBO}. For BHC systems and lower luminosity NS LMXRBs the correlation involves either two QPOs, or a QPO and a peaked noise component. In all cases the frequency separation is about a decade and an approximate linear relationship ($\nu_{HBO} \sim \nu_1^{0.95}$) holds. The QPO frequencies from the peculiar NS system Cir X-1 varies over nearly a decade while closely following the PBV correlation and bridging its low and high frequency ends. Psaltis, Belloni & van der Klis (1999) noted also the that ν_2 vs. ν_1 relations of different Atoll and Z-sources line-up with good accuracy (see also Psaltis et al. 1998).

We show here that the frequencies predicted by the RPM match precisely the PBV correlation, without resorting to any additional assumption (see Stella, Vietri & Morsink 1999). We assume that in all QPO sources, including BHCs, $\nu_{HBO} \simeq 2\nu_{nod}$ as in 4U1728-34 (see Sect.3). Fig. 3A shows $2\nu_{nod}$ and ν_ϕ obtained from Eqs. 1-3 as a function of ν_{per} for corotating orbits and selected values of M and a/M. The high frequency end of each line is dictated by the orbital radius reaching the marginally stable orbit at r_{ms} (where $\nu_\phi = \nu_{per}$, i.e. $\nu_r = 0$, one of the defining characteristics of the marginally stable orbit). It is apparent that r_{ms} depends mainly on M and to a lesser extent on a/M.

The separation of the lines in Fig. 3A testifies that while ν_{nod} depends weakly on the mass and more strongly on a/M, the opposite is true for ν_ϕ. By taking the weak field ($M/r \ll 1$) and slow rotation ($a/M \ll 1$) limit of Eqs. (1)-(3) the relevant first order dependence is made explicit,

$$\nu_\phi \simeq (2\pi)^{-2/5} 3^{-3/5} M^{-2/5} \nu_{per}^{3/5} \simeq 33\ m^{-2/5} \nu_{per}^{3/5}\ \text{Hz}\ , \tag{6}$$

$$\nu_{nod} \simeq (2/3)^{6/5} \pi^{1/5} (a/M) M^{1/5} \nu_{per}^{6/5} \simeq 6.7 \times 10^{-2}\ (a/M) m^{1/5} \nu_{per}^{6/5}\ \text{Hz}\ . \tag{7}$$

For the case of rotating NSs where the effects of stellar oblateness introduces modifications relative to a Kerr spacetime, we adopt the numerical approach outlined in Sect. 3. Results are shown in Fig. 3B for a NS mass of 1.95 $M_\odot$, EOS AU and $\nu_s = 300$, 600, 900 and 1200 Hz (corresponding to $a/M = 0.11$, 0.22, 0.34 and 0.47, respectively). While also in this case ν_ϕ depends only very weakly on ν_s, the approximately linear dependence of ν_{nod} on ν_s is apparent. Note that the approximate scalings in Eqs. 6-7 remain valid over a wide range of frequencies. Only for the largest values of ν_{per} and ν_s, ν_{nod} departs substantially from the $\sim \nu_{per}^{6/5}$ dependence.

The measured QPO and peaked noise frequencies giving rise to the PBV correlation are also plotted in Fig. 3B. Higher kHz QPO frequencies from

NS systems (ν_2) are included (for the sake of clarity NBOs and FBOs were excluded). The agreement over the range of frequencies spanned by each kHz QPO NS system should not be surprising: together with the accurate matching of the corresponding $\nu_1 - \nu_2$ relationship in Z-sources, this is indeed part of the evidence on which the RPM model was proposed. However the fact that the dependence of ν_{nod} on ν_{per} matches the observed $\nu_{HBO} - \nu_1$ correlation to a good accuracy over ~ 3 decades in frequency (down to ν_1 of a few Hz), encompassing both NS and BHC systems, provides additional independent evidence in favor of the RPM. The observed variation of ν_{HBO} and ν_1 in individual sources (Cir X-1 is the most striking example, see Fig. 3B) further supports the scaling predicted by the RPM. The matching of the observed ν_2 vs. ν_1 relation in terms of ν_ϕ vs. ν_{per} is also quite accurate.

In the case of NS LMXRBs the best matching is obtained for relatively stiff EsOS (*e.g.* AU, as in Fig. 3B, and UU), which can also achieve the relatively high masses ($m \sim 1.8 - 2.0$) that are required to match the observed ν_1 and ν_2 relationship. For EOS AU and $m = 1.95$, the ν_{HBO} vs. ν_1 values of most NS LMXRBs are best matched for ν_s in the ~ 600 to 900 Hz range. It is apparent from Fig. 3B that ν_s as low as ~ 300 Hz are required for the Atoll sources with ν_{HBO} somewhat below the main PBV correlation: these are 4U1728-34 (see Sect. 3) and 4U1608-52 (from which burst oscillations have not been detected yet). The values above are close to the range of ν_s inferred from ν_{burst} in a number of other Atoll sources (van der Klis 1999). Z-type LMXRBs, appear to require ν_s in the ~ 600 to 900 Hz range, a possibility that is still open since for none of these sources there exists yet a ν_s measurement based on burst oscillations or coherent pulsations (see also Stella, Vietri & Morsink 1999b). Note that the upper HBO branch of 4U1728-34 matches well the main PBV correlation. In the interpretation of Sect. 3 the lower branch corresponds to $2\nu_{nod}$ and the upper branch to $4\nu_{nod}$. One could further speculate that sources following the main PBV correlation, Z-sources in particular, are also in the upper HBO branch at $4\nu_{nod}$; in this case their ν_s might be expected in the $\sim 300 - 400$ Hz range.

For BHC LMXRBs the scatter around the PBV correlation implies values of a/M of $\sim 0.1 - 0.3$ (see Fig. 3B). Note that so far no evidence has been found for a BHC QPO (or peaked noise) signal that is the equivalent of ν_2 in NS systems. The points from XTE J1550-564, while inconsistent with any single value of a/M, might lie along two distinct branches separated by a factor of ~ 2 in ν_{HBO}, similar to the case of 4U1728-34. Interpreting the PBV correlation within the RPM implies that both NS and BH systems share relatively high values of their specific angular momentum ($a/M \sim 0.1 - 0.3$ for BHCs or $\nu_s = 300 - 900$ Hz for relatively stiff EOS NSs). Population

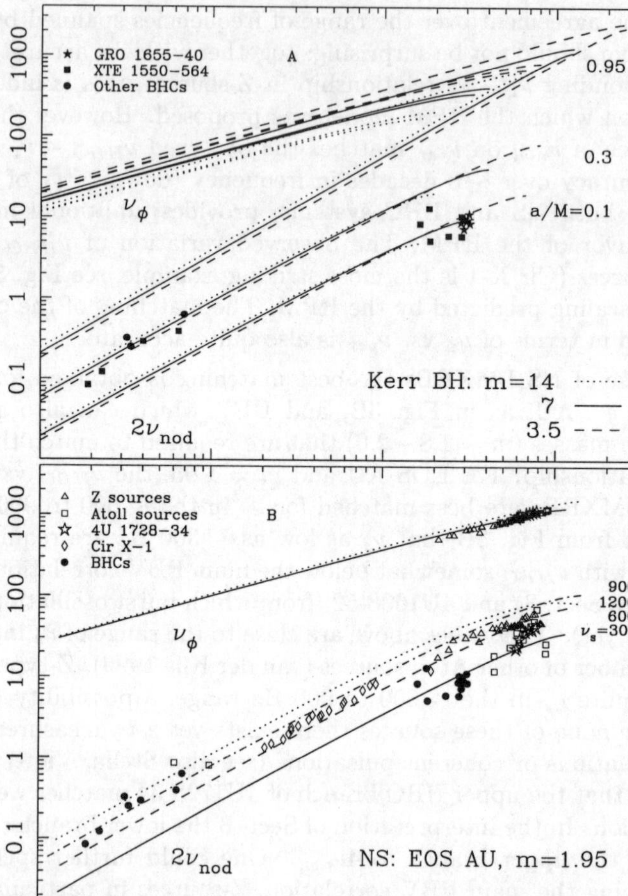

Figure 3: Twice the nodal precession frequency, $2\nu_{nod}$, and ϕ-frequency, ν_ϕ, vs. periastron precession frequency, ν_{per}, for black hole candidates of various masses and angular momenta (panel A) and rotating neutron star models (EOS AU, $m = 1.95$) with selected spin frequencies (panel B). The measured QPO (or peaked noise) frequencies ν_1, ν_2 and ν_{HBO} giving rise to the PBV correlation are also shown in panel B for both BHC and NS LMXRBs and in panel A for BHC LMXRBs only; errors bars are not plotted (see Psaltis, Belloni & van der Klis 1999 for a complete list of references). We included only those cases in which QPOs at ν_1 were unambiguously detected. NBO and FBO frequencies are not plotted.

synthesis calculations show that a significant fraction of low magnetic field, relatively stiff EOS NSs in LMXRBs can be spunup by accretion torques to $\nu_s \sim 1$ kHz, provided that $> 0.3 - 0.4$ $M_\odot$ are accreted (Burderi et al. 1999). The combination of the relatively high NS mass spin frequency required by the RPM is at least qualitatively in agreement with evolutionary scenarios for LMXRBs. In the Kerr metric, the increase in ν_{nod} for increasing values of a/M shows no signs of saturation (Fig. 3A). BHCs can in principle achieve extreme values of $a/M \simeq 1$ (NSs are instead limited to $a/M < 0.6 - 0.7$, cf. Salgado et al. 1994). In practice spinup to high values of a/M is unlikely to occur in BHC LMXRBs, because of the limited mass that can be accreted from the mass donor star (King & Kolb 1999). In the RPM the high frequency BHC QPOs (*e.g.* the $\sim$ 300 Hz QPOs of GRO1655-40) are interpreted terms of $\nu_1 = \nu_{per}$ (while the lower frequency QPOs $\nu_{HBO} = 2\nu_{nod}$) implying moderate values of $a/M \sim 0.1 - 0.3$. This is at variance with the $\nu_1 = \nu_{nod}$ interpretation of Cui et al. (1998), which requires high values of a/M (~ 0.95 in GRO1655-40), in contrast with BHC accretion-driven spinup scenarios. From Fig. 3A it is apparent that the $\sim$ 300 Hz QPOs from GRO 1655-40 lie close to the high frequency end of the $a/M = 0.1$, $m = 7$ line. Since the mass of the BHC in GRO 1655-40 determined through optical observations is ~ 7 $M_\odot$ (Shahbaz et al. 1999), we conclude that, according to the RPM, $\nu_{per} \simeq 300$ Hz close to the marginally stable orbit, where by definition $\nu_\phi = \nu_{per}$. This suggests that any additional QPO signal at $\nu_2 = \nu_\phi$ would be very close to or even blended with the QPO peak at ν_1. The detection of two closeby or even partially overlapping QPO peaks close to $\sim$ 300 Hz in GRO 1655-40 would therefore provide further evidence in favor of the RPM interpretation. Similar considerations might apply to the $\sim$ 284 Hz QPOs of XTE J1550-564 (Homan et al. 1999), provided its BHC mass is in the ~ 7 $M_\odot$ range.

5 DISCUSSION

The RPM makes several predictions that can be tested through future analyses and/or observations:

(a) The frequency difference $\Delta\nu = \nu_r$ is expected to decrease also for low values of $\nu_2 = \nu_\phi$ (see Eq. 4 and Fig.1). Moreover if the highest $\nu_2 = \nu_\phi$ frequencies do originate from nearly circular orbits (see Fig. 1A) then $\Delta\nu$ should quickly decrease as ν_2 increases further and the orbital radius approaches the marginally stable orbit.

(b) $\nu_2 = \nu_\phi$ is expected to scale as $\nu_1^{3/5} = \nu_{per}^{3/5}$. Extending the ν_2 vs. ν_1 correlation in NS systems toward lower frequencies and detecting the signal at ν_2 in BHC systems would provide important new tests.

In the RPM the QPO signals are produced at $r/M \sim 100\ m^{-2/5}\nu_{per}^{-2/5}$, which, for individual sources, must decrease for increasing mass accretion rates (as $\dot{M}$ is positively correlated with *e.g.* ν_2). The inferred radii range from close to the marginally stable ($r/M \sim 6$) to $r/M \sim 30$ over the frequency span covered by the PBV correlation. Many NS and BHC LMXRBs display two component X-ray spectra consisting of a soft thermal component, usually interpreted in terms of emission from an optically thick accretion disk, and a harder, often power-law like, component, likely due to a hot inner disk region. The QPOs might originate at the transition radius between the optically thick disk and the hot inner region as a result of occultation by orbiting blobs or modulated emission by self-luminous blobs (Stella, Vietri & Morsink 1999; see also Di Matteo & Psaltis 1999). For kHz QPO NS systems the inferred radii are remarkably close to the marginally stable orbit. A non-negligible eccentricity would also alter these frequencies slightly (see Sect. 2).

Mechanisms that can induce a finite (though small) eccentricity and tilt in the blobs' orbits, are currently under investigation. In the case of NSs, some kind of resonance between the star spin and the motion of the blobs (see *e.g.* Vietri & Stella 1998) might be responsible for the close commensurability of ν_{burst} and $\Delta\nu$ (or $2\Delta\nu$) observed in a few sources. These issues will be addressed elsewhere. Note that in the approach explored by Psaltis & Norman (1999) (see also Psaltis 2000), the same frequencies as in the RPM (plus an additional frequency at $\nu_\phi + \nu_r$) are selected by the response of the innermost annulus of the disk, when this is subject to a wide-band input noise; in this case no blobs would be required.

If confirmed, the RPM will provide an unprecedented opportunity to test GR effects in the strong field regime, such as the periastron precession in the vicinity of the marginally stable orbit and the radial dependence of the Lense-Thirring nodal precession frequency. In principle, accurately measured kHz QPO and HBO frequencies would yield crucial information on the compact object such as its mass and angular momentum (*e.g.* by solving Eqs. 1-3 for m, a/M and r). Should suitable, additional observables be found, it might become possible to obtain a self-consistency check of the RPM, together with tests of GR vs. alternative theories of gravity in the strong field regime.

Acknowledgements

I am grateful to my collaborators in this project, S. Morsink and M. Vietri.

References

1. Alpar, A. & Shaham, J. 1985, Nature, 316, 239

2. Andersson, N., Kokkotas, K.D. & Stergioulas, N. 1999, ApJ, 516, 307
3. Bardeen J.M., Press W.H. & Teukolski S.A. 1972, ApJ, 178, 347
4. Bildsten, L. 1998, ApJ, 501, L89
5. Burderi, L., Possenti, A., Colpi, M., Di Salvo, T. & D'Amico, N. 1999, ApJ, 519, 285
6. Chakrabarty D. & Morgan E.H. 1998, Nature, 394, 346
7. Cook, G.B., Shapiro, S.L., & Teukolsky, S.A. 1992, ApJ, 398, 203
8. Cui, W., Zhang, S.N., Chen, W. 1998, ApJ, 492, L53
9. de Felice F. & Usseglio-Tomasset S. 1996, Gen. Rel. Grav., 28, 179
10. Di Matteo, T. & Psaltis, D. 1999, ApJ, 526, L101
11. Ford, E.C. & van der Klis, M. 1998, ApJ, 506, L39
12. Ford, E.C. et al. 1998, ApJ, 508, L155
13. Friedman, J.L, Ipser, J.R. & Parker, L. 1986, ApJ, 304, 115
14. Homan, J., van der Klis, M., Wijnands, R., Vaughan, B, & Kuulkers, E. 1998, ApJ, 499, L41
15. Homan, J., Wijnands, R. & van der Klis, M. 1999, IAU Circ. 7121
16. Kato S. 1990, PASJ, 42, 99
17. King, A.R. & Kolb, U. 1999, MNRAS, 305, 654
18. Lamb, F.K., Shibazaki, N., Alpar, A. & Shaham, J. 1985, Nature 317, 681
19. Mendez, M. et al. 1998a ApJ, 494, L65
20. ——— 1998b ApJ, 505, L23
21. Mendez, M. & van der Klis, M. 1999, ApJ, 517, L51
22. Mendez, M., van der Klis, M., Ford, E.C., Wijnands, R. & van Paradijs, J. 1999, ApJ, 511, L49
23. Miller, M.C., Lamb, F.K. & Psaltis, D. 1998, ApJ, 508, 791
24. Morsink, S. 1999, this volume
25. Morsink, S. & Stella, L. 1999, ApJ, 513, 827
26. Perez C.A., Silbergleit A.S., Wagoner P.V. & Lehr D.E. 1997, ApJ, 476, 589
27. Orosz, J.A. & Kuulkers, E. 1999, MNRAS, 305, 132
28. Psaltis, D. 2000, in "X-ray Astronomy '99: Stellar Endpoints, AGNs and Diffuse Background", Ap. Lett. & Comm., in press
29. Psaltis, D. & Norman, C. 2000, ApJ, in press.
30. Psaltis, D., et al. 1998, ApJ, 501, L95
31. ——— 1999, ApJ, 520, 763
32. Psaltis, D., Belloni, T., & van der Klis, M. 1999, ApJ, 520, 262
33. Salgado, M., Bonazzola, S., Gourgoulhon, E. & Haensel, P. 1994, A&A, 291, 155.
34. Shahbaz, T. et al. 1999, MNRAS, 306, 89

35. Smith, D.A., Morgan, E.H. & Bradt, H. 1997, ApJ, 479, L137
36. Stella, L. & Vietri, M. 1998a, ApJ, 492, L59
37. ———— 1998b, in "The active X-ray sky", Nucl Phys B (Proc. Suppl.), 69, 135
38. ———— 1999, Phys. Rev. Lett, 82, 17
39. Stella, L., Vietri, M. & Morsink, S.M. 1999, ApJ, 524, L63
40. Stergioulas, N. 1995, code available via ftp from `pauli.phys.uwm.edu` in directory `/pub/rns`
41. Stergioulas, N., & Friedman, J.L. 1995, ApJ 444, 306
42. Strohmayer, T.E. et al. 1996, ApJ, 469, L9
43. Strohmayer, T.E et al. 1998, ApJ, 503, L147
44. Strohmayer, T.E. 2000, in "X-ray Astronomy '99: Stellar Endpoints, AGNs and Diffuse Background", Ap. Lett. & Comm., in press
45. van der Klis, M. 1995,in "X-ray Binaries", Eds. W. H. G. Lewin, J. van Paradijs & E. P. J. van den Heuvel (Cambridge University Press), p. 252
46. ———— 1998, in: Proc. NATO ASI "The many faces of neutron stars", Series C, Vol. 515, p. 337
47. ———— 1999, this volume.
48. ———— 2000, in "X-ray Astronomy '99: Stellar Endpoints, AGNs and Diffuse Background", Ap. Lett. & Comm., in press
49. van der Klis, M. et al. 1997 ApJ, 481, L97
50. Vietri, M. & Stella, L. 1998b, ApJ, 503, 350
51. White, N.E. & Zhang, W. 1997, ApJ, 490, L87
52. Wijnands R.A.D. & van der Klis M. 1997, ApJ, 482, L65
53. ———— 1998, Nature, 394, 344
54. Wijnands, R., van der Klis, M. & Rijkhorst, E.J. 1999, ApJ, 512, L39
55. Wiringa, R.B., Fiks, V. & Fabrocini, A. 1988, Phys. Rev. C, 38, 1010

LESSONS FROM X-RAY ASTRONOMY APPLIED TO HST

ETHAN J. SCHREIER and RODGER DOXSEY
Space Telescope Science Institute
3700 San Martin Drive
Baltimore, Maryland 21218, USA
E-mail: schreier@stsci.edu

Riccardo Giacconi, probably more than any other single individual, established x-ray astronomy as an essential sub-discipline of astronomy. Its incorporation into the mainstream of astronomy was substantially completed with the Einstein Observatory which, with its imaging capabilities and its Guest Observer program, invited non-x-ray astronomers to use the facility. It was therefore perhaps fitting that when optical astronomy moved into space, with the Hubble Space Telescope, it called on Riccardo to oversee the transition. He brought with him lessons about building and operating space observatories, experience working with NASA on large science projects, a business-like approach to attacking tasks, and his unique vision and abilities. Among the guiding principles he brought to HST were: involvement of a strong scientific research staff in all aspects of the program; establishment of a vital, active research environment; attention to "science system engineering" and applying a rational scientific approach to problems; creation of an atmosphere of "ruthless intellectual honesty;" and maintenance of the highest regard for process. These formed the basis both for attacking the problems of HST, and for building an Institute to do so.

1 A Personal Preamble

It was 16 years ago today (the time of the writing of this paper), just before Memorial Day in May 1981, that Riccardo Giacconi called together his senior staff of the High Energy Astrophysics Division at the Center for Astrophysics to announce that he had accepted a position as first Director of the Space Telescope Science Institute. This close-knit group of x-ray astronomers, many of whom had worked with Riccardo since the 1960s, was shocked. Not only was the head of their division, the scientific leader of their research group, and the mentor of several, leaving, but the "father" of x-ray astronomy was moving into optical astronomy. In retrospect, it was an obvious move, both for the optical-UV community that selected him, and for the individual who pioneered a "new way of doing business" in astronomy. But at the time, it was quite a surprise.

Riccardo's long discussion with his staff concentrated on leaving the high energy group intact, especially necessary to oversee the long-awaited development of AXAF. However, it was clear to him that he needed to bring some of x-ray astronomy's accumulated expertise in space science operations and data systems with him to Baltimore. And thus, one of the authors (EJS) of this paper started commuting to Baltimore in June 1981, becoming the first Chief Data and Operations Scientist, and second x-ray astronomer, at the Space Telescope Science Institute. By the end of that summer, having assessed the true need for science operations expertise in the space telescope project, yet a third x-ray astronomer, the other author (RD) of this paper, was recruited as his deputy.

2 Introduction

Riccardo, probably more than any other single individual, established x-ray astronomy as an essential sub-discipline of astronomy. Coming from high energy physics, he became convinced that this wavelength domain, characteristic of high energy processes in nature, was of essential importance to the physical understanding of many areas of astronomy. He participated in the pioneering rocket flights of discovery, convinced NASA of the importance of x-ray astronomy, forged collaborations in the expanding field, and conceived of and sold a program of research and technology development which, even in the mid-1960s, envisioned an imaging x-ray capability on the scale of AXAF, just now on the verge of being realized. By 1970, he and his group had built and started using UHURU, the first satellite observatory dedicated to x-ray astronomy, whose discoveries, documented elsewhere in this volume, not only revolutionized the field, but showed the close inter-dependence between x- ray observations and the existing branches of astronomy. From the discovery of the binary x-ray systems to the measurements of masses of clusters of galaxies, UHURU's x-ray observations significantly affected most fields of astronomy, bringing new understandings impossible without combined x-ray, optical and radio observations. The merging of x-ray astronomy into the mainstream of astronomy, which started with UHURU, was substantially completed with the Einstein Observatory which, with its imaging capabilities and its Guest Observer program, invited non-x-ray astronomers to use the facility.

It is therefore perhaps fitting that when optical astronomy moved into space in a "big way," with the Hubble Space Telescope, it called on Riccardo to oversee the transition. But it was the lessons about building and operating space observatories learned in x-ray astronomy, the experience of working with NASA on big science projects, Riccardo's business like approach to attacking tasks, and most of all, his unique vision and abilities, which were being recruited.

In addition to, or perhaps because of, the experience accumulated by Riccardo in developing, operating, and doing research with x-ray astronomy's pioneering observatories, there were a number of guiding concepts which Ric-

New Scientist 8 June 1972

Interactions

A business approach to modern astronomy

American Science and Engineering Inc, of Cambridge, Massachusetts, supplies scientific payloads for many NASA astronomy experiments. The company has achieved great fame for its success with the Uhuru X-ray satellite. What is it like to do pure research in a company rather than a university?

Dr Simon Mitton

Left to right: Ethan Schreier, Riccardo Giacconi, Terry Matilsei, Harvey Tannenbaum, Edwin Kellog

Figure 1: A New Way of Doing Business — Riccardo with some of the UHURU group in his office at American Science and Engineering.

cardo brought to the Space Telescope Science Institute. Foremost among these was the need for a strong scientific research staff, involved in all aspects of the program, in order to apply a scientific, rational approach to problems, and to provide the vitality of an active research environment. Along with this came attention to what we came to call "science system engineering;" the creation of an atmosphere of "ruthless intellectual honesty;" and the maintaining of the highest regard for process. These formed the basis both for attacking the problems of HST, and for building an Institute to do so.

3 The Beginning — Building A Staff

One of the most exciting, sometimes daunting, aspects of being on the staff of the STScI in 1981 was the realization that we were creating, essentially from scratch and nearly overnight, what needed to become one of the premier astronomy institutions in the world. The task of operating a breakthrough facility such as HST required a first rate, active research staff as well as an excellent technical and operational staff. There wasn't the slightest doubt in Riccardo's mind concerning the nature of the organization he was creating, clearly beyond the purely operational organization NASA was expecting, and necessarily beyond the scope envisioned in the Hornig report. The range of issues which Riccardo focused on even in the first year was guided by his experience in x-ray astronomy, both in building and operating space hardware, and in dealing with NASA. It set the groundwork for the successful development and operation of the STScI.

The first year saw the clear definition of the role for research astronomers at the STScI, and the start of the process of building the science staff. It included establishing the STScI as an independent voice representing the astronomy community in project deliberations. During the first year, the staff of the STScI took true ownership of the technical problems associated with HST science operations, began strongly influencing project development of the ground system, and defined the visions for the Guide Star Selection and Science Data Analysis Systems. And not least, the organization began creating it's own infra-structure, including construction of a building, for long term operations.

In the chaos of the first year, with a very small staff and with launch then looming only a few years away, Riccardo kept his, and his staff's, attention focused on essential, long range goals. The very first staffing plan and budget created by the STScI itself, in December 1981, included essential aspects such as the research component for the science staff, research support facilities, first realistic estimates of the operations staffs, staff to handle important functions such as observing proposal solicitation and review, and graduate student programs. Riccardo understood that the outline of the eventual Institute had to be reflected in the very first budget he submitted, even if it was guaranteed to be controversial. The knowledge of the importance of the functions included was firmly based in two decades of experience with x-ray astronomy facility development and operations.

At times it seemed a little incongruous, writing about and budgeting for an organization of several hundred staff, when we could hold "all hands" meetings in an old Johns Hopkins physics department reading room, seated around one table. Meetings would get interrupted to send someone over to the new building site to quickly make some decision, as they were about to pour concrete. More than one procedure or process was invented on the fly the first time we found we needed it, but based on logic and extrapolated from experience. Many early official project meetings and reviews drew standing room only audiences, to see how the new guys on the program did (or to hear what outrageous idea they would suggest next). The sense of invention, and the enjoyment of participating in an adventure, came out frequently in one of Riccardo's common questions: "So what do you really want to do when you grow up?"

Figure 2: The STScI Under Construction — Riccardo (center) in a hardhat walking with Frank Martin (then NASA Associate Administrator for Space Science) with other early STScI staff and NASA personnel at the building site.

4 Science System Engineering

Science system engineering involved looking at the entire system that was being developed — instruments, spacecraft, operations, calibrations, data systems, staffing — as a single system to do science. It was essential to look at the system from the standpoint of what the end-users, both the operations staff and especially the astronomers, would really need. Thus, if HST was supposed to be able to carry out planetary observations, it was necessary to ask what kind of observations would have to be carried out, and then to ask how each piece of the entire system — hardware, software, and people — would do its part. In a project as big as HST, it was perhaps not surprising that some sub-systems would be built without full knowledge of how other subsystems would work. But it certainly was a surprise to find out that, although HST was expected to do first class planetary science, the initially planned on-board software and the planning and scheduling system only knew about linear motion — HST would not be able to track real solar system objects!

A prime example of the science system engineering methodology brought from x-ray astronomy into the Space Telescope project was the establishment of the calibration database system. In high energy physics, the original discipline of many x-ray astronomers, devoting a great deal of attention to instrument calibration was taken for granted; exploring the discovery space at the detection threshold of an instrument necessitated a detailed understanding of instrumental characteristics. Riccardo's x-ray observatories were fully in keeping with this tradition. With the Einstein Observatory especially, a very large effort was devoted to understanding mirror performance and calibrating the spatial and spectral response of the detectors, and indeed, of the observatory as a whole. Equally important, software was developed for the pre-flight calibration of the observatory, which could later be used for the calibration of the science data itself. This level of attention to calibration was not originally envisioned in the HST program, nor considered in-scope. Nonetheless, a coherent program of calibrations was implemented, adding to the instrument calibrations a set of standard star measurements, cross-calibrations, a set of software routines to be included in the data analysis system, and a database of calibrations to be included in the HST archive.

Other examples of science system engineering emerged in reviewing the plans for operations, for planning and scheduling the observatory, for selecting guide stars, and for archiving data. For example, when the initial design for the planning and scheduling system was reviewed, it became apparent that although the basic capabilities for entering observing proposals were incorporated, it would take some dozen keypunch operators working 24 hours a day to keep up. Thus, we found that although subsystems might be built to meet their individual specifications, continuing attention to the overall system was necessary to meet the spirit of the requirements, and to create an observatory which could really be used.

Perhaps a more significant and long-lasting impact of Riccardo's attention to system engineering came with the development of the guide star selection system. In reviewing the concept for target acquisition and guiding, we found that the plan called for digitizing individual areas of survey plates for each observation, in order to select guide stars for HST's fine guidance sensors to track. Having the time-intensive plate scanning and guide star selection process in-line with the planning and scheduling activities was clearly not optimal — not only was guide star selection dependent on the time of observation, difficult to know long in advance, but the scanning of individual sections of thousands of survey plates in random order would lead to a nightmare of plate handling logistics and risk. As a result of looking at the entire process from the standpoint of what was really needed, it became clear that scanning ALL the plates, in a well-defined order, and creating a source catalog of potential guide stars over the entire sky, to be referenced during the planning and scheduling process, was a far superior approach. In this case, it not only optimized science operations, but it created a science resource which extended far beyond the needs of HST. The Digitized Sky Survey, the Guide Star Catalog, and the software tools to use them, became widely used scientific products of the HST program even before any data was taken by the telescope. These products, available via the Internet and on CD-ROMs, revolutionized ground-based and space-based astronomy, contributing to science projects and telescope operations around the world.

5 Ruthless Intellectual Honesty

This phrase would not come as a surprise to anyone who worked with Riccardo for any length of time. Riccardo insisted on an atmosphere of honesty and trust in the work environment. His original x-ray group was steeped in this tradition, and he made it an underpinning of the Space Telescope Science Institute. The STScI staff was asked to impose the same questioning approach to technical and programmatic tasks as they would to a research endeavor. This often led to conflict with other elements of the ST program. But by expecting serious debate on all issues, pros and cons of technical approaches were thoroughly aired. Not questioning a questionable decision was one of the worst sins. This total respect for hearing all sides of an issue was combined with little respect for un-focused "academic" deliberations, and no respect for deceit, either of colleagues or of yourself. It was not uncommon to be told "you're not thinking straight" — meaning you were indeed being heard, and engaged in a communal decision making process. By the time a consensus was reached, everyone could be sure that all sides had been heard and duly considered, that nearly everyone owned the decision, and that there was little or no confusion about what the decision was.

6 The Strategy Panel — An Outstanding Example

Probably the ultimate and most visible example of the vision, methodology, and science system engineering which Riccardo brought to HST was the process that led to the successful correcting of HST's optical problems. Arising from the program-wide depression and despair which followed the discovery of spherical aberration, Riccardo's STScI mobilized to attack the problem. While investigating how to make the best of the aberrated HST images via software approaches to image restoration, we worked with NASA, the astronomy community, and the optics community to determine what, if anything, could be done to "fix" HST. The approach was to convene a "Strategy Panel" representing the best talent that could be assembled — astronomers, opticians, engineers, astronauts — and to solicit the broadest range of possible solutions. Nothing was too crazy to be considered. Science system engineering led to looking at all aspects of the system and the problem. Process involved including representatives of all HST's stake-holders. An atmosphere of honesty and trust was required to solicit the widest variety of solutions. The ultimate success of the approach is demonstrated with every new HST discovery and every HST image which appears in the media.

7 Conclusion

Riccardo had, and still has, a vision of how to do science, how to attack problems, and how to build a cohesive group. This vision enabled him and his x-ray group to become very successful in x-ray astronomy. He carried this same vision into optical astronomy in his building of the Space Telescope Science Institute and his oversight of the operations development for HST. His vision for the Institute involved not so much what it should look like, but what it had to do, how it had to function, and how the Space Telescope Science Institute would meet the needs of the community it had to serve. Its key mission was to conduct the science program of HST, and to do that well, it had to continually strive for excellence in both service and research.

Riccardo's deep beliefs in a scientific approach to problem solving, to process, to applying the principles of science system engineering, and to an envi-

Figure 3: Riccardo's Garibaldini — Riccardo with some of the veterans of the early years at the STScI, just before he departed. The red shirts they wear refer to one of his favorite stories, Garibaldi instructing his soldiers to wear red so that their wounds would not show.

Figure 4: The Space Telescope Science Institute — Riccardo with the staff of the Institute just before the launch of HST.

ronment of honesty and trust, were essential to ensure the strength and vitality of the Space Telescope Science Institute, producing an institution which could meet the challenges of the HST Program. He has gone on to apply his vision and the principles summarized above to the European Southern Observatory and the building of the VLT. Having helped realize the enormous scientific potential of the Hubble Space Telescope, Riccardo moved back to earth to transform an entire mountain into a precision scientific research tool. Optical astronomy will never be the same.

HST OBSERVATIONS OF EXTRAGALACTIC OPTICAL JETS

F. DUCCIO MACCHETTO[a]
Space Telescope Science Institute - 3700 San Martin Drive,
Balt., MD 21218, USA
E-mail: macchetto@stsci.edu

The HST has become one of the most important instruments to investigate the physical nature of the optical counterparts to radio-jets. High-spatial and spectral resolution observations of a variety of sources has been carried-out. HST observations of the complete 3CR sample of radio sources has led to the discovery of several new optical counterparts to radio jets, more than doubling the sample of known objects. HST observations have revealed, for the first time at optical wavelengths, proper motions in the M87 jet, with doppler-boosted velocities, as high as 2.5c. Also relevant to the understanding of the physical properties of these jets are the numerous HST observations of the central regions of AGN. These include LINERS, Seyferts 1 and 2, as well as powerful radio sources. The HST results are in broad agreement with the unified theories for AGN. However, these results have also made important modifications to the basic theory, such as the need for additional sources of ionization (e.g., shocks) to explain the observed filamentary structure in the Narrow-Line Region.

"To study optical jets is as much fun as sucking a dry nail". Riccardo Giacconi (Circa 1986)

1 Introduction

The investigation of the physical properties of the nuclear regions of active galaxies has been the subject of many HST observing programs.

A central problem is the study of the optical counterparts to the radio jets. We know that these jets play a fundamental role in transporting energy from the central source to the extended radio lobes. Observations of optical and ultraviolet wavelengths with the HST are essential to obtain spatial resolutions similar to, or better than, those achieved in the radio band and, thus, provide the possibility of directly comparing the sites and mechanisms responsible for the emission at these different wavelengths.

In all cases to date, the emission has been attributed to the synchrontron mechanism, and since the electron lifetime is a strong function of the observed frequency, observations at optical and ultraviolet wavelengths offer the possibility to determine the precise location where particle acceleration occurs.

[a] On assignment from the Space Science Department of ESA

Comparison of the radio and optical morphologies further allows the study of the confinement mechanisms and diffusion processes within the jet. A number of important discoveries and observations that place the theoretical models on firmer observational grounds have been published based on the HST data.

However, to fully understand the physics underlying the jet phenomenon, we must explore the whole range of properties of the active galactic nuclei.

In the now standard paradigm for AGN, the basic differences between the different classes of objects is simply explained as a result of different orientations to the line-of-sight. In the unified picture, a central energy source, generally assumed to be a massive black-hole, is surrounded by two spatially and kinematically distant regions. The Broad-Line Region (BLR) has scales of the order of a parsec in diameter and emits broad permitted emission lines with widths up to 10000 km s^{-1}. The Narrow Line Region (NLR) emits narrow (a few hundred km s^{-1}) permitted and forbidden lines and has sizes of up to a kiloparsec in diameter. The orientation effects and, therefore, the classes of AGN, are determined by an optically thick torus composed of dense molecular clouds, whose inner diameter is comparable to the size of the BLR and can extend for tens of parsecs. The symmetry axis of the torus determines the direction of the kiloparsec scale radio jet and is independent of the orientation of the galactic disc. In the standard model if the torus is seen face on, we can see the continuum source and BLR directly and the galaxy is classified as a Seyfert 1. If, instead, the torus obscures the central regions, the galaxy is classifed as a Seyfert 2.

While HST observations to date seem to confirm the broad validity of this picture, these high spatial resolution observations as well as the important new imaging polarimetry results have shed new light on the fundamental physical phenomona that are at play in the nuclear regions of active galaxies.

In this review, I will discuss statistical studies of a sample of Seyfert galaxies and then discuss in detail some important examples for each of the main categories of Seyfert galaxies. In the main section, I will highlight important HST discoveries related to the optical counterparts to radio jets.

2 Statistical Studies of AGN

Nelson et al (1995) have carried out a SNAP program to observe 52 Markarian Seyfert and 48 Markarian non-Seyfert galaxies with WFPC 1. They find that the nuclei of type 1-1.5 Seyfert galaxies are dominated by strong point sources, while those of Seyfert 2 and non-Seyfert Markarian galaxies tend to be resolved, less distinguished, and similar in shape to the luminosity profiles of normal galaxies. This result confirms earlier FOC observations (Boksenberg et al.

1992) of the UV properties of Seyfert nuclei which showed that only Seyfert 1-1.5 nuclei were detectable at UV wavelengths in line with the predictions of the unified model.

There are two possible interpretations of this result for type 2 Seyfert galaxies. The first is that their nuclear continuum sources are undetected, contributing less than 10% of the nuclear light (within 0.5" radius) in all cases. The second is that their nuclear components are resolved and blend in smoothly with the brightness profile of the host galaxy's bulge. Since spectroscopic studies support typical nuclear continuum fractions distinctly greater than 10%, the latter conclusion is clearly preferable.

If the continua observed in Seyfert 2 galaxies orginate as nuclear light which is redirected into the line-of-sight by scattering, as predicted by unified models of AGN, then the scattering regions must be extended. Simple simulations suggest that these regions must cover several tens of parsecs, or more, in agreement with estimates of the sizes of the scattering "mirrors" in other Seyfert 2 galaxies. However, the similarity of the profiles of non-Seyfert Markarian and type 2 Seyfert nuclei suggests that circumnuclear star formation may also be an important component in the nuclear profiles of the latter.

2.1 NGC 4151

NGC 4151 is the nearest (~30 Mpc) example of a (sometime) type 1 Seyfert galaxy although it is rather a low-luminosity example of the class. The broad emission component of Hβ, which is a characteristic of Seyfert 1 galaxies, varies dramatically and, in a low state, can almost disappear. This led to NGC 4151 being reclassified as Seyfert 1.5. Both the permitted and the continuum emission show variations on time scales as short as days. FOC observations in [0 III] and nearby continuum have shown that the NLR is resolved into a number of emission-line clouds (sizes 10~ pc) with elongated morphology. These are distributed in a biconical structure with apices coincident with the central point source and a cone opening-angle projected on the sky of $\sim 75^o \pm 10^o$. The cone position angle of $60^o/240^o \pm 5^o$ is aligned with the extension of the nuclear VLBI radio source, suggesting that the same mechanism may align both the optical ionizing radiation field and the parsec scale radio structure.

Long-slit spectroscopic observations were carried out with the (uncorrected) FOC (Boksenberg et al. 1995). They found a high and a low radial velocity component within the narrow emission lines and identify the low-velocity component with the bright, extended, knotty structure within the cones, and the high velocity component with more confined diffuse emission. Also present are strong continuum emission and broad Balmer emission-line components, which are attributed to the extended point spread function arising from the intense nuclear emission.

Adopting the geometry pointed out by Pedlar et al. (1993) to explain the observed misalignment of the radio jets and the main optical structure they model an ionizing radiation bicone, originating within a galactic disk, with apex at the active nucleus and axis centered on the extended radio jets, and confirm that through density bounding the gross spatial structure of the emission-line region can be reproduced with a wide opening angle that includes the line-of-sight consistent with the presence of a simple opaque torus allowing direct view of the nucleus. In particular, the modelling reproduces the observed decrease in position angle with distance from the nucleus, progessing initially from the direction of the extended radio jet, through the optical structure, and on to the extended narrow-line region.

More recently with Winge et al (1997), we have carried out Faint Object Camera long-slit spectroscopy of the inner 8" of the Narrow Line Region of NGC 4151 at a spatial resolution of 0".029. The emission gas in characterized by an underlying general orderly behaviour consistent with galactic rotation over which are superposed kinematically distinct and strongly localized emission structures. High velocity components shifted up to $\sim 1500 kms^{-1}$ from the systemic velocity are seen, associated with individual clouds located preferentially along the edges of the radio knots. Off-nuclear blue continuum emission is also observed associated with the brightest emission-line clouds. Emission line ratios of key diagnostic lines vary substantially between individual clouds. The high spatial resolution long-slit spectroscopy of the nuclear region of NGC 4151 by Winge et al (1997), Fig. 1, clearly shows that the popular picture of anisotropic illumination by the nuclear source is overly simplistic to explain the complex morphology and kinematics observed in the NLR. Comparison of the radio and optical data shows that the line emission is enhanced along the edges of the radio knots, as would be expected from the interaction between the jet and the surrounding medium. This clear association implies that the interaction with the radio jet plasma, and not illumination effects, is the dominant mechanism in determining the morphology , as well as the physical conditions in the individual clouds.

2.2 NGC 1068

A number of observations of the prototypical Seyfert 2 galaxy NGC 1068 (D $\sim$ 22.7 Mpc, 0.1" = 11 pc) have been carried with the FOC and WFPC2 (Macchetto et al. 1994, Capetti et al. 1995b, 1996a, Bower et al. 1995, Capetti, Axon and Macchetto, 1997, Marconi et al. 1997). These include visible and

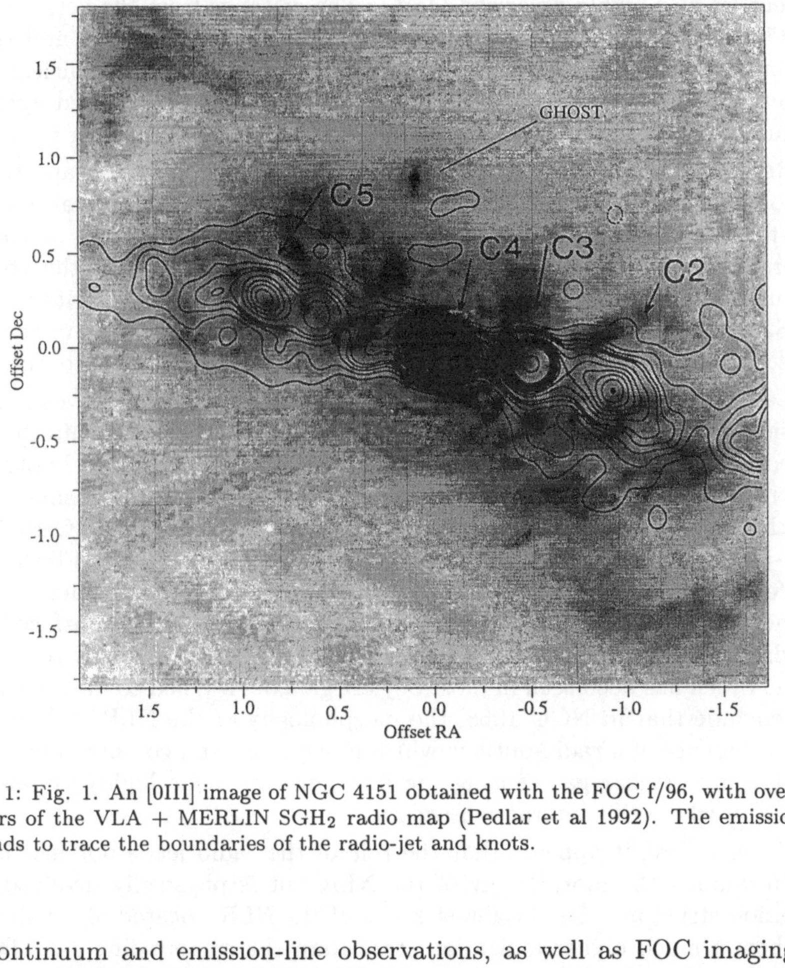

Figure 1: Fig. 1. An [0III] image of NGC 4151 obtained with the FOC f/96, with overlayed contours of the VLA + MERLIN SGH$_2$ radio map (Pedlar et al 1992). The emission-line gas tends to trace the boundaries of the radio-jet and knots.

UV continuum and emission-line observations, as well as FOC imaging polarimetry (F253M, F372M & F501N). These observations show that the inner morphology is very complex. In first approximation, it appears as a "bi-cone", with the axis of symmetry changing with distance from the nucleus. This may be due to the different location of the gas as it streams around the inner bar. Alternately, the ionization and scattering cone is rather narrow and tracks the radio jet more closely. The brightest feature in the continuum and line emission images is "Cloud-B" which is highly polarized (65%) and contrary to previous claims is not the location of the nucleus. The polarization data shows that the true nucleus is located some 0.6" South of Cloud B and is obscured. This is very consistent with the unified scheme of AGNs. A strange feature, the "twin crescent" is also highly polarized (45%). The distance from the active nucleus $\sim$ 0.1" makes it unlikely that it has any physical relations to it and thus, it remains a mystery. The high-degree of polarization in the circumnuclear region implies that most, if not all, of the observed UV light is scattered light from the nucleus, with either dust or electrons providing the scattering medium.

Recent high precision ground-based and HST astrometry (Capetti, Macchetto & Lattanzi, 1997) has allowed the alignment of the optical and radio data to be carried-out. The maps have been registered with a precision of better than $\pm$ 0".030. With this determination the position of the obscured nucleus, as derived from the polarization measurements, falls within the S1 and S2 peaks of the radio data, but are not coincident with any of these two peaks, whereas it is within 1 σ of the position of the H_20 maser source.

Capetti et al.(1997b) have used HST/WFPC2 imaging to investigate the emission-line structure of the NLR of NGC 1068 and its relationship to the extended radio emission. Previous works showed that the brightest knots of the NLR lie along the opposite edges of the radio-jet. These WFPC2 images indicate that a similar association also holds on a larger scale. In NGC 1068 both radio-jets and lobes are present and the NLR also displays a quasi-linear emission feature associated with the radio-jet and emission-line filaments around the radio lobe (Fig.2). Outside the radio lobe, the emission-line surface brightness drops dramatically. This NLR structure is, therefore, strongly reminiscent of that which has been seen in most Seyfert galaxies studied to date with HST. We conclude that in NGC 1068, the morphologhy of the NLR is determined by the presence of a radio-outflow which is sweeping and compressing the surrounding interstellar gas causing the line-emission to be highly enhanced in the region where this interaction occurs.

Furthermore, it appears that the role of the radio jet is not just limited to determining the morphology of the NLR but is physically involved in its ionization structure. The brightest knots of the NLR, located along the radio jet, show a much higher ionization state than the surrounding gas. Density measurements of individual knots derived from archival FOS/HST spectra and compared with ground based data indicate that they are high density condensations within the NLR. Their higher ionization and higher density imply that these knots are illuminated by an incident ionizing flux greater by at least one order of magnitude than the rest of the NLR. Shocks can produce large ionizing flux as soon as their velocities exceed a few hundred km s^{-1}. This source of ionizing photons can prevail locally over the nuclear radiation field and dominates the ionization conditions in these regions of the NLR. While there are hints of such complex NLR ionization structure changes in the HST observations of other more distant Seyferts, only in NGC 1068 can we really

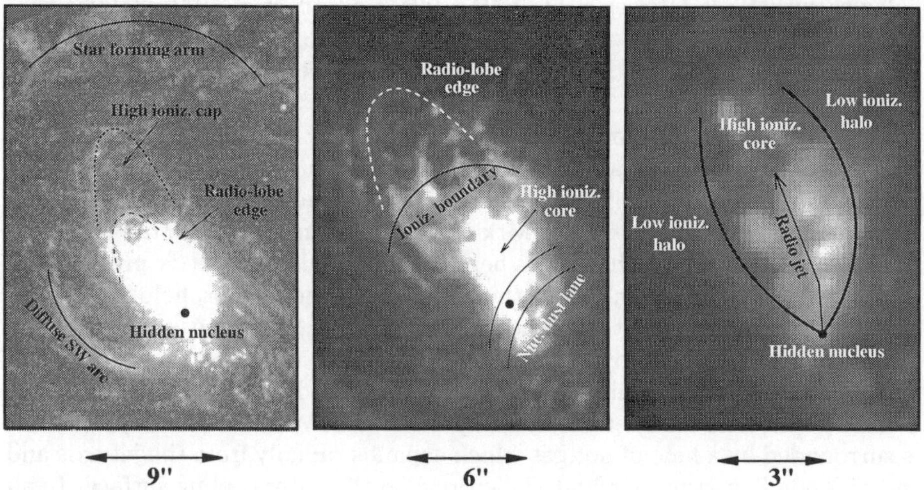

Figure 2: Images of NGC 1068 in the Hα + [NII] lines obtained with HST at different spatial scales. Overlaids on these images are the schematic representations of the various components of the VLR and other features of interest. The black dot shows the location of the hidden nucleus.

resolve it.

At a distance of $\sim$ 4" ($\sim$ 300 pc) from the nucleus, the ionization structure of the NLR shows a sharp and well-defined boundary between an inner low-ionization zone and an outer higher zone in correspondence to the transition in the radio structure from jet-like to lobe-like. This ionization change can best be explained with a density drop where the jet enters the lobe which we interpret as evidence for backflowing jet material.

2.3 Radio Outflow and NLR Structure

For many years, it was generally accepted that in the NLR of Seyfert galaxies, the gas is photoionized by nuclear radiation. The discovery of the NLR with "conductive" morphology on initial HST observations of some nearby Seyfert galaxies (NGC 1068 Evans et al. 1991, NGC 4151 Evans et al. 1993, NGC 5728, Wilson et al. 1993) seemed to give further support to that view. However, ground-based studies have shown that the NLR is spatial with the radio emission and its kinematics clearly shows signs of the effect of interactions with the ejected radio-plasma. This association, now clearly confirmed by the observations of NGC 4151 and NGC 1068 described in the previous sections has been given even stronger support by the observations of several other Seyfert galaxies with HST. (Bower et al. 1994, 1995, Capetti et al. 1995c, 1996a,b).

These and other HST observations of the nuclear regions of Seyfert 2 have been discussed by Capetti et al. (1996b). They show that in all cases the physical structure of the NLR in Seyfert 2's is closely related to the radio-emission. The optical morphology is dominated by the interaction with the radio ejecta. The NLR appears to take a different form depending on the structure of the radio emission. Where there are radio lobes, there are shell-like emission-line structures. For Mrk 573, Mrk 78 and Mrk 348, the emission-line structures are bow-shocks. Where a collimated radio jet is present, the morphology is different. In Mrk 3, the NLR follows the jet morphology. Capetti et al. (1996b) show that this dichotomy implies that bow-shock emission-line structures are produced by the sweeping-up of gas at the advancing working surface of the ejected radio plasma. The corrugated structure indicates that instabilities have developed in the compressed gas. Where a jet is apparent, it is surrounded by a halo of hot gas which expands radially from the jet axis and the emission line forms a cylindrical cocoon on the outer cooling surface. In all cases, the ionization conditions, as determined for example from the emission-line ratios, are such that the ionization parameter increases with distance from the nucleus. This requires a source of ionization in addition to the unclear ionizing flux.

Capetti et al. (1996b) show that shock mechanisms are the best candidates to produce the relatively small amount of locally produced excess radiation. They are fully consistent with both the measured ionization conditions and the observed filamentary morphology.

3 Optical Counterparts to Radio Jets

To date, nine optical synchrotron jets have been identified and confirmed with radio maps. The jets are located in the radio galaxies M 87 (3C 274), 3C15 (Martel et al. 1997), 3C 66B, 3C 78 (Sparks et al. 1995), 3C 200, 3C 264, 3C 273, 3C 346, and PKS 0521-36. Of these 3C78 was discovered with the FOC (Crane et al. 1993) and four (3C15, 3C 200, 3C 264 and 3C 346) were discovered in the 3CR Snapshot survey, which uses the WFPC camera to image in two filters all the 3CR souces.

In the standard model of jet formation, material accretes on a supermassive black-hole leading to the ejection of a collimated, relativistic hot plasma. The jet becomes visible in the optical if the Doppler boosted synchrotron radiation is "beamed" towards the observer. High-resolution optical imaging with HST has revealed common properties among the jets and their nuclei, consistent with this hypothesis: the jet originate from a bright, compact, unresolved nucleus, characteristic of an AGN, no counter jet is observed, the jets are curved and possess a distorted morphology, and they are smaller and narrower in appearance than their radio counterparts (Sparks et al. 1995). Presently, the only known exception to this relativistic beaming picture is 3C 66B whose counterjet has been imaged in both the radio (Hardcastle et al. 1996) and the optical (Fraix-Burnet 1997), suggesting that the observed synchrotron optical emission is due to environmental effects. On the other hand, despite a superficial consistency with the relativistic beaming picture, Sparks et al. (1995) showed that qualitatively, the statistics of optical jet sizes and power favor, instead, an intrinsic difference between the optically emitting jets and those radio jets which show no optical emission.

More important issues regarding the formation, collimation, acceleration, evolution and lifetime of optical jets still need to be addressed comprehensively. For example, the acceleration mechanism responsible for the emission of optical synchrotron radiation is still a matter of debate, especially when the radio and optical structures are broadly similar (Meisenheimer 1996). Either the particle acceleration is *in situ* to the jet through processes such as shocks or magnetic reconnection or the electrons are accelerated in the nucleus and are transported along the jet in a channel having low magnetic field and consequently low radiation losses as suggested for M 87 (Owen, Hardee, & Cornwell 1989). Eilek & Arendt (1996) have recently shown that the observed synchrotron spectrum can be attributed not only to power-law particle distribution functions, but also to power-law magnetic field distributions, instead of the uniform magnetic field assumed in the standard model. This has important ramification for spectral aging studies since observed synchrotron souces can be significantlyu older than predicted by the standard model. Ultimately, the properties of optical jets need to be understood in context of their host galaxies and their environment. Since so few optical jets have been identified, it is important to carefully study the few that are known to understand their physical properties and improve the theoretical models.

3.1 PKS 0521-36

PKS 0521-36 is a relatively isolated radio galaxy at a redshift z=0.055 which also harbours a bright V = 16 BL Lac nucleus and extended optical line emission. Sparks et al. (1990) reported optical polarization measurements of the jet and nucleus, which confirmed that the emission is due to synchrotron radiation.

The FOC images obtained in 1990 with the FOC (Macchetto et al. 1991a) show a bright knot located ≈ 1.8" to the NE and clearly resolved as in the

VLA data (Keel 1986). The width of the knot is $\approx 0.8"$. Beyond this bright knot, the jet has approximately constant surface brightness and a morphology similar to the VLA image with a total length of 6.5". The jet is also resolved in width, 0.6" wide in the fainter regions of the jet, with little or no evidence of structure on a scale of $\leq 0.1"$. The FOC data appears to show more flux than the VLA data in the region at slightly larger radius from the nucleus but close to the southern tip of the knot. A large bright knot further along the jet is a clear counterpart to the radio knot. The bright knot is an important site where particle acceleration is occurring.

Using standard formulae and values for the relevant parameters, we derive a mean lifetime for the electrons of $t_{1/2} \sim 600$ yrs. This implies that there must be continuous acceleration along the jet of the electrons responsible for the optical emission, since electron diffusion from the bright knot could not account for the observed optical extent.

3.2 3C 66B

3C 66B is a relatively nearby bright radio source associated with a 13th magnitude galaxy at a distance of 86 Mpc. At this distance an angular scale of 0.1" corresponds to a projected linear size of 41 pc. Images of 3C 66B were obtained with the FOC and compared with the best VLA map of 3C 66B (Macchetto et al. 1991b). Several conclusions can be drawn from these observations. On the scale of the HST resolution, the jet of 3C 66B is filamentary. Two distinct "strands" can be traced from $\geq 3.7"$ (1.5 kpc) from the nucleus out to a distance of 7.6" (3 kpc), where they disappear into the noise. The separation between the strands varies between about 0.3" and 0.4", that is about 150pc, and they appear to undergo sharp "kinks" at distances of 2.5" (1.0 kpc) and 6.2" (2.5 kpc) from the nucleus.

The origin of these kinks is unclear. The fact that they are mimicked in more than one filament suggests that they are not due to an instability mode in an individual filament. They may well trace out irregularities in the ISM of the galaxy and/or be due to time-dependent variations in the power of the nuclear source responsible for producing the jet.

3.3 3C 273

3C 273 is one of the nearest and brightest quasars known. Its jet has been extensively studied at radio and optical wavelengths. Using the FOC, Thomson, Mackay & Wright (1993) have carried out high-resolution imaging polarimetric observations of the jet. More recently, Bahcall et al. (1995) have obtained

WFPC2 as well as Merlin observations with matching resolution. The projected jet length is more than 70 kpc ($H_o = 50kms^{-1}Mpc^{-1}; q_o = 0.5$). The width is only a few tenths of an arcsecond ($\sim 0.5'' = 1kpc$). The optical emission is highly confined to the core of the radio jet. It runs along the ridge of the radio emission and is asymmetric compared to the radio. All oblique radio features coincide with optical knots, though there are some optical features without radio counterparts.

Bahcall et al. (1995) suggest that the 'radio jet' consists of two components. The first is the fast-moving jet, shown by the coincident oblique radio and optical features. The second consists of the emission from a surrounding, slow-moving "cocoon". The radio data suggests that the oblique radio features coincident with the optical knots may be in the form of a helix. If the velocity is relativistic, the emission will appear brightest where the velocity vector is closest to the line-of-sight, the enhancement being independent of wavelength. This would explain the close correspondence between the optical and radio knots. The helical form may arise from a driven Kelvin-Helmholtz instability.

3.4 3C 264

NGC 3862 (3C 264) at a distance of 86.2 Mpc is an FR1 source. This galaxy was observed in two bands with the FOC and resulted in the first discovery of an optical jet with HST. The images showed a prominent jet-like feature emanating from the nucleus (Crane et al. 1993) morphologically similar to the jets seen in M87 and 3C 66B. Indeed, the bifurcation seen at the end of the jet is reminiscent of the "double-stranded" feature seen in the HST image of 3C 66B (Macchetto et al. 1991b).

Broad-band (F702W) imaging data taken with the WFPC2 shows a very intriguing feature in this galaxy (Baum et al. 1997). In addition to the very evident optical jet, an almost perfectly circular ring is observed, which is the manifestation of a nearly face-on circumnuclear dust disk or dense gaseous region. The jet collimation, brightness, and polarization properties change dramatically as it crosses the outer boundary of the 'ring' suggesting an interaction between the jet and a dense gaseous region interior to the apparent ring. They suggest that the jet may be entraining disk material and decelerating, and derive equations for the brightness evolution of a relativistic jet which expands adiabatically and apply this model to the jet in 3C264, assuming it is initially relativistic and that the subsequent changes in jet brightening are due to a combination of Doppler boosting, and adiabatic and radiative losses. This model is roughly consistent with the observed properties of the jet in 3C264 and suggest we are viewing an intially highly relativistic (v $\sim$ 0.98c) jet at roughly ~ 50 degrees inclination angle. The unique properties of 3C264 provide a laboratory for the study of relativistic entraining jets and may help to quantify unifying models for BL Lac and Fanaroff and Riley Class I radio galaxies.

3.5 M87

The giant elliptical galaxy M87 contains the closest extragalactic jet, which makes it a prime target for studies of jet structure and kinematics. Optical and ultraviolet observations of M87 have been carried out with HST and have been extensively reported (Macchetto 1991, Boksenberg et al. 1992; Macchetto, Biretta & Sparks 1992).

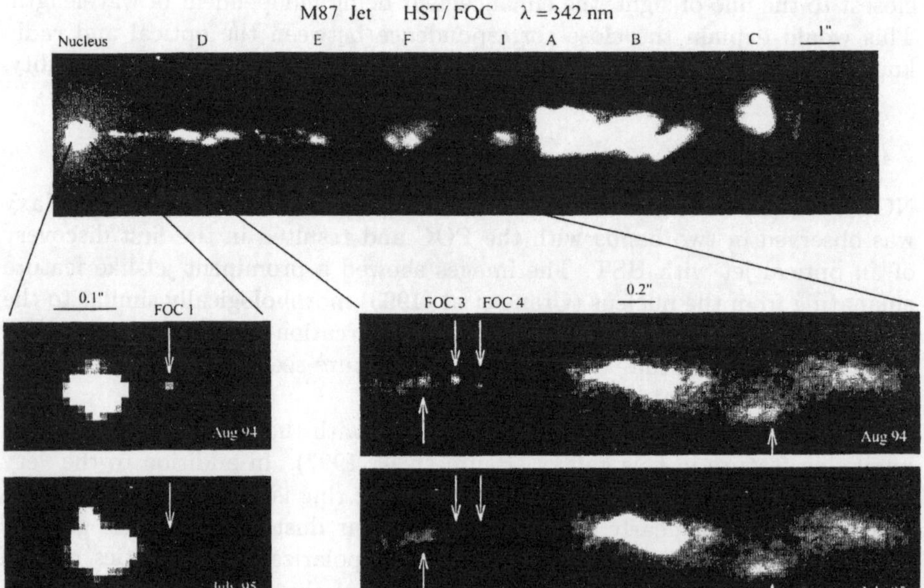

Figure 3: Comparison of M87 FOC images from August 1994 and July 1995. Feature FOC 1 near the nucleus (lower left panel) appears to move outward at 2.5c. Features FOC 2 and DM (lower right panel) appear to move outwards at 3 to 5c, while features FOC 3 and FOC 4 fade from view. The images have been aligned at the bright point source in the nucleus. Dark spots are rescue marks on the photo cathode.

While the radio and optical images present a remarkable degree of similarity, there are nevertheless significant differences. The optical/UV images show intrinsically higher contrast than the radio, with compact regions of emission localized within the knots. The jet is narrower in the opitcal/UV, and more concentrated to the jet center in the optical/UV than in the radio. The radio-to-optical spectral index of the inter-knot regions is steeper than that of the knots themselves. There are also differences in the detailed knot structure of the optical emission compared to the radio, and there is a weak overall spectral steepening with distance from the nucleus beyond knot A.

Capetti et al. (1997c) have analyzed polarization observation of the M87 jet taken with the FOC in the ultraviolet and with the WFPC1 in the visual. The degree of polarization is typically 30% over most of the jet. At the edges of the jet the polarization is as high as 60%, requiring a highly ordered magnetic field. In the center of the jet the small scale structure of the magnetic field produces significant cancellation reducing the polarization to $\sim$ 10%. The degree of polarization and the polarization pattern are very similar at radio and optical wavelengths. No significant depolarization or Faraday rotation have been detected, in agreement with previous radio determinations. However, the morphology of knot D is considerably different in the VLA observations by Owen et al. (1989) and these HST observations. Knot D1 appears to be relatively brighter and closer to the nucleus in the optical than in the radio images. Capetti et al. (1997c) conclude that this component is associated with a shock front. At the location of a shock front acceleration of relativistic electrons occurs, enhancing the synchrotron emission at shorter wavelengths, and the transverse component of the magnetic field is amplified by the compression produced by the shock.

Figure 3 shows FOC images of the nucleus obtained in August 1994 and July 1995 by Biretta et al. (1997). The images, which have been aligned at the bright point source, show a feature approximately 150 milliarcseconds (12 pc) from the core, which corresponds roughly to knot M in the 18 cm VLBI image of Reid et al. (1989). Comparison of the two images shows that this feature moves outward by 8.5 mas or 0.66pc, for a velocity of 2.3c+0.3c. From this result it is clear that the jet is already relativistic within the core, and that acceleration is probably unimportant on these scales. Velocity measurements for three features in the region of knot D were also obtained. We find a speed of 2.6c+0.4c for feature DW, and a somewhat larger 3.7c+0.5c for DM. A third feature, FOC1, which was not seen in the radio data, has a speed of 2.1c+0.3c. Two additional features, labeled FOC2 and FOC3, appear to fade from view between 1994 and 1995. This fading rate is much higher than typical synchrotron lifetimes in the jet (100 yrs) suggesting that the adiabatic expansion of the emitting region is likely to play an important role in the evolution.

In a related investigation, Macchetto et al. (1997) and Marconi et al. 1997 have obtained long-slit observations of the circumnuclear region of M87 at three different locations, with a spatial sampling of 0".028 using the Faint Object Camera f/48 spectrograph on board HST. These data allow us to determine the rotation curve of the inner $\sim$ 1" of the ionized gas disk in [0II]λ3727 to a distance as close as 0".07 ($\simeq$5pc) to the dynamic center, thereby significantly improving on both the spatial resolution and coverage of previous FOS observations. We have modeled the kinematics of the gas under the assumption of the existence of both a central black-hole and an extended central mass distribution, taking into account the effects of the instrumental PSF, the intrinsic luminosity distribution of the line, and the finite size of the slit. We find that the central mass must be concentrated within a sphere whose maximum radius is 0".05 ($\simeq$ 3.5pc) and show that both the observed rotation curve and line profiles are consistent with a thin-disk in keplerian motion. We conclude that the most likely explanation for the observed motions is the presence of a super-massive black-hole and derive a value of $M_{BH} = (3.2 \pm 0.9) \times 10^9 M\odot$ for its mass.

3.6 3C 78

As part of the HST "snapshot" program to acquire broad-band images with high spatial resolution of 3CR radio sources with the WFPC2 F702W filter, Sparks et al. (1995) reported the discovery of an optical synchrontron jet emanating from the nucleus of NGC 1218, the galaxy hosting the radio source 3C 78. (Fig. 4). The jet is similar to that of M87, although smaller in projected length.

The visible jet has a total length of 1.37 arcsec which is 0.75 kpc in projection for $H_o = 75$ km.s^{-1} Mpc^{-1}. Within $\approx$ 0.5 arcsec of the nucleus, the jet is essentially unresolved across its width even at PC resolution, although it is slightly curved and there are distinct knots within that portion of the jet. The gap between the brightest part of the jet and the nucleus appears to be real. At $\approx$ 0.5 arcsec, the jet fades dramatically, and fans out into a broad plateau of emission. There are also two discrete knots, and a possible third fainter one, within the plateau portion of the jet, seen in the optical data.

3.7 3C 15

The radio galaxy 3C 15 was imaged with the WFPC2 in February 1997 in the F555W filters (Martel et al. 1997). After subtraction of the underlying stellar light, the most striking morphological feature of 3C 15 is its extended, broad,

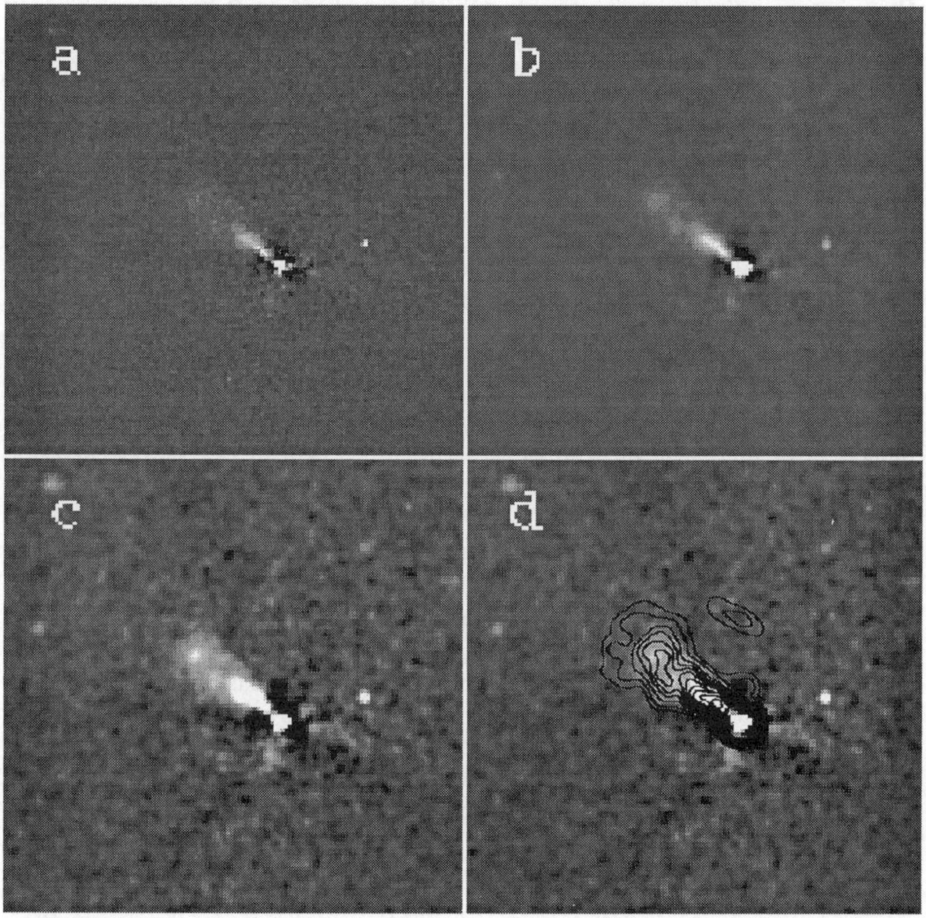

Figure 4: Optical jet in 3C 78 after subtraction of the underlying galaxy. (a) Low-contrast image at full resolution. (b) same as (a), but smoothed with a Gaussian (σ =0.8 pixels) (c) High-constract version of the lighly smoothed image. (d) Optical image shown in (b) with overlaid radio contours.

and well-defined optical jet at PA $\sim 30^o$. In Figure 5, they have labeled three dominant knots in the jet. The innermost knot, Knot A, is broad and diffuse. It extends from 2.6 kpc (2".1) to 3.5 kpc (2.".8) from the nucleus. Knots B and C are comparatively compact and well-delineated. Knot B lies at a projected distance of 3.9 kpc (3".1) from the nucleus, while the outermost knot, Knot C, is located 5.1 kpc (4".1) away. For comparison, the optical jet of M 87 spans a distance of only $\sim$ 1.6 kpc while the jet of 3C 273 extends out to about $\sim$ 100 kpc. Its radio-optical spectral index, $\alpha_{ro} \sim 1$, is the softest yet measured. The host elliptical galaxy of 3C 15 is located in a loose galaxy group and is quite ordinary in appearance and color. On average, the optical jet is bluer than its host by approximately 0.4 mag.

The optical morphological disturbances observed in the circumnuclear region suggest that 3C 15 is the result of a merger event, perhaps from the capture of a single or multiple satellite galaxies. The merger led to the deposition of dust and the formation of emission filaments or arms in the circumnuclear region as well as the episodic ejection of energetic particles along the jet. Emission filaments or arms are also detected in the central 2" of the V and R band images, previous work on similar structures observed in the centers of other ellipticals has shown that photoionization by post-AGB stars, perhaps combined with heating by hot electrons from a coronal X-ray halo, represent the likely ionization mechanisms (Macchetto et al. 1996). The absence of a bright, point-like nucleus implies that the central AGN is either presently inactive or heavily obscured from our line-of-sight by copious amounts of dust produced in the merger. In the first case, the nucleus of 3C 15 may undergo periodic episodes of activity, when material is infalled towards the central supermassive black hole. Material is ejected into the ISM of the host galaxy, thus accounting for the knots and bow shocks observed in the outflowing jet. In the second case, an extinction of $A_R \sim 20$ mag would be requirted to mask an AGN of absolute magnitude $M_R = 1$ mag, a typical value for the AGN nuclei of galaxies with optical synchrotron jets. Multi-wavelength, broad-band imaging, in particular in the near-IR and ultraviolet, is necessary to determine the location of the synchrotron break frequency and will greatly improve our knowledge of the particle acceleration mechanisms and lifetimes in synchrotron optical jets.

4 Conclusions

The HST observations conducted so far have already led to the discovery of five new optical jets and the identification of other candidates, which will be further studied in the future.

Although few in number, there are several common features shared by the

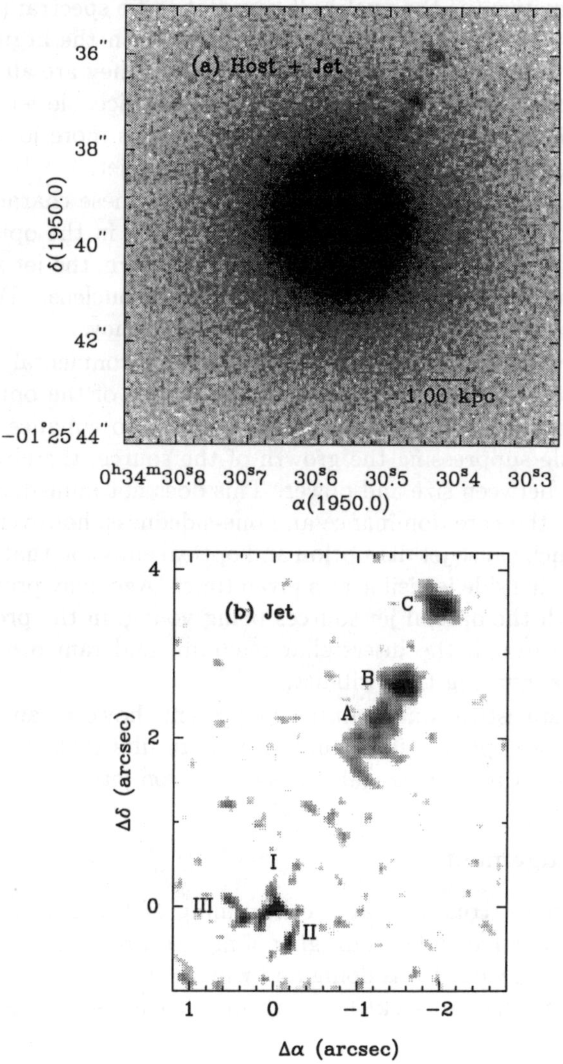

Figure 5: (a) V-band image of 3C15. The optical jet is clearly visible at PA of -30°. (b) The optical jet after subtraction of the host galaxy light. The dominant knots are labeled A,B and C.

optical jet radio sources: (i) all have relatively prominent nuclei in both radio and optical domains; (ii) the nuclei all have flat radio spectra; (iii) the jets are small compared to typical radio jet dimensions, with the arguable exception of 3C 273; (iv) there are no two-sided optical jets - they are all one-sided with a large jet to counterjet lower limit; (v) there is noticeable jet curvature; (vi) in addition, where measured, the optical emission is more localized than the radio and the optical jet is narrower than the radio jet.

An obvious candidate explanation for most of these characteristics is relativistic 'beaming', in which the jet becomes visible in the optical only when pointing towards the observer. In the beaming picture, the jet appears brightened, foreshortened and with a prominent active nucleus. Beaming is also expected to blueshift the synchrotron 'break' frequency.

As an alternative to relativistic beaming, environmental effects may be considered. If the pressure is higher in the vicinity of the optically emitting sources, then the additional confinement may act to enhance their radiative luminosity while suppressing the growth of the source, thereby giving rise to the correlation between size and power. This does not immediately suggest an explanation for the core dominance and one-sidedness; however, there may be instabilities which cause jet disruption and optical emission that are sufficiently rapid that only one side is visible at a given time. 'Age' may provide yet a third alternative, with the optical jet sources being young, in the process of forcing their way out through the interstellar medium, and ram pressure playing a similar role in enhancing the visibility.

There are statistical uncertainties at present, however an extensive analysis of many more optical jets should provide results that will be essential in elucidating the nature of extragalactic synchrotron jets.

5 Acknowledgement

Despite the (true!) comment at the beginning of this review, Riccardo was always very supportive and encouraging of my research work. In his inimitable way, he questioned what I was doing and why so that I would be able to really understand that I had to seek the truth in science, as well as in every other endeavor.

This is what a great teacher does. Grazie, Riccardo!

6 References

Bahcall, J.N., Kirhakos, S., Schneider, D.P., Davis,R.J., Muxlow, T.W.B., Garrington, S.T., Conway, R.G., & Unwin, S.C., 1995, ApJL, 452, L91

Baum, S.A., de Koff, S., Sparks, W.B., Biretta, J., Golombek, D., Macchetto, F.D., McCarthy, P., & Miley, G.K., 1997. ApJS, in press

Biretta, J.A., Sparks, W.B., Macchetto, F., & Capetti, A. 1997 ApJ, submitted

Boksenberg, A., et al. 1992, A&A, 261, 393

Boksenberg, A., Catchpole, R.M., Macchetto, F., et al. 1995, ApJ, 440, 151

Bower, G.A., Wilson, A.S., Mulchaey, J.S., Miley, G.K., Heckman, T.M. & Krolik, J.H., 1994, AJ. 107, 1686

Bower, G.A., Wilson, A.S., Morse, J.A., Gelderman, R., Whittle, M., & Mulchaey, J. 1995, ApJ, 454, 106

Capetti, A., Axon, D.J., Kukula, M., Macchetto, F., Pedlar, A., Sparks, W.B., & Boksenberg, A. 1995a, ApJ, 454, 185

Capetti, A., Axon, D.J., Macchetto, F., Sparks, W.B., & Boksenberg, A. 1995b, ApJ, 446, 155

Capetti, A., Macchetto, F., Sparks, W.B., & Boksenberg, A. 1995c, ApJ, 448, 600

Capetti, A., Macchetto, F., Axon, D.J., Sparks, W.B., & Boksenberg, A. 1996a, ApJ, 466, 169

Capetti, A., Axon, D.J., Macchetto, F., Sparks, W.B., & Boksenberg, A. 1996b, ApJ, 469, 554

Capetti, A., Macchetto, F., & Lattanzi, M.G. 1997a, ApJ, 476, 67

Capetti, A., Axon, D.J., & Macchetto, F.D., 1997b, ApJ, in press

Capetti, A., Macchetto, F., Sparks, W.B., & Biretta, J.A. 1997c, ApJ, submitted

Crane, P., et al. 1993, ApJL, 402, L37

Eilek, J.A., & Arendt, P.N., 1996, ApJ, 457, 150

Evans, I.N., Ford, H.C., Kinney, A.L., Antonucci, R.R.J., Armus, L. & Caganott, S, 1991, ApJL, 369, L27

Evans, I.N., Tsvetanov, Z., Kriss, G.A., Ford, H.C., Caganott, S., & Koratkar, A.P., 1993, ApJ, 417, 82

Fraix-Burnet, D., 1997, MNRAS, 284,911

Hardcastle, M.J., Alexander, P., Pooley, G.G., & Riley, J.M., 1996, MNRAS, 278, 273

Keel, W.C. 1986, ApJ, 302, 296

Korista, K., et al. 1995, AJ, 97, 285

Leahy, J.P., Blade, A.R.S., Dennet-Thorpe, Hardcastle, M.J., Komissarov, S., Perley, R.A., Riley, J.M. & Schener, P.A.G., 1997, MNRAS, in press

Macchetto, F. 1991, Proceedings, Physics of AGN, ed. S.J. Wagner, W.J. Duschl, Springer-Verlag: Berlin, 325

Macchetto, F., et al. 1991a, ApJL, 369, L55

Macchetto, F., et al. 1991b, ApJL, 373, L55

Macchetto, F., Biretta, J.A., & Sparks, W.B. 1992, Proceedings 182nd AAS Meeting, 24, 1183
Macchetto, F., Capetti, A., Sparks, W.B., Axon, D.J. & Boksenberg, A. 1994, ApJL, 435, L15
Macchetto, F.D., Pastoriza, M., Caon, N., Sparks, W.B., Giavalisco, M., Bender, R., & Capaccioli, M., 1996, A&A Supp, 120,463
Macchetto, F.D., Marconi, A., Axon, D.J., Capetti, A., Sparks, W.B., & Crane, P., 1997, ApJ, in press
Marconi, A., Axon, D.J., Macchetto, F.D., Capetti, A., Sparks, W.B., & Crane, P., 1997, MNRAS, in press
Martel, A.R., Sparks, W.B., Macchetto, F.D., Biretta, J.B., Baum, S.A., Golombek, D., McCarthy, P.J., de Koff, S., & Miley, G.K., 1997, ApJ, submitted
Meisenheimer, K., 1996, in Jets from Stars and Galactic Nuclei, ed. W. Kundt (New York: Springer)
Nelson, C.H., Mackenty, J.W., Simkin, S.A., & Griffiths, R. E. 1996, ApJ, 466, 713
Owen, F.N., Hardee, P.E. & Cornwell, T.J. 1989, ApJ, 340, 698
Pedlar, A., Howley, P., Axon, D.J., & Unger, S.W. 1993, MNRAS, 259, 369
Reid, M.J., Biretta, J.A., Junor, W., Spencer, R., & Muxlow, T. 1989, ApJ, 336, 125
Sparks, W.B., Miley, G., & Macchetto, F. 1990, ApJL, 361, L41
Sparks, W.B., Biretta, J.A., & Macchetto, F.D., 1996, ApJ, 473, 254
Sparks, W.B., Golombek, D., Baum, S.A., Biretta, J., deKoff, S., Macchetto, F.D., McCarthy, P., & Miley, G.K., 1995, ApJ, 450,L95
Thomson, R.C., Mackay, C.D. & Wright, A.E. 1993, Nature, 365,133
Wilson, A.S., Braatz, J.A., Heckman, T.M., Krolik, J.H., & Miley, G.K., 1993, ApJ, 419, L61
Winge, C., Axon, D.J., Macchetto, F.D. & Capetti, A., 1997, ApJL, in press

HIGH REDSHIFT RADIO GALAXIES: FROM HST to VLT–A PROTO-CLUSTER AT z=2.2?

GEORGE MILEY, LAURA PENTERICCI, HUUB RÖTTGERING
Leiden Observatory, P.O. Box 9513, 2300 RA, Leiden, The Netherlands

1 Introduction

It is a privilege to contribute to this festschrift in celebration of Riccardo Giacconi's 65th birthday. Riccardo is one of the giants of twentieth century astronomy. He has played a leading role in the construction of at least four facilities, Uhuru, Einstein, Hubble and the VLT that have or will revolutionise astronomy.
High redshift radio galaxies are a fitting topic to include in this tribute because they impinge on all the various facilities with which Riccardo has been associated. They have been an important topic for study with the Hubble Space Telescope, will be prime targets for the VLT and may well allow us to study X–ray emitting gas in the early Universe.
Studying high redshift radio galaxies (HZRGs) can contribute substantially to our understanding of galaxy formation and evolution in the early universe. HZRGs are of exceptional interest as potential signposts for pinpointing high-redshift X-ray clusters of galaxies. It has been shown that high luminosity radio sources associated with quasars and radio galaxies in the range $0.5 < z < 1$ are located in rich clusters (e.g. Hill and Lilly, 1991). At $z > 2$ the existence of clusters around HZRGs has not been established. but there are many indications that HZRGs might be in clusters (Dickinson, 1996 and references therein). The host galaxies of HZRGs are also the most massive galaxies known at high–redshifts, and since they are probably located in forming clusters of galaxies they are likely to be the ancestors of nearby cD galaxies. Here we shall describe recent observations of one of the most fascinating objects for such studies, the radio galaxy 1138−262 at redshift 2.2.
The work we present was carried out in collaboration with Chris Carilli, Pat McCarthy, Hy Spinrad, Wil van Breugel and Duccio Macchetto.

2 The remarkable properties of 1138−262

We have observed 1138−262 using the HST, the VLA and the NTT.
The data reduction and results are extensively discussed in Pentericci *et al.*

(1997a and 1997b). 1138−262 has several characteristics which are extreme amongst all known high redshift radio galaxies :

(i) The optical continuum morphology of 1138−262 is one of the clumpiest of all HZRGs, as evident from Figure 1. The optical color index B-R of the external regions is systematically bluer than the central part.

(ii) While the optical continuum is very clumpy, near infrared imaging shows an extremely bright and compact object. The K-band magnitude is 16.07, i.e. almost 2 magnitudes brighter that the average $z = 2$ radio galaxy in the Hubble K-z diagram (McCarthy, 1996).

(iii) The radio morphology (Figure 2) is the most distorted of all known powerful HZRGs, consisting of a series of knots and showing a double bend towards south in the eastern lobe (Carilli *et al.*, 1997).

(iv) The rotation measure in the western lobe (6250 rad m^{-1}) is the highest measured yet for $z > 2$ radio galaxies and the gradient of RM across the source is also one of the highest (Carilli *et al.*, 1997).

(v) There is a small companion object in a direction perpendicular to the radio axis, which is gas rich and shows clumpiness on the same scale as the main optical galaxy.

(vi) The distribution of the Lyα emission gas (as shown in Figure 2) does not follow the optical continuum. The peak of the line emission does not coincide spatially with the central part of the optical galaxy, but with the location at which the radio jet bends. The gas shows no sign of ordered motion, such as rotation.

3 Is 1138−262 at the center of a cluster?

A recent result from the study of powerful radio galaxies at the centers of X-ray clusters is a correlation between the magnitude of Faraday rotation observed towards the radio source and the density of the X-ray gas in the surrounding cluster (Taylor *et al.*, 1994). This indicates that the Faraday rotation is mostly due to the cluster gas which is substantially magnetized. Drawing the analogy between low redshift radio galaxies and 1138−262, we conclude that this galaxy is also likely to be situated at the center of a dense, X-ray cluster gas, possibly with a cooling flow as large as $500 M_{\odot} \mathrm{yr}^{-1}$. Nearby radio sources located in strong cooling flows often show amorphous radio morphologies, due to the collimation of the jet being disrupted by the dense ICM (Taylor *et al.* 1994). If 1138−262 is located in a rich cluster, this scenario could also explain its extremely distorted radio morphology. Finally, the Lyα narrow band image (Figure 2) suggests that the object labelled F in Figure 1 is an emission line galaxy: if the line is Lyα, then F is at the same redshift as 1138−262, i.e. a

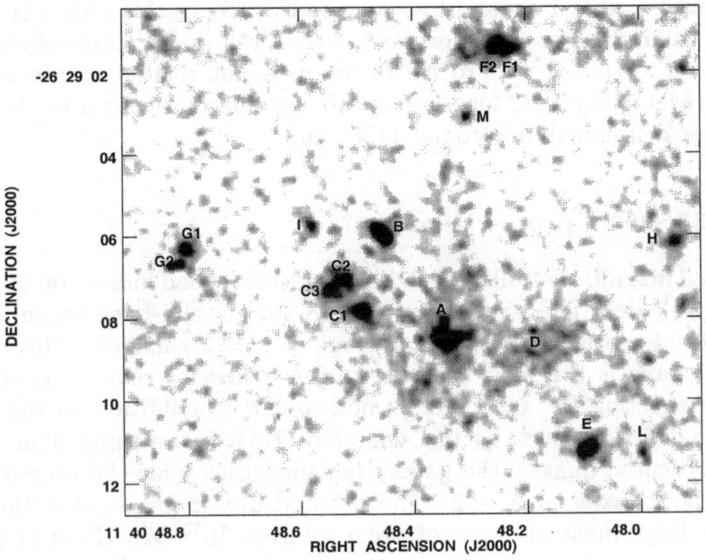

Figure 1: HST image of 1138-262 in the F6060W filter. Labels indicate the components of the galaxy and the companion object is designated F.

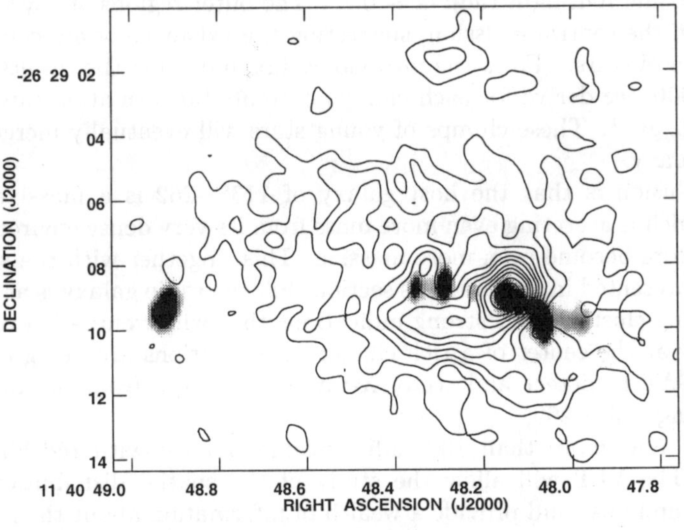

Figure 2: Lyα image of 1138-262 in a contour representation with the 8.2 GHz VLA map in grey-scale superimposed.

companion in a cluster around the radio source. Since this object is observed not along the radio axis direction (as in other well known cases of companion galaxies) but perpendicular to it, its origin is not related to the alignment effect (see McCarthy 1993 for a review on this effect), but is a further sign of a truly over-dense medium around 1138−262.

4 Is 1138−262 a proto-cD galaxy?

The K-band magnitude of distant radio galaxies is a good indication of the mass of hosting galaxies. The host galaxy of 1138-262 has a K-band magnitude that is 2 magnitudes brighter that the average of $z = 2$ galaxies. This indicates that this galaxy is the most massive known galaxy at redshift greater than 2. Even correcting for any possible non-stellar contribution to the infrared continuum emission, such as line emission (Hα) or scattered light from an optically obscured quasar, the galaxy remains still by far the brightest of all known radio galaxies at $z \sim 2$. The infrared flux indicates that this object contains a large mass of stars, of order at least 10^{12} $M_\odot$ (Pentericci *et al.*, 1997a). Moreover the surface brightness distribution is well represented by a classical de Vaucouleurs profile with effective radius ~ 10 kpc, indicating that the galaxy is a well formed elliptical.

Contrary to the K-band morphology the UV/optical broad band morphology is very clumpy, consisting of a central region and a number of clumps of sizes of few kpc and half-light radii of $\sim 0.2''$. The outer regions are systematically bluer than the central emission, suggesting the existence of a recently formed population of stars. ¿From the ultraviolet flux and using the results of Steidel *et al.*, (1996) we derive for each clumps current star formation rates between 2 and $5 M_\odot yr^{-1}$. These clumps of young stars will eventually merge with the central galaxy.

Our conclusion is that the host galaxy of 1138−262 is a massive elliptical galaxy which is accreting even more mass from its very dense environment and will therefore become even more massive. This, together with the substantial evidence presented in the previous section that the radio galaxy is at the center of a forming cluster, gives strong indications that will eventually evolve into a cD galaxy at the center of a rich cluster. Observations are being carried out with ROSAT to detect and study X-ray emitting gas from the protocluster surrounding 1138−262.

There are now more than 100 radio galaxies known with redshifts greater than 2. The VLT will allow the study of the spatial distribution of their stars, gas and dust and provide a wealth of information about the role of such forming galaxies in the early history of the Universe. We look forward to report considerable progress by the time we celebrate Riccardo's 75th birthday.

References

1. Carilli, C. L., Röttgering, H., van Ojik, R., Miley, G. K., and van Breugel, W.: 1997, APJ: in press
2. Dickinson, M. :1996 in 'HST and the High Redshift Universe,", proceedings of the 37th Herstmonceux Meeting, Cambridge, UK; N. Tanvir, A. Aragon-Salamanca, and J.V. Wall (eds.).
3. Hill, G. and Lilly, S.: 1991, *ApJ* **367**, 1
4. McCarthy, P.: 1996, in M. Bremer,P. van der Werf, H. Röttgering and C. Carilli, (eds.), *Cold Gas at High Redshifts*, Kluwer
5. McCarthy, P. J.: 1993, *ARA&A* **31**, 639
6. Pentericci, L., Röttgering, H.,Miley, G. K.,Carilli, C. L., and McCarthy P.: 1997,*A&AS* in press
7. Pentericci, L., Röttgering, H.,Miley, G. K., McCarthy P., van Breugel W., Macchetto D and Spinrad H.: 1997,APJ submitted
8. Röttgering, H. J. A., West, M., Miley, G., and Chambers, K.: 1996b, *A&AS* **307**, 376
9. Steidel, C. C., Giavalisco, M., Pettini, M., Dickinson, M., and Aldebergerw, K. L.: 1996, *ApJ* **462**, 17
10. Taylor, G., Barton, E., and Ge, G.: 1994,*AJ* **107**, 1942

PROBING THE COSMIC "DARK AGE" WITH THE VLT

MARTIN J. REES
Institute of Astronomy, Madingley Road, Cambridge, CB3 OHA, UK

This paper considers how the first subgalactic structures produced the UV radiation that ionized the intergalactic medium before $z = 5$. These 'pre-galaxies' may eventually be detectable.

1 Introduction

Riccardo Giacconi's uniquely distinguished and versatile career has kept him at the leading edge of cosmic discovery for the last 40 years. He remains so, now that he is spearheading the VLT project. Rather than focus on his past achievements, I shall use these few pages to look forward to the coming decade. I shall describe why there is good reason to hope that the unparalleled sensitivity of the VLT, complemented, perhaps by a New Generation Space Telescope (NGST), will open up the exploration of the pregalactic era, at ultrahigh redshifts lying beyond the range of current telescopes.

The structures in our present universe are the outcome of more than 10 billion years of evolution. Slight irregularities imprinted at very early eras led to increasing contrasts in the density from place to place, until overdense regions evolved into bound structures. Any acceptable theory must account for the present clustering properties of galaxies and dark matter; it must also match the actual universe at all past eras that can be probed observationally. We have for some years known about quasars with redshifts up to 5. Quasars may themselves be associated with atypical (even exceptional) galaxies, so their intrinsic properties are hard to relate to the general trend of galaxy formation What has been especially exciting about recent developments is that the morphology and clustering of ordinary galaxies can now be probed out to similar redshifts: the powerful combination of HST and the Keck Telescope has revealed many galaxies at $z > 3$. Also, the absorption features in quasar spectra (the Lyman forest, etc), probe the history of the clumping and temperature of a typical sample of the universe on galactic (and smaller) scales.

The mystery lies at still higher redshifts, between (in round numbers) a million years ($z = 1000$) and a billion years ($z = 5$). When the primordial radiation cooled below a few thousand degrees, it shifted into the infrared. The universe then entered a dark age, which continued until the first bound structures formed, releasing gravitational or nuclear energy that lit up the universe again. How long did the 'dark age' last? We now know that at least some galaxies and quasars had already formed by a billion years. But how much earlier did structures form, and what were they like?

In most cosmological theories, especially those that postulate adiabatic gaussian irregularities in the early universe, quasars and large galaxies should thin out beyond $z = 5$, but subgalactic structures may exist even at redshifts exceeding 10. I shall discuss the effects of the earliest stars and supernovae – production of UV radiation, non-uniform reheating of the IGM, and the production of the first heavy elements – and the implications for observations at ultra-high redshifts.

2 The very earlist galaxies

2.1 Arguments that $\sim$ 10 UV photons were generated for each baryon

According to most 'hierarchical' cosmological models, the UV that reheated and photoionized the IGM before $z = 5$ came from stars that formed in dark matter clumps with total mass greater than 10^8 solar masses, and internal velocity dispersion greater than about 15 km/sec. What is the chance of detecting these ancient stellar systems?

It is straightforward to estimate the cumulative amount of activity at high redshifts. The total integrated UV production beyond $z = 5$ must have been enough to ionize the IGM and build up the UV background whose strength can be inferred directly out to $z = 5$ from models of the Lyman forest, etc. We therefore have a firm lower limit to the ionizing UV generated. The total amount could be substantially above this limit, because much could 'go to waste' through reprocessing in dense clouds, local absorption in the sources, etc.

A separate estimate of the number of O and B stars at these early eras comes from the heavy element abundance in high-z absorption clouds. The diffuse gas in the Lyman α forest was already enriched to a heavy element abundance $Z \sim 10^{-2} Z_\odot$ by a redshift $z = 3$. The evidence comes from absorption lines of CIV and other species associated with Lyman α forest lines with $N_{\rm HI} \gtrsim 10^{14}$ cm^{-2} (Songaila & Cowie[1] and references therein). Detailed calculations of the expected column densities of the observed absorption lines, using hydrodynamic simulations of the Lyman α forest and realistic models for the spectrum of the ionizing background have shown that the carbon abundance needed to reproduce the observations is [C/H] $= -2.5$. However, the metal abundances are similar to those of Population II stars, where oxygen is the most abundant element and is overabundant by a factor ~ 2 relative to carbon. With $Z_\odot = 0.02$, the metallicity of the Lyman α forest is then

$Z \simeq 10^{-4}$. This metallicity should be approximately the same as the mean metallicity of the universe at $z = 3$ if the Lyman α forest contains most of the baryons, as is found to be the case in models of structure formation for the Lyman α forest similar to those analyzed in refs 2 and 3. The mean metallicity of the universe could be significantly lower only if most of the baryons were metal-free and in $N_{\rm HI} < 10^{14}\,{\rm cm}^{-2}$ absorption systems, or in a very diffuse, unobserved intergalactic medium.

The ratio of the mass of heavy elements ejected by a star to the energy in ionizing photons emitted over the lifetime of the star, as derived from models of stellar evolution and supernova explosions, turns out to be about constant over the relevant mass range $10 M_\odot \lesssim M \lesssim 50 M_\odot$, so given a mean metallicity $\bar{Z}$ we can predict the energy in ionizing photons that was emitted for each baryon in the universe. According to Madau & Shull,[4] this energy is $0.002\bar{Z} m_p c^2$ per baryon. For $\bar{Z} = 10^{-4}$, 10 ionizing photons were emitted per baryon by the stars that produced the Lyman α forest heavy elements (assuming a mean energy of 20 eV per ionizing photon).

Most of these photons must also have been absorbed by neutral hydrogen. A fraction f_i will be absorbed internally, in the galaxies where the emitting stars were formed; the rest should be absorbed after having escaped their original galaxies, in other dense absorbing systems where protons can recombine many times during the reionization epoch (similar to the observed Lyman limit systems). Of the photons that were absorbed internally, about 70% resulted in the production of a Lyman α photons, will then be scattered many times in the host galaxy, until they move to the wings of the Lyman α line and escape. We define f_d as the fraction of Lyman α photons that can eventually escape without being absorbed by dust. The mean comoving number density of these Lyman α photons is then $7\, f_i f_d (\bar{Z}_{-4}) n_b$, where n_b is the comoving baryon density and $\bar{Z}_{-4} = \bar{Z}/10^{-4}$. The mean surface brightness of the Lyman α emission line from the galaxies is simply obtained by multiplying the photon comoving number density by $c/4\pi$, which yields $10^{-7}\, f_i\, f_d\, \bar{Z}_{-4} (\Omega_b h^2/0.023)$ photons $\mathrm{cm}^{-2}\,\mathrm{sec}^{-1}\,\mathrm{arcsec}^{-2}$. The fiducial value of $\Omega_b h^2$ used here is close to that implied by a primordial deuterium abundance $D/H \simeq 3 \times 10^{-5}$.

2.2 The likely luminosity of individual 'galaxies'

The above arguments determine the average surface brightness of the high-redshift sky in the Lyman α emission line, or in other words, the product of the number of galaxies per solid angle and their individual, mean Lyman α flux. But the ease of detection obviously depends crucially on whether the UV background comes from huge numbers of individually ultra-faint systems, or (more optimistically) at least in part from systems of higher luminosity. The uncertainty here is much greater – it is essentially tied to the poorly-understood efficiency of forming massive stars in low-mass dark matter halos with shallow potential wells.

If star formation were highly efficient in all collapsed halos with velocity dispersion $V \gtrsim 15\,\mathrm{km\,s^{-1}}$, corresponding to the virial temperature $T = \mu V^2/k \simeq 10^4$ K , then reionization would take place as soon as the highest density peaks on the scales of these low velocity dispersions collapse. For a top-hat spherical region of mass M that turns around when the age of the universe is $t_f/2$, the relation between mass and velocity dispersion is $M = 2^{3/2}/(2\pi G)V^3 t_f \sim 10^9 M_\odot (V/20\,\mathrm{km\,s^{-1}})^3 (t_f/10^9\,\mathrm{yr})$. If all the baryons turned into stars over a timescale of order the free-fall time of the halo, the star formation rate would then be $0.3 M_\odot\,\mathrm{yr}^{-1}(V/20\,\mathrm{km\,s^{-1}})^3)$. For a normal IMF, this can yield a luminosity close to the peak luminosity of a supernova. The Lyman α luminosity from this efficient, rapid starburst would be $10^{52}\, f_i\, f_d\, (V/20\,\mathrm{km\,s^{-1}})^3 (f_b/0.1)$, photons sec^{-1}. The corresponding Lyman α flux at present (for $\Omega = 1$ and $H_0 = 70\,\mathrm{km\,s^{-1}\,Mpc^{-1}}$) is $10^{-6}\, f_i\, f_d\, (V/20\,\mathrm{km\,s^{-1}})^3$ $(f_b/0.1)\left[1-(1+z_f)^{-1/2}\right]^{-2}(1+z_f)^{-1}$photons $\mathrm{cm^{-2}\,sec^{-1}}$. If the sources that reionized the universe have these characteristics, then we can use the mean Lyman α surface brightness we obtained earlier to conclude that the number density of these sources in the sky should be ~ 1 per arcsec^2 for the fiducial numbers we have chosen, and a formation redshift $z_f \sim 10$. The redshift z_f should of course depend on the detailed model for the amplitude of the primordial fluctuations on small scales.

The main problem with this scenario is that supernova explosions in galaxies of such low velocity dispersions may well expel the gas before more than a small fraction has turned to stars. This leaves us with two other options regarding the sources of the reionization photons:

(i) If just a small fraction of the gas in each object turns to stars before the remainder is expelled in a wind, then each galaxy emits fewer photons. More therefore have to form, from lower amplitude peaks. The luminosity of each galaxy would be reduced.

Or

(ii) Star formation may be so inefficient in the low-escape-velocity halos that reionization has to await the formation of more massive galaxies with $V \gtrsim 100\,\mathrm{km\,s^{-1}}$.

Some moderately massive galaxies are in any case expected to form before reionization, because in cold dark matter the amplitude of fluctuations decreases very slowly with scale, implying that at the same epoch, halos with

$V = 70\,\mathrm{km\,s^{-1}}$ (with a mass ~ 40 times larger than halos with $V = 20\,\mathrm{km\,s^{-1}}$) would be collapsing from $3-\sigma$.peaks on this larger scale. In a gaussian theory, the $3-\sigma$ peaks should contain $\sim 10\%$ as much mass as the $2-\sigma$ peaks at a fixed epoch so we see that the mass distribution of these first-generation galaxies should probably be broad.

2.3 Detectability of galaxies beyond the reionization redshift z_i

What is the chance of detecting these galaxies? Given the mean surface brightness of Lyman α photons derived previously, the surface brightness in the UV continuum follows with little additional uncertainty, because the stars emitting most of the light in the range $1216\text{Å} < \lambda \lesssim 2500$ Å are also mostly very young stars. For every Lyman α photon produced from ionizing photons there should be ~ 10 UV continuum photons. The mean surface brightness from all galaxies in the rest-frame UV continuum is $10^{-6}\mathrm{photons\,cm^{-2}\,sec^{-1}\,arcsec^{-2}}$. The main spectral feature that should identify any such galaxies at $z \gtrsim 5$ is a sharp break of the UV continuum at the Lyman α wavelength, due to the Gunn-Peterson trough; in addition, the Lyman α emission line may be present, depending on dust absorption and scattering of the Lyman α photons. Notice that the redshift at which the IGM was reionized is only marginally relevant to the presence of the Gunn-Peterson edge, because even if the medium was reionized at $z \gg 5$, we know that the flux decrement caused by the Lyman α forest reaches a factor of 2 at $z \simeq 4$ and grows rapidly with redshift. Thus, the technique of identifying galaxies at $z \simeq 3$ from the Lyman continuum break[5,6] should be replaced by the Gunn-Peterson trough at $z \gtrsim 5$.

The mean surface brightness in the rest-frame UV continuum from these galaxies can also be expressed as $S_\nu = 6\times10^{-33}\,\mathrm{erg\,cm^{-2}\,sec^{-1}\,Hz^{-1}\,arcsec^{-2}} = 32\mathrm{AB\,arcsec^{-2}}$ (where AB denotes AB magnitudes in the band where the UV continuum is observed). If there is one galaxy per arcsec^2 at redshifts $z_f \sim 10$, each galaxy would have an AB magnitude of 32, i.e., about 3 magnitudes fainter than the detection limit in the *Hubble Deep Field* (HDF) for the I-band.[7] However, as mentioned above, galaxies with a mass 40 times larger would be ~ 400 times rarer, which still implies a number density of a few galaxies per square arc minute with AB magnitudes of ~ 28 [with the optimistic assumption of a rapid and efficient starburst in these more massive galaxies].

Detection with ground-based telescopes should be possible at $z < 6$. The Keck telescope can detect point sources to AB magnitudes $R < 28$, and $I < 27$, in a night of observing time, and VLT should do substantially better. Galaxies up to $z = 6$ can be detected in the I band; for higher redshift, the magnitude limit from a ground-based telescope degrades rapidly due to the high atmospheric background at $\lambda > 8500$ Å. The magnitude limit could be improved significantly with adaptive optics. And the HST detects galaxies in the HDF down to $I \simeq 28.5$. The NGST could image galaxies to AB magnitudes ~ 31 in the near-infrared, and should be able to detect galaxies to much higher redshift[8] with a much higher number-density than has been seen so far.

In general, the lensing magnification in rich lensing clusters may be used here to stretch the magnitude limit (see also ref 9). As an example, a lensing cluster with an Einstein ring radius $b = 30''$ should magnify to $A > 10$ an area of ~ 30 arcsec^2 in the source plane. In the example used above, about 30 galaxies with $AB = 32$ could be in this area, which would be magnified to $AB = 29.5$. Magnified images of high-redshift galaxies should characteristically appear in pairs around the critical lines, in a region that can be predicted from lensing models,[9,10] so this should help in their identification. These numbers indicate that a new deep field (similar to the HDF) imaged with HST in a rich cluster, adding also the H and J filters in the near-infrared, might well identify several galaxies at $z > 5$. In fact, the largest redshift object known at present (at $z = 4.92$) is already a gravitationally lensed galaxy.[11]

The foregoing discussion has assumed that these high-redshift galaxies would be sufficiently small to remain unresolved. Resolved objects would need to have higher fluxes to be detected, since the detection is limited by the sky background. The likely dimensions of the star-forming regions are ~ 100 pc, corresponding to angular sizes ~ 0.01 arcsec– not resolvable even with NGST.

2.4 The Lyman α emission line

Another possible way to detect the faint galaxies is to search directly for the Lyman α emission line. As we discussed before, for a normal starburst spectrum we expect $\sim 10\%$ of the UV photons to be in the Lyman α emission line if dust absorption is not important. Therefore, if the sensitivity for detecting galaxies is still limited by the sky background for emission-line searches, the width of the line should be $\Delta\lambda/\lambda < 0.01$ to allow detection on a shorter time than for the UV continuum. The width of the Lyman α line in emission from a region of neutral gas with column density $N_{\rm HI} = 10^{22} N_{22}\,{\rm cm}^{-2}$, and velocity dispersion $V = 10 V_6\,{\rm km\,s}^{-1}$ is $\Delta\lambda/\lambda \simeq 2 \times 10^{-3} (N_{22} V_6)^{1/3}$. Thus, the sensitivity of emission-line searches might be at best similar to searches for the UV continuum, and of course a much smaller range of redshift is searched since a very narrow band must be used.

The Lyman α emission line may be suppressed due to internal dust extinction. In addition, any photon emitted in the blue side of the Lyman α line will be scattered in the surrounding IGM, reducing the strength of the line by a factor of 2. For a galaxy observed when the surrounding IGM is still mostly neutral, even photons on the red side can be scattered due to the damping wing of the Gunn-Peterson trough.[12] This width of the damped absorption from the IGM is broader than the emission lines from starburst galaxies, so any Lyman α emission line should be suppressed by the IGM in any galaxy observed before the reionization. The edge of the Gunn-Peterson trough should be 'rounded off' into a characteristic shape, except for the fact that the inhomogeneity of the IGM may alter the profile (in particular, the presence of a halo of gas accreting on the galaxy may substantially increase the column density contributing to the damped profile from gas at a redshift close to that of the galaxy).

Thus, even though the Lyman α emission line might not be as good a spectral signature as the Gunn-Peterson trough as a technique to search for high redshift galaxies, its presence or absence should be an important diagnostic of the ionization state of the IGM in the neighborhood of any detected galaxy.

2.5 A note on AGNs and non-thermal UV background

AGN formation requires virialised systems with larger masses and deeper potential wells[13] AGNs may 'take over' as the dominant UV source at redshifts below 5 (and the second ionization of He may be delayed until AGNs can provide a power-law contribution to the spectrum), but at higher z, when the H itself was reheated and ionized, OB stars would dominate.

It is worth noting also that the diffuse IGM at redshifts of 3-5 has such a long thermal timescale that it retains a 'memory' of its thermal history at even higher redshifts. The adiabat this gas is on at $z = 5$ depends on the value of z_i, and also on the slope of the UV spectrum. This adiabat can be constrained by observations of weak and narrow lines in the Lyman forest.[14,15]

3 Very high redshift supernovae (and gamma-ray bursts?)

The intergalactic gas was already highly photoionized by $z = 5$; also, the mean abundance of heavy elements had attained a level about 0.01 solar by that time, this degree of contamination being about what would be expected if the reheating and ionization were due to OB stars.[16,17]

It is straightforward to calculate how many supernovae would have gone off, in each comoving volume, as a direct consequence of this output of UV and heavy elements: there would be one, or maybe several, per year in each square arc minute of sky.[18] These would be primarily of Type 2. The typical observed light curve has a flat maximum lasting 80 days. One would therefore (taking the time dilation into account) expect each supernova to be near its maximum for nearly a year. It is possible that the explosions proceed differently when the stellar envelope is essentially metal-free, yielding different light curves, so any estimates of detectability are tentative. However, taking a standard Type 2 light curve (which may of course be pessimistic), one calculates[38] that these object should be approx 27th magnitude in J and K bands even out beyond $z = 5$. The detection of such objects would be an easy task with the NGST. With existing facilities it is marginal. The best hope would be that observations of clusters of galaxies might serendipitously detect a magnified gravitationally-lensed image from far behind the cluster.

As a speculative addendum, note that a few percent of observed gamma-ray bursts may come from redshifts as large as 5: this would be expected if the burst rate, as a function of cosmic epoch, tracks the star formation rate. At the time of writing, data on optical and X-ray afterglows are still very sparse, but it is at least an exciting possibility that there may be occasional flashes, far brighter than supernovae, from very large redshifts.

4 The first stars: some uncertainties

The gravitational aspects of clustering can all be modeled convincingly by computer simulations. So also, now, can the dynamics of the baryonic (gaseous) component – including shocks and radiative cooling. But the nature of the simulation changes as soon as the first stars (or other compact objects) form. The first stars exert crucial feedback – the remaining gas is heated by ionizing radiation, and perhaps also by an injection of kinetic energy via winds and even supernova explosions. The three uncertainties here are:

(i) What is the IMF of the first stellar population? The high-mass stars are the ones that provide efficient (and relatively prompt) feedback. It plainly makes a big difference whether these are the dominant type of stars, or whether the initial IMF rises steeply towards low masses, so that very many faint stars form before there is a significant feedback.

(ii) The influence of the early stars depends on whether their energy is deposited locally or penetrates into the all-pervasive medium that is not yet in contracting systems. The UV radiation could, for instance, be mainly absorbed in the gas immediately surrounding the first stars, so that it exerts no feedback on the condensation of further clumps – the total number of massive stars needed to build up the UV background, and the concomitant contamination by heavy elements, would then be greater.

(iii) Quite apart from the uncertainty in the IMF, it is also unclear what fraction of the baryons that fall into a clump would actually be incorporated into stars before being re-ejected. The retained fraction would almost certainly depend on the virial velocity: gas more readily escapes from the shallow potential wells relevant to dwarf galaxies. Ejection is even easier in potential wells so shallow that they cannot confine gas at the photoionization temperature: this effect may lead to a reduced efficiency during the second, as compared to the third of the three cosmogonic stages discussed in section 2.

All these three uncertainties would, for a given fluctuation spectrum, affect the redshift at which molecules were destroyed, and at which full ionization occurred.

5 Where are the oldest stars?

The efficiency of early mixing is important for the interpretation of stars in our own galaxy that have ultra-low metallicity[38] – lower than the mean metallicity that would have been generated in association with the UV background at $z > 5$. If the heavy elements were efficiently mixed, then these stars would themselves need to have formed before galaxies were assembled. To a first approximation they would cluster non-dissipatively; they would therefore be distributed in halos (including the halo of our own Galaxy) like the dark matter itself. More careful estimates slightly weaken this inference, This is because the subgalaxies would tend, during the subsequent mergers, to sink via dynamical friction towards the centres of.the merged systems. There would nevertheless be a tendency for the most extreme metal-poor stars to have a more extended distribution in our Galactic Halo, and to have a bigger spread of motions.

The number of such stars depends on the IMF. If this were flatter, there would be fewer low-mass stars formed concurrently with those that produced the UV background. If, on the other hand, the IMF were initially steeper, there could in principle be a lot of very low mass (macho) objects produced at high redshift. These could provide a few percent of the halo if omega were 1 (and a higher proportion of the $\lesssim 30$ kpc region probed by lensing searches); a larger proportion in a low-density universe.

Most of the small first-generation galaxies should by now have merged into more massive systems (forming part of the halo population of stars in normal galaxies like the Milky Way), but some could survive until today in galactic halos, or even as isolated objects. This may[12] be the explanation of present-day dwarf spheroidal galaxies.

Acknowledgments

I am especially grateful to Jordi-Miralda Escude for collaboration , and would also like to thank Zoltan Haiman, Avi Loeb, and Max Tegmark, for discussion. This is also an opportunity to express appreciation for the stimulus provided by many discussions with Riccardo Giacconi, dating back to the exciting discoveries of UHURU.

1. Songaila, A., & Cowie, L. L. 1996, AJ, 112, 335
2. Hernquist, L., Katz, N., Weinberg, D. H., & Miralda-Escudé, J. 1996, ApJ, 457, L51
3. Miralda-Escudé, J., Cen, R., Ostriker, J. P., & Rauch, M. 1996, ApJ, 471, 582
4. Madau, P., & Shull, J. M. 1996, ApJ, 457, 551
5. Guhathakurta, P, Tyson, J. A., & Majewski, S. R. 1990, ApJ, 357, L9
6. Steidel, C. C., Giavalisco, M., Pettini, M., Dickinson, M., & Adelberger, K. L. 1996, ApJ, 462, L17
7. Williams, R. E., *et al.* 1996, AJ, 112, 1335
8. Mather, J., & Stockman, H. 1996, NASA Report.
9. Miralda-Escudé, J., & Fort, B. 1993, ApJ, 417, L5
10. Kneib, J. P., Ellis, R. S., Smail, I. R., Couch, W. J., & Sharples, R. 1996, ApJ, 471, 643
11. Franx, M., Illingworth, G. D., Kelson, D. D., van Dokkum, P. G., & Tran, K.-V. 1997, submitted to ApJL (astro-ph 9704090)
12. Miralda-Escudé, J. & Rees, M.J. 1997 ApJ (submitted)
13. Haehnelt, M. & Rees, M.J. 1993 MNRAS, 263, 168
14. Haehnelt, M. & Steinmetz, M. 1997, MNRAS (in press)
15. Miralda-Escudé, J. & Rees, M.J. 1994, MNRAS, 266, 343
16. Haiman, Z. & Loeb, A. 1997, ApJ (in press)
17. Ostriker, J. P., & Gnedin, N. 1996, ApJ, 472, 630
18. Miralda-Escudé, J., & Rees, M. J. 1997, ApJ, 478, L57

ON THE CRITICAL MASS: THE CASE OF WHITE DWARFS

REMO J. RUFFINI
Physics Department, University of Rome, La Sapienza, 00185 Rome, Italy

I.C.R.A.-International Center for Relativistic Astrophysics c/o Physics Department, University of Rome, La Sapienza, 00185 Rome, Italy
E-mail: ruffini@icra.it

We recall the generalization of the Feynman-Metropolis-Teller approximation for a compressed atom using a relativistic Fermi-Thomas model. These results within a Wigner-Seitz approximation lead to a new equation of state for white dwarfs and to a new value of their critical mass, smaller than the one obtained by Chandrasekhar. The possible observations of these effects in binary neutron stars are outlined.

1 Introduction

The problem of critical mass against gravitational collapse is one of the most important in relativistic astrophysics and has been the subject of a very large number of theoretical investigations with many observational consequences, the most important one being the paradigm for the identification of black holes in binary X-ray sources. The concept of critical mass appears in a large variety of astrophysical settings: white dwarfs, neutron stars, galactic nuclei. The occurrence of gravitational collapse in galactic nuclei is still matter of great debate both from the observational and theoretical point of view: the physical conditions leading to the onset of gravitational collapse, even the matter constituents of such a process, are still far from settled although the existence of extremely massive relativistic systems in the range of $10^6 - 10^8 M_\odot$ in galactic nuclei is certain. The occurrence of gravitational collapse in neutron stars has obtained grandiose observational verification in binary X-ray sources although, from a theoretical point, it still presents problems of unsurpassed difficulty due to theoretical and observational unknowns in the behavior of matter compressed at supra-nuclear densities. In this case only a firm upper limit on the numerical value of the critical mass has been obtained from first principles: the validity of general relativity, the causality non-violation principle and the behavior of matter at a fiducial nuclear density.[10]

In this respect the case of white dwarfs appears most interesting: it is amenable to a complete theoretical treatment which is also testable by refined observational measurements thanks to the superb observations of binary pulsars. The critical mass for white dwarfs occurs at densities relatively low in comparison with nuclear densities and many of the intricacies of the behavior of matter in bulk at supra-nuclear densities can be completely neglected. On the other hand the theoretical analysis of the white dwarf critical mass can become the arena for a detailed treatment of gravitational and electromagnetic effects at relativistic regimes in the fully tested region of validity of these field theories: general relativity, quantum theory and electromagnetism. I will devote this contribution to this topic which has reached complete maturity and in principle can give important observational clues for the understanding of astrophysical high-precision measurements.

2 Historical Background

That white dwarfs cannot have an arbitrary large mass and that a critical mass against gravitational collapse must exist in their configuration of equilibrium reachable by increasing their central densities was clearly understood in the thirties independently by Subranaman Chandrasekhar[1,2] and by Lev Davidovich Landau.[3] Their treatments differ enormously in style.

After publishing his result in an article, Chandrasekhar gave his derivation in a 343 page classic book[13] which summarized: a) some of the results originally presented in German in the classic textbook *Gaskugeln* by Emden, b) basic aspects of quantum mechanics, developed in the classic book of R.H. Fowler[18] and in the work of P.A.M. Dirac, and 3) the detailed computations of the mass-radius relation for white dwarfs and their agreement with observations with the clear statement of the existence of a critical mass equal to $1.45 M_{\odot}$.

The work of Lev Landau is totally different in its essential approach: Landau also uses all the expertise which could be gained from the classic work of Emden and quantum theory, at that time just formulated, and the corresponding relativistic corrections. Similarly he reaches the conclusion of the existence of a critical mass for a white dwarf, but now, everything, in a dramatic paper of no more then *three* pages. There by the stroke of crucial shortcuts in the physics of self-gravitating degenerate matter in relativistic regimes, Landau reaches that unexpected conclusion. Landau's paper is not only beautiful for the masterful interplay of physics and astronomy, but can also be considered the first paper introducing a new style of work in the scientific analysis of astronomical phenomena.

Another non-negligible difference must be mentioned distinguishing the work of Landau and Chandrasekhar: Landau *did not* believe the possible existence in nature of the new concept of a critical mass he had just obtained, while Chandrasekhar never revealed such a critical attitude towards it.

At the time of presentation of his work in Cambridge, Chandra met with very strong resistance and criticism from Arthur Eddington, so much so that his paper published in the proceedings of the Royal Astronomical Society was preceded by an article of strong criticism by Eddington himself which clearly stated, with reference to Chandra's work, "... but its physical foundation does not inspire confidence, since it is a combination of relativistic mechanics with non-relativistic quantum theory."

I was always interested in understanding some of the reasons which motivated Eddington in his criticism. After all, at the time, Eddington was one of the greatest experts in the theory of general relativity and was the author of fundamental textbooks both in astronomy[15] and general relativity:[16] for me more than his expressed criticism toward Chandrasekhar's work, it was interesting to understand whether there was any fundamental theoretical issue adopted by Chandrasekhar which could be disturbing to the eyes of a relativist.

The answer to Eddington's criticism came finally in 1974 from Chandrasekhar himself in his outstanding article "Why are the stars as they are" presented at the Varenna Summer school on the Physics and Astrophysics of Neutron Stars and Black Holes.[14]

Nevertheless reconsidering recently the issue of the white dwarf critical mass, I have found a basic unproven assumption in Chandrasekhar's work. The distribution of the degenerate electrons is assumed to be uniform in the stars and unaffected by Coulomb interaction with the nuclei all the way to infinite densities. Indeed in Chandrasekhar's work the value of the critical mass is reached for an infinite central density of the star. On this issue Landau had made two important remarks:

1. In 1938, following the earlier suggestions by Hund,[22] Landau[21] pointed out *"it is well known that matter consists of nuclei and electrons. Nevertheless it can be shown that in bodies of very large mass, this usual "electronic" state of matter can become unstable. The reason for this lies in the fact that the "electronic" state of matter does not lead to extremely great densities, because at such densities electrons form a Fermi gas having an immense pressure. On the other hand, it is easy to see that matter can go into another state which is much more compressible—the state where all the nuclei and electrons have combined to form neutrons."*. Later Yakov Borisovich Zel'dovich[23] pointed out, quantitatively, that this "neutron condensation" for inverse beta decay starts to occur for hydrogen already at $10^7 \frac{g}{cm^3}$. Then Harrison et al,[24] by analyzing the pulsational modes of white dwarfs, pointed out that all configurations of white dwarfs after the

onset of the inverse beta decay are indeed unstable and the critical mass is reached not at an infinite density but at precisely the finite density marked by the onset of inverse beta decay.

2. The second contribution of Landau is contained in the treatise with Yevgeny Lifshitz, where Landau[17] gives an estimate of the effects of the electromagnetic interaction between nuclei and electrons and explicitly shows how these interactions can be neglected only in the asymptotic regime for very large densities. The fiducial densities where indeed the Coulomb interactions can be neglected occur at $\rho > 10^7 g/cm^3$ for Z=56, namely at densities too high if compared to the density regime at which the stable white dwarfs and their critical mass occur. The interactions between nuclei and degenerate electrons therefore are not negligible for white dwarfs: they were certainly difficult to evaluate at the time of the Chandrasekahr's original work.

Actually an interesting alternative approach, with respect to the evaluation of the interaction term between the electrons and the nuclei, was advanced two years earlier than the work of Chandrasekhar and Landau by the Soviet physicist Yacov Ilich Frenkel.[19] This paper, quoted by Chandrasekhar in his classic book,[13] has been in the past and is still today surprisingly ignored. Frenkel proposed to use a relativistic Thomas-Fermi model within a Wigner-Seitz approximation in order to describe stellar matter. The astrophysical motivations were not clear to Frenkel, though the theoretical formulation he proposed would have deserved a much more thorough examination.

An important turning point came in this field due to the fundamental work of Feynman, Metropolis and Teller. While at los Alamos, they considered the equation of state of compressed matter.[4]

It was Ed Salpeter[7,8] who first applied the Frenkel equations and the Feynman-Metropolis-Teller[4] approach to the treatment of white dwarfs: Salpeter's treatment was somewhat affected by the intricacies of the numerical integrations and the drastic approximations he adopted in the numerical solutions.

We ourselves rediscovered the Frenkel Equations and the relativistic Fermi-Thomas treatment applied to an appropriate Wigner-Seitz problem following the study of self-gravitating systems [25] and the riformulation of the treatment of these systems by using a relativistic generalization of gravitational Fermi-Thomas like equations [9 12 26]. Today with relatively simple numerical computations, we can reconsider the problem and present a comprehensive treatment by first considering the issue of the electromagnetic interactions between nuclei and electrons and turning later to the issue of inverse beta decays.[11] The new treatment[20] leads to the disappearance of all scaling laws with the chemical composition of the stars, which have been commonly assumed in all treatments up to now. A new value of the critical mass of white dwarfs is obtained, smaller than the one originally proposed by Chandrasekhar and by Salpeter, and one which strongly depends on the chemical composition of the star. These results appear in principle to be observationally testable. Particularly attractive is the possibility of explaining the observed masses of binary pulsars: these systems are clearly neutron stars which could be formed by the onset of instability of stellar cores very close to the critical mass of white dwarfs as presented by Chandrasekhar.[14] The evaluation of their masses will be clearly affected by our computations.

3 The Relativistically Compressed Thomas-Fermi Atom

In the well known Feynman-Metropolis-Teller treatment of the classical Fermi-Thomas model[5,6] an accurate numerical analysis is made of atomic configurations submitted to external pressure. Following Ferreirinho *et al.* [9] and Ruffini and Stella *et al.*[12] we reconsider such a treatment in the framework of a relativistic generalization of the Fermi-Thomas model, which is here briefly derived and solved for different atomic species in different states of compression. Results so obtained are applied to determine the equation of state of cold degenerate stars (white dwarfs) to show how the relativistic and electromagnetic effects acting on microscopic scales affect the overall structure of the star. Finally, the equation of state obtained is used to determine the equilibrium configurations of white dwarfs. A comparison is made with the classic work of Chandrasekhar and Salpeter to show the differences introduced by the use of the new equation of state, and to show how those treatments correspond respectively to a zero-order and first-order approximation to the one we present here.

Let us consider the spherically symmetric problem of a nucleus with Z protons and A nucleons interacting with a fully degenerate gas of Z electrons. The equation of electrostatics for this problem reads, with usual meanings for the symbols appearing in it,

$$\Delta V(r) = 4\pi e n_e(r) - 4\pi e n_p(r) \ , \tag{1}$$

where we have introduced the number density of electrons $n_e(r)$ and of protons $n_p(r)$.

In Eq. (1) the quantities $V(r)$ and $n(r)$ are of course not independent, being related by the equilibrium condition for a relativistic gas in a Coulomb external potential (see [17])

$$c\sqrt{p_F^2 + m^2c^2} - eV(r) = \text{const} \equiv E_F\ , \tag{2}$$

where E_F stands for *Fermi energy* of the electrons.

To put the Eq. (1) in a simple and dimensionless form we introduce the new function $\Phi(r)$, related to the Coulomb potential by

$$\Phi(r) = V(r) + E_F/e \tag{3}$$

and the corresponding dimensionless function $\chi(r)$, implicitly defined by

$$\Phi(r) = \frac{Ze\chi}{r} \tag{4}$$

Similarly we have the dimensionless expression for the radius $r = bx$, where

$$b = (3\pi)^{2/3}\frac{\hbar^2}{me^2}\frac{1}{2^{7/3}}\frac{1}{Z^{1/3}}\ . \tag{5}$$

We then have

$$p_F = 2mc\left(\frac{Z}{Z_{cr}}\right)^{2/3}\left(\frac{\chi}{x}\right)^{1/2}\left[1+\left(\frac{Z}{Z_{cr}}\right)^{4/3}\frac{\chi}{x}\right]^{1/2}, \tag{6}$$

where

$$Z_{cr} = \left(\frac{3\pi}{4}\right)^{1/2}\left(\frac{\hbar c}{e^2}\right)^{3/2} \approx 2462.4\ . \tag{7}$$

From

$$n_e = \frac{p_F^3}{3\pi^2\hbar^3} \tag{8}$$

we then obtain

$$n_e = \frac{Z}{4\pi b^3}\left(\frac{\chi}{x}\right)^{3/2}\left[1+\left(\frac{Z}{Z_{cr}}\right)^{4/3}\frac{\chi}{x}\right]^{3/2}\ . \tag{9}$$

Let us express the second term of the source of Eq. (1) in terms of dimensionless quantities: we assume a homogeneous nucleus, with a radius given by the approximate formula

$$r_{nuc} = 1.2A^{1/3}\ \text{Fermi} \tag{10}$$

The number density of protons is therefore

$$n_p = \frac{3Ze}{4\pi r_{nuc}^3}\Theta\left(x_{nuc} - x\right) \tag{11}$$

Finally we can write Eq. (1) in the form

$$\frac{d^2\chi}{dx^2} = \frac{\chi^{3/2}}{x^{1/2}}\left[1+\left(\frac{Z}{Z_{cr}}\right)^{4/3}\frac{\chi}{x}\right]^{3/2} - \frac{3x}{{x_{nuc}}^3}\Theta\left(x_{nuc}-x\right) \tag{12}$$

where x_{nuc} is the dimensionless size of the nucleus ($r_{nuc} = bx_{nuc}$). Eq. (12) is what we call the "*generalised dimensionless Fermi-Thomas equation.*"

The first boundary condition follows from $\chi \propto r\Phi$ and therefore $\chi \xrightarrow{r\to 0} 0$, and so

$$\chi(0) = 0\ . \tag{13}$$

The second condition comes from the normalization condition

$$N = \int_0^{r_0} 4\pi n_e r^2 dr = Z\int_0^{x_0}\frac{\chi^{3/2}}{x^{1/2}}\left[1+\left(\frac{Z}{Z_{cr}}\right)^{4/3}\frac{\chi}{x}\right]^{3/2} x\ dx \tag{14}$$

with atom size $r_0 = bx_0$. Using Eq. (12) we obtain

$$N = Z\int_0^{x_{nuc}} x\chi''\ dx + \frac{3Z}{x_{nuc}^3}\int_0^{x_{nuc}} x^2\ dx + Z\int_{x_{nuc}}^{x_0} x\chi''\ dx \tag{15}$$

which gives the simple relation

$$N = Z\left[x_0\chi'(x_0) - \chi(x_0) + 1\right]\ . \tag{16}$$

For a neutral atom we have $N = Z$ and the Eq. (16) reads simply

$$x_0\chi'(x_0) = \chi(x_0) \tag{17}$$

Note that the physical quantities of interest, such as the Coulomb potential and the density of electrons do not show any singularity in the center, nor on the surface of the nucleus, since they only depend on the function χ and his first derivative. The only discontinuity appears in the second derivative of χ due to our rough assumption of a homogeneous nucleus.

4 The equation of state

Under these hypothesis, we determine the equation of state of compressed matter simply by computing for each value of the atomic compression parameter x_0 the corresponding values of pressure and mass density: the first is generated by the degenerate gas of electrons and the second is given by the mass density of nuclei. We assume that the nuclei are arranged in a Wigner-Seitz lattice, each cell being filled by a relativistic gas of degenerate electrons. The shape of the lattice cell is unimportant in the lowest approximation,[7] so we will assume a spherical cell, with a radius equal to the atomic radius. The average density of cells is thus

$$\overline{\rho}(x_0) = \frac{Am_0}{V_{cell}} = \frac{3Am_0}{4\pi(bx_0)^3} , \tag{18}$$

where A and m_0 are respectively the number and mass of the nucleons.

In this scheme the pressure is also easily determined, being simply the pressure generated by a Fermi gas of number density equal to the number density of electrons at the surface of the atom $n_e(x_0)$, which can be expressed in terms of dimensionless quantities as follows

$$n_e(x_0) = \frac{Z}{4\pi b^3}\left(\frac{\chi(x_0)}{x_0}\right)^{3/2}\left[1+\left(\frac{Z}{Z_{cr}}\right)^{4/3}\frac{\chi(x_0)}{x_0}\right]^{3/2} . \tag{19}$$

Our treatment differs from the zero-order approximation of Chandrasekhar (uniformly distributed electrons) and also from the first-order approximation of Salpeter in which the Thomas-Fermi equation for each value of Z and x_0 was not solved and instead an expansion of the number density of electrons was introduced

$$n(x) = n_0 + \epsilon(x) . \tag{20}$$

This expansion applies only when the electrostatic interaction within nuclei and electrons is small with respect to the Fermi energy of electrons, i.e. in the limit of very high densities.

5 Equilibrium Configurations for White Dwarfs

We can now compute the equilibrium configurations for white dwarfs. The results are shown in figures 1 and 2. We obtain, for the same central density, less massive configurations than those obtained by both Chandrasekhar and Salpeter. Namely for a fixed mass density, i.e. for a fixed value of A and of the compression parameter, our pressure is systematically lower than the one generated by a uniform shell of electrons.

Numerical integration of the models is performed here for stars of helium and iron in the framework of a Newtonian theory of gravitation. We thus deal with the usual equations of hydrostatic equilibrium and add to these equations the numerical equation of state obtained in the previous section.

Results of integrations show that we recover Chandrasekhar's results in the limit of nuclear density, where effectively the density of electrons reaches

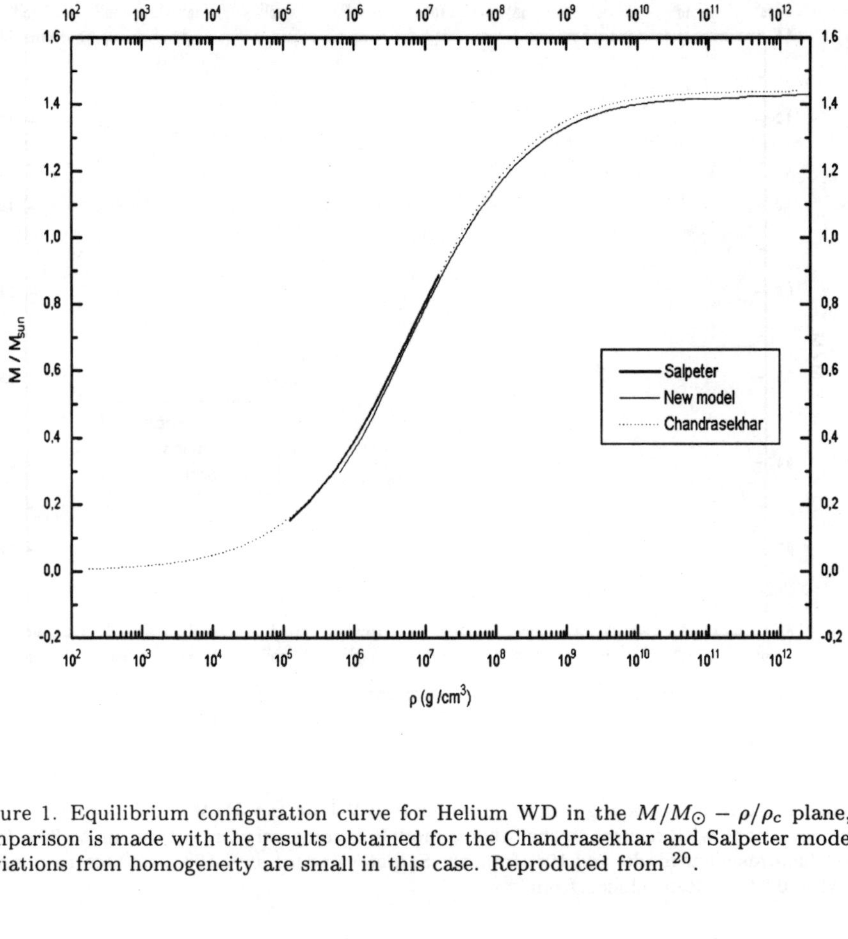

Figure 1. Equilibrium configuration curve for Helium WD in the $M/M_{\odot} - \rho/\rho_c$ plane, a comparison is made with the results obtained for the Chandrasekhar and Salpeter models; deviations from homogeneity are small in this case. Reproduced from [20].

homogeneity. The deviations from classical models becomes very important at the lower densities.

The mass for an iron star corresponding to a central density of the order of $10^6 g/cm^3$ is about 30% smaller than that predicted from the Chandrasekhar model and about 10% smaller than Salpeter's one. Of course deviations are the greatest for atomic specie with high Z, being related, as seen, to Coulomb corrections to the equation of state.

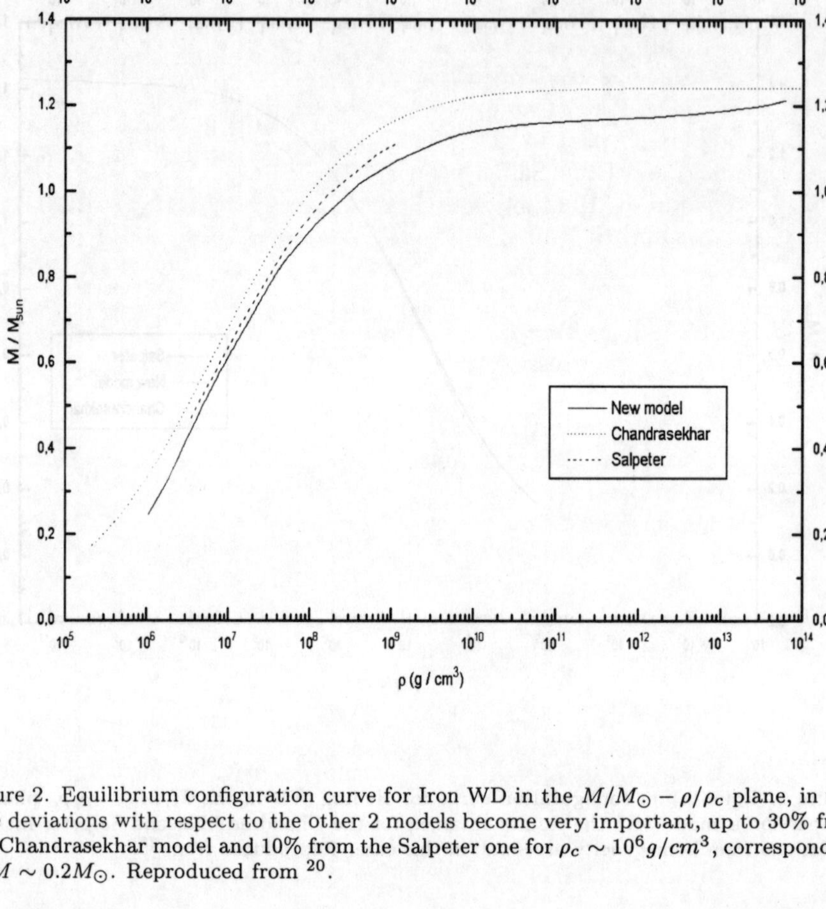

Figure 2. Equilibrium configuration curve for Iron WD in the $M/M_{\odot} - \rho/\rho_c$ plane, in this case deviations with respect to the other 2 models become very important, up to 30% from the Chandrasekhar model and 10% from the Salpeter one for $\rho_c \sim 10^6 g/cm^3$, corresponding to $M \sim 0.2 M_{\odot}$. Reproduced from [20].

These results and differences will apply as well when the effects of inverse beta decay are taken into account and in the general relativistic treatment [20] where corresponding differences in the values of the critical mass are found.

References

1. S. Chandrasekhar, Phyl. Mag. **7**, 63 (1929)

2. S. Chandrasekhar, Mont. Not. Roy. Astronom. Soc. **91**, 456 (1931)
3. L. Landau, Phys. Zeits. Sov. **1**, 285 (1932)
4. R. P. Feynman, N. Metropolis and E. Teller, Phys. Rev. **75**, 61 (1949)
5. E. Fermi, Rend. Acc. Lincei **6**, 602 (1928)
6. L. H. Thomas, Proc. Cambr. Phil. Soc. **33**, 542 (1927)
7. E. E. Salpeter, ApJ **134**, 669 (1961)
8. T. Hamada and E. E. Salpeter, ApJ **134**, 683 (1961)
9. J. Ferreirinho, R. Ruffini and L. Stella, Phys.Lett. **91B**, 314 (1980)
10. C. Rhoades and R. Ruffini, Phys.Rev.Lett. **32**, 324 (1974)
11. A. Merloni, R. Ruffini, Int.Jour. of Mod. Phys. **5**, 507 (1996)
12. R. Ruffini and L. Stella, Phys. Lett. **B102**, 6 (1981)
13. S. Chandrasekhar, *Introduction to the Study of Stellar Strucuture*, University of Chicago Press, Chicago (1939)
14. R. Giacconi and R. Ruffini, *Physics and Astrophysics of Neutron Stars*, North Holland, Amsterdam(1978)
15. A. Eddington, *The Internal Constitution of Stars*, Cambridge University Press, Cambridge (19)
16. A. Eddington, *The Mathematical Theory of Relativity*, Cambridge University Press, Cambridge (1923)
17. L. D. Landau and E. M. Lifshitz, *Statistical Physics*, Addison Wesley, Reading, Massachussets (1969), chapter XI
18. R. H. Fowler, *Statistical Mechanics*, Cambridge University Press, (1929,1936,1955)
19. Y. I. Frenkel, Zeit. fur Phys. **50**, 234 (1928)
20. G. Bertone and R. Ruffini, *Proceedings of the 3rd Icra Network Workshop, C.Cherubini and R.Ruffini eds., SIF 2000.*
21. L. D. Landau, Nature **19**, 333 (1938)
22. F. Hund, Erg. d. exacten Natwis. **15**, 189 (1936)
23. Ya. B. Zel'dovich, JETP **6**, 760 (1958)
24. Harrison,
25. R. Ruffini, S. Bonazzola, Phys. Rev. **187**, 1767 (1969)
26. A. Merloni, R. Ruffini, V. Torroni, Il Nuovo Cimento **113B**, 255 (1998)

